Transformation of Agri-Food Systems

K. C. Bansal • W. S. Lakra • Himanshu Pathak
Editors

Transformation of Agri-Food Systems

 Springer

Editors
K. C. Bansal
Formerly at ICAR-National Bureau
of Plant Genetic Resources
New Delhi, India

W. S. Lakra
Formerly at ICAR-CIFE
Mumbai, India

Himanshu Pathak
ICAR- DARE
Government of India
New Delhi, India

The opinions expressed in Chapters 6 and 15 are those of the author(s) and do not necessarily reflect the views of the Food and Agriculture Organization of the United Nations, its Board of Directors, or the countries they represent.

ISBN 978-981-99-8013-0   ISBN 978-981-99-8014-7   (eBook)
https://doi.org/10.1007/978-981-99-8014-7

This Springer imprint is published by the registered company Springer Nature Singapore Pte Ltd.
The registered company address is: 152 Beach Road, #21-01/04 Gateway East, Singapore 189721, Singapore

Paper in this product is recyclable.

Preface

The current challenges in agricultural production and food systems including climate change, shrinking of natural resources—cultivable land and water, and loss of biodiversity need to be addressed while meeting the ever-increasing global food demand. The available practices to improve agriculture rely on finite resources such as land, water, and energy. However, it has become essential in view of the fast degrading agro-ecosystems that solutions be nature-based with a focus on the sustainable development of agriculture, leading to enhanced climate resilience, water conservation, improved soil health, and halting the loss of biodiversity. This calls for a transformation of agri-food systems with an all-inclusive approach.

The National Academy of Agricultural Sciences (NAAS), a credible think tank of agricultural science experts, organized its XVI Agricultural Science Congress from October 10 to 13, 2023, with the theme **"Transformation of Agri-Food Systems."** The aim of the Congress was to bring together leading academicians, researchers, policy makers, students, farmers, and entrepreneurs to exchange their research findings, ideas, and experiences on all aspects of agri-food systems and allied sectors. The book is organized to cover different themes as discussed in the Congress with national and international speakers as lead authors. The chapters include major areas such as Innovations for Reconfiguring Food Systems; Transforming High-Value Food Commodities; Demand-Supply of Agri-food Commodities; Balancing Human Demand and Ecological Supply; International Partnership for Transformation of Agri-food Systems; Transforming Animal Health and Aquatic Food Systems; Climate-Resilient Agriculture; Natural Resource Management; Improving Water Productivity; Combating Micronutrient Deficiencies and Nutritional Security; Plant Genetic Resources for Food Security and Nutrition, Genome Editing for Crop Improvement; and Biosafety and Socioeconomic Considerations.

Key actions and strategies that will revolve around making agriculture more resilient, sustainable, and equitable will help transform the agri-food systems. On the food production side, optimum management of natural resources for sustainable agro-ecosystems, diversification to high-value climate-resilient crops to increase economic water productivity, and income under normal climatic conditions are the major driving factors. Implementation of sustainable soil management practices across different landscapes will lead to safeguarding food and nutritional security. Enhanced supply and demand of millets through promotion and policy initiatives and bringing millets to the mainstream food chain will aid in achieving sustained nutritional security. Improving food safety and popularizing nutrition education will boost the enhancement of dietary diversity and the reduction of post-harvest food losses. Additionally, ensuring animal health, resilient value chain, and environmental sustainability in the overall livestock sector will drive the transformation of agri-food systems in a holistic way. However, concerted collaborative efforts at the global, regional, and local levels and the application of scientific and technological advanced methodologies are needed to ensure achieve the SDGs by 2030.

The book provides readers with updated knowledge and is expected to contribute to formulating the way forward in transforming the global agri-food systems to meet the Sustainable Development Goals (SDGs-2030) of the United Nations, particularly in achieving zero hunger and building an action strategy for a sustainable future for all. The book focuses on the use of latest science-based technologies and policies for sustainable agri-food systems, highlights solutions for addressing issues of climate change and sustainability, and covers agri-horticultural crops, livestock and fisheries, field-based technologies, and agriculture policies and institutions. We are confident that this edited book will serve as a valuable resource for academicians, policymakers, and professionals and help contribute to the ongoing global efforts to transform agriculture to a more sustainable, resilient, and equitable food system.

We thankfully acknowledge the distinguished contributors of various chapters, and Ms Rashmi Singh, Editorial Manager at NAAS, for her able academic assistance.

New Delhi, India	K. C. Bansal
Mumbai, India	W. S. Lakra
New Delhi, India	Himanshu Pathak

The original version of this book has been revised: Copyrights HolderName updated in all the Chapters. The correction to this book can be found at
https://doi.org/10.1007/978-981-99-8014-7_26

Contents

Editors and Contributors

About the Editors

K. C. Bansal is Secretary, National Academy of Agricultural Sciences and is a former Professor of Plant Biotechnology at IARI. He served as the Director of the National Bureau of Plant Genetic Resources (ICAR), New Delhi. He served as Vice-Chair (Asia), Commission on Genetic Resources for Food and Agriculture, United Nations (2013–2015). Currently, he is on the Board of Directors of the Global Plant Council and Board of Trustees of MS Swaminathan Research Foundation. He has 40 years of research experience and played a leadership role in research management, human resources development, and policy making in the area of plant genome engineering in India. His unprecedented initiative to evaluate the entire wheat germplasm (~22,000 accessions) conserved in the Indian National Genebank was recognized by the Limca Book of Records. He is Rafi Ahmed Kidwai awardee of ICAR and Haryana Vigyan Ratna awardee of the State of Haryana. He is a fellow of the National Academy of Agricultural Sciences and the National Academy of Sciences, India.

W. S. Lakra is Secretary, National Academy of Agricultural Sciences. After graduating from the University of Delhi, he obtained his Ph.D. and D.SC. degrees from HNB Garhwal University, UK and post-doc research and training from CSIRO, Australia, and Harvard University, USA. Dr Lakra has over 40 years of experience in fisheries research and education including Director of ICAR's National Bureau of Fish Genetic Resources, Lucknow, and Central Institute of Fisheries Education, Mumbai. He has been an external expert for the WorldFish Centre, Malayasia, and FAO and has been an invited speaker at national and internal conferences. His area of expertise includes aquatic biodiversity management, aquaculture, genetics, and biotechnology. He has guided 29 Ph.D. students, published over 350 research papers, and authored several books in the field of aquaculture and fisheries. He is a fellow of the National Academy of Agricultural Sciences and the National Academy of Sciences.

Himanshu Pathak is the Secretary, Department of Agricultural Research and Education, Govt. of India, and Director General, Indian Council of Agricultural Research, New Delhi. He provides leadership to one of the largest agricultural research and development organizations of the world. Earlier, he served as Director of ICAR-National Rice Research Institute, Odisha, and Director of ICAR-National Institute of Abiotic Stress Management, Maharashtra, India. He has over three decades of experience in the areas of agricultural research, education, and extension. He has published about 250 research papers with an h-index of 76, i10-index of 238, and more than 22,700 citations. He is a fellow of Indian National Science Academy; National Academy of Sciences, India; National Academy of Agricultural Sciences; and Humboldt Foundation. He is also Rafi Ahmed Kidwai Awardee of the Indian Council of Agricultural Research. He is the President of the National Academy of Agricultural Sciences.

Contributors

R. C. Agrawal Indian Council of Agricultural Research (ICAR), New Delhi, India

Upali A. Amarsinghe International Water Management Institute, Colombo, Sri Lanka

K. C. Bansal Formerly at ICAR-National Bureau of Plant Genetic Resources, New Delhi, India

Rutwik Barmukh WA State Agricultural Biotechnology Centre, Centre for Crop and Food Innovation, Murdoch University, Murdoch, WA, Australia

Vinay Bhardwaj ICAR-National Research Centre on Seed Spices, Ajmer, Rajasthan, India

B. Venkatesh Bhat ICAR-Indian Institute of Millets Research, Hyderabad, India

Pratap S. Birthal ICAR-National Institute of Agricultural Economics and Policy Research, New Delhi, India

K. M. Bujarbaruah Assam Agricultural University, Jorhat, Assam, India

Kamal Malla Bujarbaruah ICAR, Assam Agricultural University, Jorhat, Assam, India

Prem Chand ICAR-National Institute of Agricultural Economics and Policy Research, New Delhi, India

Suresh Kumar Chaudhari Krishi Anushandan Bhawan-II, Pusa, New Delhi, India

Viswanathan Chinnusamy ICAR-Indian Agricultural Research Institute, New Delhi, India

Shaun J. Curtin Department of Agronomy and Plant Genetics, University of Minnesota, Saint Paul, MN, USA
Center for Plant Precision Genomics, University of Minnesota, St. Paul, MN, USA
Plant Science Research Unit, United States Department of Agriculture, Saint Paul, MN, USA

P. Das ICAR, New Delhi, India

Ndeye Ndack Diop Plant Production and Protection Division, Food and Agriculture Organization of the United Nations, Rome, Italy

Stefano Diulgheroff Plant Production and Protection Division, Food and Agriculture Organization of the United Nations, Rome, Italy

K. Elavarasan Fish Processing Division, ICAR-Central Institute of Fisheries Technology, Cochin, India

Bonnie Furman Plant Production and Protection Division, Food and Agriculture Organization of the United Nations, Rome, Italy

R. Hemalatha ICMR-National Institute of Nutrition, Hyderabad, Telangana, India

Jagadish B. Hiremath ICAR–National Institute of Veterinary Epidemiology and Disease Informatics, Bengaluru, Karnataka, India

Wilson Hugo Plant Production and Protection Division, Food and Agriculture Organization of the United Nations, Rome, Italy

Seema Jaggi Indian Council of Agricultural Research (ICAR), New Delhi, India

Michael G. K. Jones Western Australian State Agricultural Biotechnology Centre, Food Futures Institute, School of Agricultural Sciences, Murdoch University, Perth, WA, Australia

P. K. Joshi International Food Policy Research Institute, New Delhi, India

Chandra Bhushan Kumar ICAR-National Bureau of Fish Genetic Resources, Lucknow, India

M. K. Mahatma ICAR-National Research Centre on Seed Spices, Ajmer, Rajasthan, India

B. Mandal ICAR, New Delhi, India

Satendra K. Mangrauthia ICAR-Indian Institute of Rice Research, Hyderabad, India

Chikelu Mba Plant Production and Protection Division, Food and Agriculture Organization of the United Nations, Rome, Italy

Shawn McGuire Plant Production and Protection Division, Food and Agriculture Organization of the United Nations, Rome, Italy

Anupam Mishra Central Agricultural University, Imphal, Manipur, India

Abhijit Mitra Department of Animal Husbandry and Dairying (DAHD), Ministry of Fisheries, Animal Husbandry and Dairying, Government of India, New Delhi, India

Kutubuddin A. Molla ICAR-National Rice Research Institute, Cuttack, India

K. Madhavan Nair Former Scientist-F, ICMR-National Institute of Nutrition, Hyderabad, Telangana, India

Arshiya Noorani Plant Production and Protection Division, Food and Agriculture Organization of the United Nations, Rome, Italy

Lucio Olivero Plant Production and Protection Division, Food and Agriculture Organization of the United Nations, Rome, Italy

Ram Krishna Pal ICAR-National Research Centre on Pomegranate, Solapur, Maharashtra, India

Anutosh Paria ICAR-National Bureau of Fish Genetic Resources, Lucknow, India

Madhusha Perera International Water Management Institute, Colombo, Sri Lanka

Pravata Kumar Pradhan ICAR-National Bureau of Fish Genetic Resources, Lucknow, India

Habibar Rahman ICAR, New Delhi and Regional Representative for South Asia, International Livestock Research Institute (ILRI), South Asia Office, New Delhi, India

R. Ramakumar School of Development Studies, Tata Institute of Social Sciences, Mumbai, India

Ravindra ICAR-National Bureau of Fish Genetic Resources, Lucknow, India

C. N. Ravishankar ICAR- Central Institute of Fisheries Education, Mumbai, India

Uttam Kumar Sarkar ICAR-National Bureau of Fish Genetic Resources, Lucknow, India

C. Tara Satyavathi ICAR-Indian Institute of Millets Research, Hyderabad, India

S. N. Saxena ICAR-National Research Centre on Seed Spices, Ajmer, Rajasthan, India

Purushottam Sharma ICAR-National Institute of Agricultural Economics and Policy Research, New Delhi, India

Rishi Sharma Fisheries and Aquaculture Division, Food and Agriculture Organization of the United Nations Viale delle Terme di Caracalla, Rome, Italy

S. K. Sharma ICAR, New Delhi, India

Alok K. Sikka International Water Management Institute, India Regional Office, New Delhi, India

R. B. Singh FAO Regional Office for Asia and the Pacific, Bangkok, Thailand

Raj Kumar Singh One Health Support Unit, DAHD, GoI, New Delhi, India

Shoba Sivasankar Plant Breeding and Genetics Section, Joint FAO/IAEA Centre of Nuclear Techniques in Food and Agriculture, Department of Nuclear Sciences and Applications, International Atomic Energy Agency, Vienna, Austria

Neeraj Sood ICAR-National Bureau of Fish Genetic Resources, Lucknow, India

Robert M. Stupar Department of Agronomy and Plant Genetics, University of Minnesota, Saint Paul, MN, USA
Center for Plant Precision Genomics, University of Minnesota, St. Paul, MN, USA

Raman M. Sundaram ICAR-Indian Institute of Rice Research, Hyderabad, India

Philip Thornton Clim-Eat, Wageningen, The Netherlands

Amit Kumar Tripathy One Health Support Unit, DAHD, GoI, New Delhi, India

Rajeev K. Varshney WA State Agricultural Biotechnology Centre, Centre for Crop and Food Innovation, Murdoch University, Murdoch, WA, Australia

Kennady Vijayalakshmy International Livestock Research Institute (ILRI), South Asia Office, New Delhi, India

Innovations for Reconfiguring Food Systems

Philip Thornton

Abstract

Reconfiguring food systems to be more resilient, sustainable and equitable presents considerable challenges: food systems are extremely diverse, they are based on biological systems that provide many key functions other than food provision that need to be safeguarded, and large-scale disruption has the potential to be catastrophic. Innovation in several parts of the food system will be necessary, consisting of a mixture of small, incremental shifts to existing systems and practices as well as radical and disruptive changes. Many technological options are already being piloted or in the pipeline that may help to shift food system practices and behaviour in ways that can benefit food security, equity and the environment. Uptake of innovations at scale needs a range of enablers to be in place, such as new policies and regulatory frameworks, innovative funding mechanisms, monitoring and assessment to help avoid unintended consequences, and market incentives to help spread risk. We also need to be innovative in doing agricultural research for development and in monitoring the impacts of change on human and natural systems.

Keywords

Food system · Innovation · Technology · Accelerators · Equity

Clim-Eat, Agro Business Park 10, Office 2.03, 6708 PW Wageningen, The Netherlands. This paper is a shortened (and only lightly modified) version of Thornton P (2023), *Tweaks and disruptions: Reconfiguring food systems through innovation*. Discussion Starter 3. Utrecht, the Netherlands: Clim-Eat.

P. Thornton (✉)
Clim-Eat, Wageningen, The Netherlands
e-mail: pthornton@cgiar.org

1.1   Introduction

As currently configured, food systems pose considerable threats to our ability to operate within a safe planetary space. We are already struggling: the number of people who were moderately or severely food insecure in 2021 was some 2.3 billion (FAO 2022). Many recent reports address the question of how current food systems need to change in a wide range of different ways, but one common thread running through them all is the need for innovation—be it technical, financial, socio-political, economic, or institutional. Another is the recognition that there is not one great fix to cure all ills but rather it will take many different innovations addressing different parts of the food system; some may be radical and disruptive, others will be small, incremental shifts to existing systems and practices (Loboguerrero et al. 2020).

The reconfiguration of food systems poses big challenges. The food sector is different in several key respects compared with other sectors (Hall and Dijkman 2019). First, culture, society and social licence play a very large role, given the existential importance of food to everyone on the planet. Second, the sector is dependent on biological systems that have other key functions in addition to food provision, and regulation and safeguarding are of critical importance to maintaining their viability. Third, the sector provides livelihoods and food security directly and indirectly for large numbers of poor people, particularly in lower- and middle-income countries (LMICs), where the sector often has limited capacity to adapt. For these and other reasons, rapid, radical change in the food sector is extremely difficult.

This paper lays out a few highly promising technological innovations that may help to improve resilience to climate change, incomes and food security in LMICs. There is a long pipeline of promising innovations, some of which may take many years before they are taken up, if ever. Those described below are just a few of the many that are in the "moving to scale" stage of development.

1.2   Some High-Potential Near-Ready Technical Innovations

1.2.1   Green Ammonia

About 180 Mt of ammonia is produced every year, and of this total, 80% is used as fertiliser. The production of ammonia is a very energy-intensive process and results in nearly 2% of global carbon dioxide emissions. Ammonia can now be produced without using fossil fuels and using renewable energy sources. Several processes have been proposed (Ghavan et al. 2021). One method uses water electrolysis to provide a hydrogen supply, and nitrogen is extracted from the air. Nitrogen and hydrogen are combined to make fertiliser ammonia. Green ammonia may be an important carbon-free fuel for the future (Royal Society 2020). Europe's first commercial-scale green ammonia facility came online in 2022, and the industry is set for considerable expansion globally in the coming years. There is considerable potential for relatively small-scale green ammonia plants in LMICs (Sun et al. 2021). The supply of cheap, timely and locally-available nitrogen fertilizer is often constrained in many countries. Green ammonia technologies have real potential for overcoming these constraints, helping to raise crop yields and improving small-scale farmers' livelihoods.

1.2.2 Protein Fermentation

Several single-cell proteins are already being used as human food, such as mycoproteins (fungus-based), with others in the pipeline. Many have substantial benefits for the climate and the environment, compared with protein from some livestock species and systems. There is considerable interest in single-cell proteins made by extracting atmospheric carbon dioxide and then combining it with water, nutrients and vitamins. This protein fermentation process can be done inside a bioreactor using electricity generated from renewable sources to convert water to hydrogen gas. Hydrogen-oxidizing bacteria are then used to produce a powder that is about 65% protein, similar to soy protein, and can be added to a wide range of food products. In October 2022, the Singapore Food Agency approved the use of one such protein for human consumption, with commercial production due to start in 2024. The production and consumption of such single-cell protein at scale could have a large (though currently uncertain) impact on food production, reducing the environmental footprint of agricultural production by freeing up land, reducing water use and increasing landscape biodiversity (Newman et al. 2023). There could be substantial health, nutrition and livelihood benefits for small-scale producers too (Rastogi et al. 2022), via protein fortification of cereals and other value additions.

1.2.3 Edible and Biodegradable Food Packaging

There are various options for environmentally sustainable forms of packaging, to promote food quality and safety as well as to prolong storage and shelf life, including edible packaging and biodegradable packaging. Edible packaging typically consists of biodegradable material that is utilized by coating or wrapping food, generating no waste. Many different types of materials can add value to packaged foods (Petkoska et al. 2021). There is considerable potential to increase the functionality of edible packaging via nanotechnology. Similarly, there are many options for biodegradable packaging materials. Some materials are already on the market: Notpla in the UK make and market packaging material from brown seaweed and plants to create edible packaging (sauce sachets, food boxes) which is completely biodegradable. Apeel Science in the USA makes a tasteless, edible coating product made from plant material that can double the shelf life of fruit. There is only limited literature on the potential application of edible and biodegradable food packaging in LMICs, though consumer awareness may be a challenge (Agossou et al. 2021).

1.2.4   Low-Tech Urban Food Production

About 70% of the world's population will be urbanised by 2050. Already, 20–25% of Nairobi's milk and meat is supplied by livestock kept in the city, and 40–50% of Kolkata's vegetables are urban produced. Data are scarce on the extent and productivity of urban agriculture, but its expansion in some situations has considerable potential to improve local food security while reducing greenhouse gas emissions (Pye-Smith et al. 2022), as well as contributing to increasing the resilience of cities (Drescher et al. 2021). In higher-income countries, urban agriculture, including high-tech Controlled Environment Agriculture (CEA), is currently attracting major investment. Further expansion of this sector seems likely. In LMICs, urban crop, animal and fish production often takes place in informal settings, using traditional methods of production, and receives little if any financial or policy support. CEA in LMICs does not have to be technologically advanced: utilising renewable energy sources to power high-efficiency LED lights, recycling of water and nutrients, using greenhouses rather than vertical structures, for example. Currently there is little policy support for urban agriculture in LMICs; providing secure land tenure, turning vacant spaces into areas that can be used for food production, and providing would-be urban farmers with grants and low-interest loans as well as with the incentives to turn organic waste into fertiliser, could facilitate uptake (Pye-Smith et al. 2022).

1.2.5   Off-Grid Renewable Energy Storage

Access to reliable energy sources in the more remote rural areas in many in LMICs continues to be a significant challenge. To meet climate and sustainable energy goals, nearly 10,000 GWh of energy storage will be required worldwide by 2040; this is 50 times the size of the present market (International Energy Agency (IEA) 2020). Currently, the energy storage market is dominated by Li-ion battery technologies, mostly for short-duration storage. There are technologies that may offer considerable potential for longer-term energy storage at local scales in areas where the electricity grid is weak or non-existent. These include different methods of thermal storage, allowing renewable energy to be stored and used when needed on-farm, to power micro-irrigation pumps and small evaporative cooling units for storing fruit and vegetables to reduce spoilage rates, for example. An example of a potentially high-impact thermal storage technology is a "sand battery", consisting of low-grade sand which is heated by renewable energy power to 500–600 °C. The energy can be stored for several months and used as heat or converted to electricity as needed. While chemical batteries can store 5–10 times more energy per unit volume than a sand battery, costs per MWh using the sand battery may be 8–10 times lower (https://polarnightenergy.fi/sand-battery). Feasibility studies on the use of sand battery technology in the context of LMIC agriculture are needed.

1.2.6 Farm-to-Fork Virtual Marketplaces

Small-scale producers often receive only a small fraction of the price consumers pay for their food, because of small volumes, long value chains and lack of marketing information. Farm-to-fork virtual marketplaces offer small-scale producers the opportunity to become connected with a wide range of food purchasers—consumers, restaurants, hotels—using online- or smartphone-based platforms that link suppliers with end-users directly (Farley et al. 2017). The benefits include more efficient and timely supply chains, increased producer profits, and decreased food loss and waste. Farm-to-fork virtual marketplaces also offer the ability to increase the traceability and accountability of food production, providing information with which food producers and consumers can make decisions. There are several examples of functioning farm-to-fork virtual marketplaces in LMICs (Ezeomah and Duncombe 2019). There are enormous opportunities for these virtual models in emerging markets, where populations are both increasing and urbanizing and incomes rising.

1.3 Facilitating and Accelerating Change

Herrero et al. (2020) proposed a set of interconnected action points to accelerate technological change and systemic innovation in food systems. These include putting in place a stable financial environment to promote innovation that addresses social and environmental objectives; designing market incentives to help spread the costs and risks associated with innovation, and to facilitate the competitiveness of new approaches; changing policies and regulations to ensure appropriate oversight of new technologies and industries while reducing bureaucratic constraints to innovation; and safeguarding against undesirable effects through monitoring impacts of innovation and making corrections where needed. For many innovations, transition pathways will be needed: plans that set out in some detail how innovations will be implemented in ways that ensure political viability of the changes being considered and their financial viability for all the food system actors affected by the change. Capacity development activities may also be important, both of the people and the institutions involved in what may be a series of shifts in food systems.

The need for food system reconfiguration has sparked new thinking about how public agricultural research for development (A4RD) could become more effective and efficient. The AR4D agenda is now more wide-ranging and complex than it was 50 years ago, involving multi- and transdisciplinary science approaches as well as taking account of cross-sectoral interests such as agriculture, food, health, energy and infrastructure. New and broadened approaches imply new tools and methods, new partnerships, new funding and incentive arrangements, and new ways to frame innovation processes themselves. This could extend to designing and piloting new ways of doing business that are more collective, decentralised and based on shared visions of what the future should look like (Körner et al. 2022).

In any case, it will be necessary to monitor future food systems for supporting decision makers at multiple scales, so that course corrections can be made where needed, as well as for accountability purposes (Fanzo et al. 2021). The risk of worsening inequity abounds when considering low-carbon innovations, as they are not automatically just or equitable (Sovacool et al. 2022). Various initiatives are underway to address this issue. For example, the Food Systems Dashboard (https://www.foodsystemsdashboard.org/) and the Innovative Food System Solutions (IFSS) portal (https://ifssportal.nutritionconnect.org/) are two initiatives that bring together many resources, tools and datasets for comparing and prioritising food system actions. Another example is the database ERA (Evidence for Resilient Agriculture, https://era.ccafs.cgiar.org/), a large collection of published information on the performance of agricultural technologies in SSA suitable for many development decisions. Efforts such as these need to be considerably expanded, so that in time it will be possible to use consistent and comparable data to monitor performance of the world's food systems (Campbell et al. 2022).

A key challenge in implementing change revolves around the design of feasible pathways to move from current to future food systems in ways that are consistent with sustainable development, poverty reduction, increased equity and increased food security. There is a growing literature on the desirable characteristics of future food systems, but analyses that address the pathways towards these desired futures are far fewer in number (Kerr et al. 2022). For example, Gerten et al. (2020) estimated that 10.2 billion people could be supported, while not exceeding planetary boundaries, by altering the places where particular crops are grown and facilitating dietary changes, among other actions (Gerten et al. 2020). But the plausible pathways for moving towards such a future in ways that are socially, economically, politically and environmentally acceptable through time are largely unknown. The difficulty of identifying and operationalising (let alone monitoring and evaluating) transition pathways at almost any scale appears to be a major bottleneck. Integrated modelling of transition pathways will be needed for improved understanding of the synergies and trade-offs that may arise in the future, and on the ways in which these may change through time. Considerable research work needed is in this area.

1.4 Conclusions

Reconfiguring food systems presents many challenges. Some of the many promising technological options may result in small tweaks to different parts of the food system, while others may be highly disruptive. At the same time, scaling up the changes needed will require a strong enabling environment. Equally important, we need better understanding of the synergies and trade-offs that may result from multiple innovations in the food system and in other sectors. This information is critical to help guide investment and decision-making in the quest to make food systems more resilient, sustainable and equitable.

References

Agossou PN et al (2021) Substitution of non-biodegradable plastic food packagings by ecological food packagings at Abomey-Calavi University (Benin): state of place and microbiological quality of packagings. Afr J Microbiol Res 15(6):295–303

Campbell BM et al (2022) Food system adaptation: upping our ambition. Nat Food. https://www.nature.com/articles/s43016-022-00656-y

Drescher AW et al (2021) Urban and peri-urban agriculture in the Global South. In: Urban ecology in the global south. Springer, Cham, pp 293–324

Ezeomah B, Duncombe R (2019) The role of digital platforms in disrupting agricultural value chains in developing countries. In: International conference on social implications of computers in developing countries. Springer, Cham, pp 231–247

Fanzo J et al (2021) Rigorous monitoring is necessary to guide food system transformation in the countdown to the 2030 global goals. Food Policy 104:102163

FAO (2022) The state of food security and nutrition in the world 2022. Repurposing food and agricultural policies to make healthy diets more affordable. FAO, Rome. https://doi.org/10.4060/cc0639en

Farley S et al (2017) Innovating the future of food systems: a global scan for the innovations needed to transform food systems in emerging markets by 2035. Global Knowledge Initiative, Washington. http://globalknowledgeinitiative.org/initiative/innovating-the-future-of-food-systems/

Gerten D et al (2020) Feeding ten billion people is possible within four terrestrial planetary boundaries. Nat Sustainability 3:200–208. https://doi.org/10.1038/s41893-019-0465-1

Ghavan S et al (2021) Sustainable ammonia production processes. Front Energy Res 9:580808

Hall A, Dijkman J (2019) Public agricultural research and development in an era of transformation independent science council of the CGIAR and Commonwealth Scientific and Industrial Research Organisation

Herrero M et al (2020) Innovation can accelerate the transition towards a sustainable food system. Nat Food 1:266–272. https://doi.org/10.1038/s43016-020-0074-1

International Energy Agency (IEA) (2020) Innovation in batteries and electricity storage: a global analysis based on patent data. https://www.iea.org/reports/innovation-in-batteries-and-electricity-storage

Kerr RB et al (2022) Chapter 5: food, fibre and other ecosystem products. In: Climate change 2022: Impacts, adaptation and vulnerability. https://www.ipcc.ch/report/ar6/wg2/

Körner J et al (2022) Perspective: how to swarm? Organizing for sustainable and equitable food systems transformation in a time of crisis. Glob Food Sec 33:100629. https://doi.org/10.1016/j.gfs.2022.100629

Loboguerrero AM et al (2020) Actions to reconfigure food systems. Glob Food Sec 26:100432. https://doi.org/10.1016/j.gfs.2020.100432

Newman L et al (2023) Cellular agriculture and the sustainable development goals. In: Genomics and the global bioeconomy. Academic, New York, pp 3–23

Petkoska AT et al (2021) Edible packaging: sustainable solutions and novel trends in food packaging. Food Res Int 140:109981

Pye-Smith C et al (2022) The future for urban agriculture. Wageningen, Clim-Eat

Rastogi YR et al (2022) Food fermentation–significance to public health and sustainability challenges of modern diet and food systems. Int J Food Microbiol 371:109666

Royal Society (2020) Ammonia: zero-carbon fertiliser, fuel and energy store. Policy brief. https://royalsociety.org/green-ammonia

Sovacool BK et al (2022) Equity, technological innovation and sustainable behaviour in a low-carbon future. Nat Hum Behav 6(3):326–337

Sun J et al (2021) A hybrid plasma electrocatalytic process for sustainable ammonia production. Energy Environ Sci 14(2):865–872

Transforming Food Systems for Higher, Sustainable, and Inclusive Agricultural Growth: Role of Policies and Institutions

2

Pratap S. Birthal, Purushottam Sharma, and Prem Chand

Abstract

Driven by the sustained income growth, increasing urbanization, changing lifestyles, and increasing participation of women in workforce, the dietary patterns in India have undergone a significant transformation in favor of high-value food commodities. This transformation is also reflected in the production portfolio at the upstream—the high-value food commodities have consolidated their share of agricultural growth, from 51% in the 1980s to over 87% in the 2010s. The factors underlying these changes in the food system have been quite robust in the recent past, and are unlikely to subside in the near future, implying an acceleration in its transformation in the future plausible socio-economic scenario. Sustaining food system transformation will require significant institutional and policy support, in terms of investment in markets, storage, food processing, and food safety and traceability systems; and information and credit support for managing the production and price risks, besides reforming the existing policies that are biased towards staple cereals. The absence of such a support may deprive the smallholder farmers, who are more engaged in high-value food production, of the benefits of the food system transformation.

Keywords

High-value food commodities · Agricultural growth · Inclusiveness · Policies · Institutions · Infrastructure

P. S. Birthal (✉) · P. Sharma · P. Chand
ICAR-National Institute of Agricultural Economics and Policy Research, New Delhi, India

2.1   Introduction

India's food system has been undergoing a gradual transformation from plate to plough. Owing to the sustained rise in per capita income, changing lifestyles, and increasing consumer preferences for the nutrient-rich foods, the consumption basket has been diversifying away from the staple cereals towards the high-value food commodities such as fruits, vegetables, spices, herbal and medicinal products, milk, meat and eggs. The shift in the food basket is likely to be more prominent in future, propelling a disproportionate growth in their demand. By 2050, India's population will cross the 1.6 billion mark, with half of it living in cities and towns. The current economic growth of around 7% is unlikely to subside, making the people more affluent than at present. According to Hamshere et al. (2014), India's demand for horticultural and animal-source foods by 2050 will be more than double of that a decade ago, i.e. in 2009.

These changes at downstream of the food system create an opportunity for farmers to diversify their product portfolio in favor of more remunerative high-value crops and animal agriculture, and for small-scale entrepreneurs to invest in food processing at local level. The smallholder farmers, who comprise bulk of the farming population, are likely to benefit more from such a transformation. The high-value agriculture, including horticultural crops and animal husbandry, is labor-intensive and generates a continuous flow of outputs. Note, the smallholders are labour surplus but capital scarce, and these characteristics of high-value agriculture match the resource endowments and cash flow requirements of the smallholders (Birthal et al. 2015). Further, the evidence shows an inverse farm size-productivity relationship for high-value crops than for staple food crops (Birthal et al. 2014); hence, the diversification of product portfolio in favour of such crops has a larger poverty-reducing effect than the widely-grown staple cereal crops (Birthal and Negi 2012; Birthal et al. 2015).

However, there is an apprehension that the speed of food system transformation, consequently its economic and social benefits and their distribution may remain shadowed in the absence of institutional and policy support. This chapter looks into some of the issues critical to the food system transformation. The next section assesses the contribution of high-value crops and animal agriculture to the growth of agricultural sector. Section 2.3 provides for the infrastructural and institutional requirements for food system transformation. A policy perspective on food system transformation is provided in Sect. 2.4. Concluding remarks are made in the last section.

2.2   Contribution of High-Value Food Commodities
        to Agricultural Growth

Table 2.1 presents the decadal growths in the value of output of different food commodities from 1980–1981 onwards. Since then, the agricultural sector has grown at an annual rate of 3%. At disaggregated level, the growth for the staple

Table 2.1 % annual growth in agricultural value of output (at 2011–2012 prices)

Commodities	1980s	1990s	2000s	2010s	Overall
Cereals	2.89	2.25	1.72	1.62	1.87
Pulses	1.54	0.83	1.94	3.95	1.54
Fruits and vegetables	2.27	6.00	3.47	4.03	4.19
Species	4.71	5.11	3.44	6.46	4.41
Milk and milk products	5.47	4.28	3.73	5.64	4.36
Meat	6.62	3.10	4.96	8.09	5.06
Egg	7.92	4.12	5.80	5.50	5.23
Food commodities	3.54	3.59	3.04	4.34	3.38
Non-food commodities	1.63	2.60	2.43	0.11	2.19
Agriculture	2.84	3.26	2.85	3.13	3.00

Source: Estimated by authors using data from the National Accounts Statistics, Government of India

Table 2.2 Changes in product portfolio (% of the value of agricultural output)

Commodities	TE 1980–1981	TE 1990–1991	TE 2000–2001	TE 2010–2011	TE 2019–2020
Cereals	26.48	26.13	23.52	20.31	17.76
Pulses	5.15	4.75	3.51	3.14	3.45
Fruits and vegetables	13.18	12.12	15.23	16.62	18.23
Spices	2.03	2.14	2.49	2.67	3.37
Milk and milk products	12.82	15.35	17.61	19.87	23.19
Meat	3.11	3.86	4.13	5.23	7.45
Egg	0.45	0.66	0.75	0.99	1.17
Food commodities	63.21	65.00	67.23	68.83	74.62
Non-food commodities	36.79	35.00	32.77	31.17	25.38

Source: As for Table 2.1

cereals has been extremely sluggish, but higher for the food commodities in general, and much higher for the high-value food commodities. In the high-value segment, the animal-source foods have experienced faster growth than the plant-based high-value foods, viz., fruits, vegetables and spices. Nevertheless, there is no definite trend for different commodity groups, except cereals, which experienced a continuous deceleration in their growth, from 2.9% in the 1980s to 1.6% in the 2010s.

The differential rates of growth for different commodities have significantly altered the production portfolio at the upstream (Table 2.2). The food commodities, as a group, consolidated their share in the gross value of agricultural output, from 63% in TE[1] 1980–1981 to 75% in TE 2019–2020. And, this occurred due to a

---

[1] TE stands for triennium ending average.

Table 2.3 % share of different commodities in agricultural growth

Food commodities	1980s	1990s	2000s	2010s	Overall
Cereals	26.05	17.10	12.84	9.58	13.48
Pulses	2.65	1.02	2.16	4.12	1.87
Fruits and vegetables	10.24	25.00	19.32	23.11	21.93
Species	3.36	3.54	3.23	6.18	3.84
Milk and milk products	27.96	22.05	24.83	38.60	27.29
Meat	8.11	4.00	8.26	16.65	8.56
Egg	1.61	0.92	1.80	1.90	1.53
Food commodities	79.64	73.21	72.11	99.46	77.05
Non-food commodities	20.77	26.84	27.68	1.04	23.09

Source: Estimated by authors using information from Table 2.1 and 2.2

significant improvement in the share of high-value food commodities, which increased from 32% in TE 1980–1981 to 52% in TE 2019–2020. Amongst the high-value food commodities, the share of animal-source foods increased faster than that of the plant-based high-value foods. On the other hand, the share of foodgrains (i.e., cereals and pulses) has fallen to 21% in TE 2019–2020 from 32% in TE 1980–1981.

The high-value food commodities have emerged as engine of agricultural growth (Table 2.3). Their share in the overall growth increased from 51.3% in the 1980s to 86.5% in the 2010s. By commodity, the contribution of animal-source foods increased from 37.7% in the 1980s to 57.2 % in the 2010s, and of the plant-based high-value foods from 13.6% to 29.3%. The foodgrains lost their share by more than half, from 28.7% in the 1980s to 13.7% in the 2010s.

2.3   Status of Markets, Infrastructure, and Institutions

Most of the high-value food commodities are perishable, and have high price volatility. Hence, these need to be sold or stored or processed into less perishable forms immediately. Further, some of the high-value commodities, such as fruits and plantations, require significantly large start-up capital. Note, the smallholder farmers, who are more engaged in production of these commodities (Birthal et al. 2014), face an acute liquidity constraint in scaling up high-value agriculture. Besides, their small-scale production limits their access to credit and remunerative markets. The marketed surplus is too small to be remuneratively traded in the distant urban markets because of the higher transaction costs (Birthal et al. 2005). This also affects their bargaining power vis-à-vis buyers. Hence, a majority of smallholders are compelled to sell their surplus produce to local traders, the price realization from whom is significantly less than from the traders in the organized markets. Hence, the transformation of high-value food system at the upstream requires different kinds of infrastructure, institutions and policies. Below we discuss the current status of the markets, infrastructure and institutions.

Markets Markets act as a catalyst in the process of food system transformation. By aggregating demand and supply they create incentives for farmers to choose producing commodities which are more remunerative. The current agri-food marketing systems and value chains in India are, however, underdeveloped and inadequate to handle the huge agricultural surpluses of high-value food commodities (Birthal et al. 2019). There are 6946 regulated markets in the country, each covering an average of area of 473 km^2. Their coverage is also uneven across states, ranging from 61 km^2 in Puducherry to 11,215 km^2 in Meghalaya. To provide farmers an adequate access to markets, the country needs about 41,000 regulated markets.

To create a national market for crops and bring transparency in market transactions, the Government of India started an electronic platform known as the National Agriculture Market or e-NAM to link the APMC markets across the country. Currently, the e-NAM has 1361 *mandis* on its platform across 27 states and union territories, facilitating transactions in 209 commodities. A total of 176 lakh farmers and 2749 FPOs are registered on the e-NAM platform. Interestingly, the trade in high-value food commodities surpasses the trade in other commodities through the e-NAM platform.

Farmer Producer Organisations (FPOs) Farmer producer organizations are one of the effective means of empowering smallholder farmers in the market place. These help improve bargaining power of farmers' vis-à-vis buyers, and help them realize better prices, reduce transaction costs, and increase local processing, besides helping them in efficient allocation of land and other resource and in adopting improved technologies, quality inputs, and food safety standards. Currently, there are 7059 FPOs in the country engaged in different value chain activities across a range of commodities. The horticultural and foodgrain crops, however, are their main focus, and the aggregation of farm produce for onward sales, and the provision of inputs and services to member-farmers are their main business activities. However, several of the FPOs are not active because of their poor paid up capital and governance, and access to finance. Nevertheless, some of the FPOs, for example, the Vasundhara Agriculture Horticulture Producer Company Ltd (VAPCOL), the Sahyadri Farmer Producer Company (SFPC) and the Abhinav Farm Club (AFC) have effectively integrated farmers on their value chains. However, the FPOs are regionally concentrated mainly in Karnataka, Maharashtra, Madhya Pradesh, Telangana, Uttar Pradesh, Tamil Nadu and West Bengal.

Cooperatives Cooperatives represent the collectives of farmers organized by themselves to improve their access to markets for both outputs and inputs, and also agricultural services. In India, cooperatives have been successful in dairying and sugarcane. The Gujarat Cooperative Milk Marketing Federation (AMUL) has established itself as the largest food company in India. The Mother Dairy Fruits and Vegetables Ltd., a subsidiary of the National Dairy Development Board, procures fruits and vegetables from farmers through the informal farmers' associations, and sells the procured produce after grading through its own retail outlets in Delhi. It also exports processed horticultural products under the brand name SAFAL. The MAHAGRAPES in Maharashtra and the HOPCOMS in Karnataka are other successful cooperatives in horticulture (Birthal et al. 2007). Such successful models of cooperatives need to be scaled up on a wider scale. The dairy cooperatives are concentrated in Gujarat, Maharashtra, Tamil Nadu and Karnataka, and need to be established or strengthened in other states, especially in the eastern and north-eastern regions.

Contract Farming Contract farming is another important institution to link farmers to markets, while reducing their production and market risks and the transaction costs associated with the acquisition of inputs and services, and the disposal of farm produce. There are several models of contract farming led by farmers and buyers including the organized retailers, processors and exporters, and facilitated by the governments and non-government organizations (Chen et al. 2015). Contract farming is more prominent in high-value food crops, dairying, and poultry. The contract farming, however, is criticized for its monopsonistic tendency, and exclusion of smallholders.

Agri Start-Ups Recently, a number of start-ups have come up in agriculture to connect farmers to markets and provide services through digital platforms. Some of these start-ups procure farm produce directly from the farmers and provide online platforms for its sale.

Direct Marketing To empower farmers in the market place, some state governments have established farmers' markets, a place where the farmers can sell their produce directly to the consumers. The *Apni mandi* in Punjab and Haryana, the *Rythu Bazaar* in Telangana, the *Uzhavar Sandhais* in Tamil Nadu, the *Krushak Bazaar* in Odisha, the *Shetkari Bazaar* in Maharashtra, and the *Raitha Santhe* in Karnataka are the examples. There are 488 such markets in different states. Under the Scheme for the Development of Agricultural Marketing Infrastructure, Grading and Standardization' of the central government 22,941 rural periodic markets/haats have been upgraded as the Gramin Agricultural Markets (GrAMS).

Another variant of the direct marketing is the procurement of produce directly from farmers' fields by the licensed processors, exporters, and retailers. A few states have issued licenses for the direct sourcing—Maharashtra tops the list with 219 licenses, followed by Karnataka (37), Himachal Pradesh (12), Punjab (11), Jammu & Kashmir, Telangana, Rajasthan and Gujarat (3 each), Andhra Pradesh (2), and Madhya Pradesh (1).

Cold Storage and Other Infrastructures Cold storages are essential for high-value food supply chains. At present, India has the cold storage capacity of 38 million tons, as against the production of 107 million tons of fruits, 205 million tons of vegetables, 221 million tons of milk, and over 9 million tons of meat. Notably, about 70% of the total cold storage capacity is concentrated in Gujarat, Punjab, Andhra Pradesh, Uttar Pradesh, Bihar, Madhya Pradesh and Maharashtra. Further, 68% of the cold storage capacity is used for the potatoes, and the rest for the other commodities, like meat, seafood, dairy products, fruits and vegetables, and pharmaceuticals. The situation regarding the refrigerated transport, pack houses, pre-cooling units, and ripening chambers is equally dismal.

2.4 Enabling Policies

The major challenges in the process of the transformation of food system as a high-value food system are the high price volatility, poor access to markets, information and credit, lack of grades and standards, poor compliance with food safety standards, and large post-harvest losses. Some of the policy requirements for food system transformation are discussed below.

Correct Policy Biases India's agri-food policy remains cereal-centric, causing distortions in the cropping pattern. The procurement of rice and wheat at minimum support prices, although is beneficial for the farmers, it acts as a disincentive for the diversification towards high-value crops that are more prone to higher production and market risks compared to the rice and wheat (Negi et al. 2021). Thus, there is a need to evolve a crop-neutral price policy, offering farmers choices of crops. Note, the farmers realize more profit from the cultivation of rice and wheat than their competing crops. To wean farmers away from cultivation of rice and wheat, the price policy reforms must be accompanied by a compensation for the loss from the cultivation of these crops.

Follow Commodity Cluster Approach Several high-value commodities have their own production niches and offer significant growth opportunities. It is, therefore, essential to identify their production regions, assess the constraints on their production, and adopt a cluster-based approach to promote them. Government of India has recently come up with a clustered-based approach for several high-value food commodities in their niche production regions. Nonetheless, the need to create an environment for the private investment in value chains, and to promote market competition to discourage tendencies of regional monopsony and collusive oligopsony cannot be undermined.

Invest in Public Infrastructure to Induce Private Investment A business firm's decision to invest in cold storage, refrigerated transport, value chains and agro-processing is guided by the availability of the supporting infrastructure, i.e., roads, electricity, communication network, and appropriate regulations and incentives. A continuous improvement in the public infrastructure is essential to foster the backward and forward linkages of high-value food commodities.

Develop Grades and Standards The price and the quality of the commodities are the two important parameters for developing successful value chains. Their absence can lead to the opportunistic behaviour. Often, the organized retailers, exporters and processors have their own grades and standards. Based on these, they can reject the produce or pay less price for it. The need is to develop the grades and standards for farm produce, and ensure their compliance by the buyers.

Manage Price Volatility The prices of high-value food commodities are more volatile compared to that of other agricultural commodities. High volatility in prices creates disincentives for farmers to allocate more resources for high value commodities despite their production being more remunerative than their competing crops. Government of India has been implementing a Market Intervention Scheme (MIS) for the horticultural commodities to protect farmers from volatile prices. Given the significant growth prospects for fruits and vegetables in the plausible economic scenario, the need for price assurance for these crops cannot be undermined.

Promote Value Chains Government should facilitate development of value chains that are considered to provide solutions to many of the constraints that farmers face. The agribusiness firms provide an assured market for the produce and the quality inputs and services, and therefore, reduce the market and price risks, and transaction costs. The organized value chains in India, however, are underdeveloped for most food commodities, except for poultry and to some extent for milk.

Contract farming often suffers from low compliance rates (Chen et al. 2015). Farmers may engage in side-selling of produce in case its market price is higher than offer price or the sponsors may not procure produce in case the offer price at the time of procurement is higher than the market price. There is a need to identify causes for the poor compliance, and accordingly evolve mechanisms to overcome the opportunistic behaviour of sellers and buyers. This also suggest the need for evolving the cost and time efficient mechanisms for dispute settlement.

Promote Agri-Start-Ups The government should aggressively promote agri-tech start-ups to provide technological and innovative solutions to all stakeholders on the supply chain. This involves assessment of the soil and water health, and provision of climate advisories, quality testing and post-harvest handling services, to bring the primary processing closer to the farm-gate, and compel the value chain participants to comply with the domestic and international food safety standards.

Encourage Collective Organizations The government and non-government organizations should encourage and facilitate formation of producers' organizations such as cooperative or self-help groups or producer companies to aggregate farm produce for its sale, and to source and deliver inputs and services to their farmer-members. Since a majority of the smallholders are liquidity constrained, there is strong need to support their collectives financially, and in capacity building in market intelligence, food safety measures, branding, compliance procedures, and conflict management to make them self-sustainable. More importantly, such collectives should be encouraged to engage in processing at the local level.

Improve Farmers' Capacity to Invest Two important factors in scaling up organized value chains include the finance and insurance. A majority of farmers lack finance for investment in production of high-value long-gestation crops and animal farming. Besides, their risk-taking capacity is also low. Thus, they need financial support in terms of easy access to institutional credit and insurance. Their financial requirements are although small, they face discrimination in the rural credit market because of the high transaction costs and lending risks associated with small loans (Birthal et al. 2017). They often lack of collateral, which is acceptable to the financial institutions against the loans. The financial institutions should appreciate the commodity market orientation of the value chain as collateral but considering its competitiveness and long-term sustainability. For this, the financial institutions must build up their capacity in identifying the potential risks, market opportunities and challenges, and develop financial products suited to different chain actors, including the farmers and small-scale entrepreneurs.

Empower Farmers to Cope with Risks The outreach of the crop insurance remains limited to about 35% of the farmers despite the nation-wide implementation of the Pradhan Mantri Fasal Bima Yojana (PMFBY). Another way of coping with risk is to promote micro-irrigation, which at present is practiced on only about 15% of the cropped area, and is concentrated in a few states. The solar-powered micro-irrigation can ensure sustainable supply of irrigation water at a lower cost. Market risks are also high in perishables. The farmer producer organizations should be encouraged to have a dedicated risk management fund created out of the profits they earned to provide financial support to farmers during the crisis.

Improve Food Safety Compliance for Exports The Agricultural Export Policy 2018 focuses on improving value chain infrastructure to improve the competitiveness and food safety compliance to boost agricultural exports. However, many a times, export consignments, especially of high-value food commodities, are rejected because of the presence of pesticides and drug residues, microbial contamination, inappropriate packaging, etc. Export potential of many horticultural products (e.g., pineapple, fresh melon, orange, desiccated coconut, and dried grapes), thus, remains underexploited due to the lack of the compliance with standards of importing countries. The adoption of good agricultural practices (GAP), good manufacturing practices (GMP), and good handling practices (GHP) along with the investment in storage and logistics will help reduce rejections and improve exports of agricultural commodities. Block chain technology and Internet of Things (IoTs) can aid in improving the traceability and transparency from farm gate to end consumers.

2.5 Conclusions

Driven by their increasing demand, the high-value food commodities have emerged the engine of agricultural growth in India. The high-value agriculture is more practiced by smallholders, and exhibits a negative size-productivity relationship;

hence, the growth in high-value agriculture has a greater role in improving smallholder farmers' income, reducing poverty, and combating malnutrition.

Smallholder farmers, however, face several constraints in transition to high-value agriculture. Although, there are prospects for transformation of food system as high-value food system, its scaling up is conditional upon the availability of appropriate infrastructure, and institutional, regulatory and policy support. The policies must target market integration, facilitating evolution of value chains and their governance, developing commodity-specific grades and standards and procedures for their compliance, and attracting private investment in food processing.

References

Birthal PS, Negi DS (2012) Livestock for higher, sustainable and inclusive growth. Econ Polit Wkly 47(26-27):89–99

Birthal PS, Joshi PK, Gulati A (2005) Vertical coordination in high-value food commodities. MTID Discussion Paper 85. IFPRI, Washington

Birthal PS, Jha AK, Singh H (2007) High-value agriculture and linking farmers to markets. Agric Econ Res Rev 20:425–440

Birthal PS, Joshi PK, Negi DS, Agarwal S (2014) Changing sources of growth in Indian agriculture: implications for regional priorities for accelerating agricultural growth. IFPRI Discussion Paper 1325, IFPRI, Washington, DC

Birthal PS, Roy D, Negi DS (2015) Assessing the impact of crop diversification on farm poverty. World Dev 72:70–92

Birthal PS, Chand R, Joshi PK et al (2017) Formal versus informal: Efficiency, inclusiveness and financing of dairy value chains in Indian Punjab. J Rural Stud 54:288–303

Birthal PS, Jaweriah H, Negi DS (2019) Diversification in Indian agriculture towards high value crops: multilevel determinants and policy implications. Land Use Policy 91(C):104427. https://doi.org/10.1016/j.landusepol.2019.104427

Chen K, Joshi PK, Cheng E, Birthal PS (2015) Innovations in financing of agri-food value chains in China and India: lessons and policies for inclusive financing. China Agric Econ Rev 7(4):1–27

Hamshere P, Sheng Y, Moir B, Gunning-Trant C, Mobsby D (2014) What India wants: analysis of India's food demand to 2050. Report No. 14.16, ABARES, Canberra

Negi DS, Birthal PS, Roy D, Hazrana J (2021) Crop choices in Indian agriculture: role of market access and price policy. Econ Bull 41(4):2249–2256

Improving Water Productivity for Transforming Agri-food Systems

Alok K. Sikka, Upali A. Amarsinghe, and Madhusha Perera

Abstract

This paper finds that many rainfed districts in India, defined as those with low access to irrigation, have the potential to harvest water and improve water productivity, where supplementary irrigation in critical periods of crop growth would increase yield and physical water productivity (PWP). There is potential for irrigated areas to increase PWP, too; the excess surface runoff under normal to moderate drought conditions, harvested in ponds and tanks, could reduce water stress in critical periods of crop growth. Rainfed and irrigated areas with water stress should consider diversifying to high-value drought-tolerant crops to increase the economic water productivity, income under normal climatic conditions, and resilience under drought conditions.

Keywords

Physical water productivity · Economic water productivity · Yield · Consumptive water use · Rainfed districts · Irrigated districts

3.1 Introduction and Context Setting

Growing water scarcity and insecurity, with the substantial increase in global water withdrawals in the last century, is threatening the food security of many countries. Water scarcity due to over-exploitation and poor water management (physical water

A. K. Sikka (✉)
International Water Management Institute, India Regional Office, New Delhi, India
e-mail: a.sikka@cgiar.org

U. A. Amarsinghe · M. Perera
International Water Management Institute, Colombo, Sri Lanka

© National Academy of Agricultural Sciences, under exclusive license to Springer Nature Singapore Pte Ltd. 2023
K. C. Bansal et al. (eds.), *Transformation of Agri-Food Systems*,
https://doi.org/10.1007/978-981-99-8014-7_3

scarcity) or due to lack of development (economic water scarcity) affects most countries in South Asia (Molden 2013). With agriculture presently accounting for 70% of freshwater consumption worldwide and exceeding 90% in many developing nations, and with rising water demands from other sectors, the agri-food system faces escalating vulnerabilities linked to water availability, as it strives to sustain the expanding global population by 2050. The looming climate change is likely to exacerbate these preexisting water challenges, especially in regions already facing water scarcity (Iglesias and Garrote 2015).

The Shared Socioeconomic Pathways (SSPs) project a substantial increase in water withdrawals by 2050, projecting an increase of 55–113% compared to current levels (Wada et al. 2016). While the irrigation water demand is expected to rise but at the same time more diversion of freshwater resources from agriculture to fulfil the increasing demands of the urban, industrial, and environmental sectors is projected to occur. In developing nations, with unregulated, inadequately priced, or subsidised abstraction of irrigation water, the expansion of irrigated agriculture is exacerbating the strain on available freshwater resources (Fishman et al. 2015). This issue is notably prevalent in several South Asian countries, including India, where agriculture remains the important source of livelihood, engaging over 50% of the Indian workforce and contributing approximately 17% to the national GDP (GOI (Government of India) 2023).

Drylands, which account for 44% of the world's cultivated area, are extremely important in transforming the global South's agri-food systems, livelihoods, and economy. Drylands/rainfed area, home to most of the world's poor and characterized by low yields and water scarcity, is witnessing accelerated risks under climate change (Rosa et al. 2020; Ahmed et al. 2022). India ranks first among the dryland/rainfed countries in the world in terms of area, with about 51% of the net sown area being rainfed, with a vast potential to close the productivity gap. Water harvesting and rainwater management are vital for augmenting local scale water and soil moisture availability to enhance land and water productivity in rainfed areas. Sharma et al. (2010), Sikka et al. (2022), AICRPDA (2019), and many more studies in India have shown the potential of runoff harvesting of rainfed districts and the opportunity for supplemental irrigation to increase the yield of rainfed crops and productivity of rainwater.

Out of India's net cultivated area of 140 million hectares (Mha), approximately 68.4 Mha (48.8%), is under irrigation (GOI (Government of India) 2023). Agriculture is the largest consumer of water in India using about 82% of the country's total freshwater withdrawals and this is projected to reduce in the future, according to some estimates this could reduce to 72% by 2025 and further 65–68% by 2050 (ADB (Asian Development Bank) 2013). The increased water requirements, coupled with an increasingly erratic and uncertain water supply, will exacerbate the challenges faced by regions already grappling with water scarcity. Furthermore, this may create water stress in areas that currently enjoy abundant water resources (UNESCO, UN-Water 2020). Consequently, the sustainable management of already stressed water resources poses a formidable challenge for ensuring the sustainability of agriculture in India and across much of South Asia.

Manifestations of widening water demand-supply gaps in India are visible with increasing water shortages, depleting groundwater, deteriorating resource quality, climate change, and reduced freshwater diversion to agriculture. Assessing the present level of water productivity in irrigated and rainfed areas and minimizing the gap by enhancing water productivity is a high priority as a strategy for transforming agri-food systems. Mainstreaming of the area and context-specific innovative agricultural water management interventions alongside innovative institutional and policy enablers to improve the water productivity of both rainfed and irrigated agriculture is essential to increase agricultural production using limited water resources. This paper assesses cereals water productivity for both rainfed and irrigated districts in India, pathways for enhancing physical and economic water productivity for improved resilience, transforming agri-food systems, and looking beyond water to enhance water productivity effectively.

3.2 Water Productivity Assessment

Multiple global studies have underscored significant opportunities for enhancing water productivity (WP) (Foley et al. 2020; Brauman et al. 2013; Zheng et al. 2018). In their water productivity analysis of rice, wheat, and corn, covering 31.9% of the world's total crop land, Foley et al. (2020) identified substantial variations in WP. Likewise, Brauman et al. (2013) identified significant prospects for improving WP of 16 major staple food crops. According to them, augmenting the WP of rainfed crops could potentially boost food production by approximately 30% while maintaining a similar level of water consumption. The assessment of WP for ten irrigated crops in India revealed considerable spatial variations (Sharma et al. 2018). For instance, rice WP ranged from 0.57 kg/m^3 in predominantly groundwater irrigation Punjab to 0.24 kg/m^3 in Karnataka, with largely surface water irrigation. This underscores the potential for enhancing WP in India. Enhancing water productivity is an important proposition for transforming agri-food systems (Sikka et al. 2022; Sharma et al. 2018). Increasing water productivity is also a measure or expression for increasing resilience and adaptation to climate change via increased water use efficiency.

Water productivity (WP) generally means the value or output of the product obtained relative to the amount of water utilized, depleted, or diverted. Increasing water productivity translates to obtaining greater benefits from each unit of water used in diverse production systems. WP is defined in various ways based on the scale and focus of the analysis. Simply put, it measures production per unit of water use (Giordano et al. 2017). It represents the ratio of physical production, denoted as physical water productivity (PWP), expressed in kg/m^3 or, in certain cases, the 'economic value' of production, termed economic water productivity (EWP), expressed in \$/m^3, in relation to water utilization (in terms of water withdrawn, applied, or consumed). The choice of the numerator and denominator hinges on the scale and focus of the analysis (Molden et al. 2010; Giordano et al. 2017; Amarasinghe et al. 2021a, b).

In this paper, water productivity has been estimated in terms of physical water productivity (PWP), kg/m^3 of consumptive water use, including both green and blue water, following a conventional water accounting protocol. WP estimates include 'green' water (effective rainfall) for rain-fed areas and both 'green' water and 'blue' water (diverted water from water systems, surface and/or groundwater) for irrigated areas. WP has been worked out for the main/predominant cereal crops at the district level based on the data published by the Ministry of Agriculture (MOA 2023). Both for irrigated and rainfed districts, classified based on irrigated area coverage, districts with a net irrigated area (NIA) of less than 25% of the net sown area (NSA) are taken as the "rainfed districts" for this study. We have looked at the WP variations across irrigated and rainfed districts of major cereal crops, which occupy 50% of the cropped area.

3.3   Water Productivity Variations Across India

The cereal area at the country level remained stagnant, around 100–103 Mha from 1999–2001 to 2018–2020, but there was a significant change in the composition of irrigated and rainfed areas (Table 3.1). Many rainfed areas have received irrigation in the last two decades, and with the addition of new irrigated areas, the total cereal area in the county increased slightly by 3 Mha (Table 3.1). There are 196 districts with a net irrigated area (NIA) of less than 25% of the net sown area (NSA), considered as the "rainfed districts" in this study, with minimal irrigation augmentation in these

Table 3.1 Cereal area and PWP growth

Factors	All districts			Major rainfed districts[a]		
	1999–2001	2018–2020	Total growth (%)	1999–2001	2018–2020	Total growth (%)
Irrigated area (Million ha)	52.6	65.8	+25	3.9	4.4	+13
Rainfed area (Million ha)	47.6	37.3	−12	17.7	15.4	−13
Total area (Million ha)	100.2	103.1	+3	21.6	19.8	−9
Irrigation CWU (Bm^3)	146.5	171.6	+17	9.7	10.7	+10
Rainfall CWU (Bm^3)	224.2	221.7	−1	58.3	54.3	−7
Total CWU (Bm^3)	370.7	393.3	+6	68.0	65.0	0.96
Yield (ton/ha)	1.34	2.41	80	1.07	1.69	58
Production (Million MT)	187.2	294.8	+57	23.1	33.2	+44
PWP (kg/m^3)	0.50	0.74	+48	0.34	0.51	+0.50

Source: authors'
[a] Rainfed districts are those with the net irrigated area (NIA area equipped for irrigation) less than or equal to 25% of the net sown area (NSA)

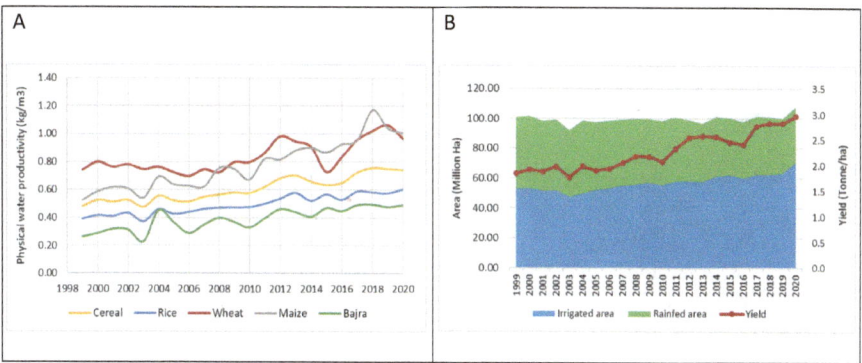

Fig. 3.1 Cereals PWP, yield, and area growth. (**a**) Cereals PWP of all districts. (**b**) Cereals yield and area. Source: authors

districts. The cereal area in the rainfed districts has declined by 9% or 2 Mha. So, much of the growth in the area under cereals occurred in non-rainfed districts.

The physical WP (PWP) of cereal crops in India (all districts included) increased significantly over the last two decades. Cereals comprise 50% of the cropped area, and its PWP increased by 48% from 1999 to 2020. Rice and wheat constitute 80% of the cereal cropped area; their PWP increased by 45% and 33%, respectively. Maize and Bajra (Pear millet) have significantly higher PWP increases of about 87% and 68%, respectively. With increased irrigated area (by 25%), the cereal irrigation consumptive water use (CWU) or evapotranspiration from irrigation increased (by 19%), while CWU from effective rainfall decreased (by 7%) (Table 3.1). Irrigation with other inputs helped cereal yield grow by 80%. The production growth (57%) helped increase PWP by 48%. Increased crop yield was one of the main drivers of cereal PWP growth (Fig. 3.1b).

Although PWP has rapidly increased, a significant spatial variation remains across districts (Fig. 3.2a). The highest PWP is observed in the northwestern and southern regions, including the States of Punjab, Haryana, western Uttar Pradesh, and Tamil Nadu. The middle to the eastern regions mainly has low to very low PWP. Many low PWP districts are rainfed or have low irrigation inputs (Fig. 3.2b).

The highest average cereal PWP is in Punjab in the northwest, with an average of 1.16 kg/m^3, which varies from 1.09 to 1.27 kg/m^3 across districts (Table 3.2). Haryana, Tamil Nadu, Karnataka, Madhya Pradesh, and Uttar Pradesh have the next highest PWP, with over 0.80 kg/m^3. The variations in large non-rainfed states, such as Tamil Nadu, Karnataka, Madhya Pradesh, and Uttar Pradesh, are also significant. Some districts in Tamil Nadu have very high PWP primarily due to the high PWP of Maize.

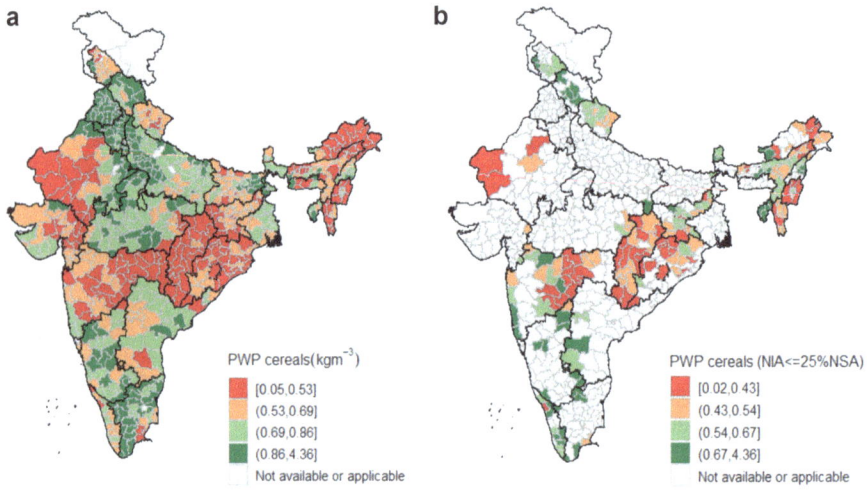

Fig. 3.2 Spatial variation of cereal PWP. (**a**) Cereal PWP of all districts. (**b**) Cereal PWP of major rainfed districts. Source: Authors

3.4   WP Improvements in Rainfed Districts

The Cereals PWP of non-rainfed and rainfed districts in the states of Assam, Chhattisgarh, Jharkhand, Kerala, and Odisha, with most rainfed districts, have no significant differences. But the average PWP is significantly lower than the districts with substantial irrigation. These states receive good rainfall, and many districts hardly require irrigation for the *Khraif* cereals. But the water deficit (rainfall—ETP) is significant for other seasons (Fig. 3.3).

The districts in these states have the potential to increase in-situ soil moisture conservation, including contour trenching, contour/compartmental bunding, and harvest rainwater in the monsoon season for subsequent use in dry periods and/or to tide over the moisture stress situation during the intervening long dry spells in the kharif season. Mountainous areas in some of the districts in these states also have scope for springs harvesting and springs shed management for augmenting water supply locally. Many districts in these states have good potential for sub-surface water harvesting, including ring wells to tap water locally. Remote areas with a lack of electricity or erratic power supply have good potential for solar irrigation pumps for which the government's developmental schemes provide subsidies for solar irrigation pumps to lift water from surface water bodies and/or sub-surface systems. In many of these districts, the 10th percentile, or a return period of one in a ten-year monthly rainfall (10% percentile) would barely meet the Kharif season water needs. However, the 25% percentile, three in four-year return period, or 50% percentile, or median rainfall have substantial surface runoff potential for water harvesting.

Maharashtra is another State with many rainfed districts. However, the NIA and NSA data used for determining rainfed districts are from 2004. Given the increased trends of irrigation development, Maharashtra's rainfed district classification may

Table 3.2 Cereals PWP

States	All districts					Rainfed districts[a]		
	No. of districts	Cereal area (Mha)	PWP (kg/m³)			No. of districts	Cereal area (Mha)	Average PWP (kg/m³)
			Average	Min.	Max.			
Andhra Pradesh	23	5.9	0.72	0.42	0.91	3	0.5	0.63
Arunachal Pradesh	15	0.2	0.45	0.36	0.77	9	0.1	0.50
Assam	23	2.5	0.54	0.35	0.92	19	2.0	0.55
Bihar	37	6.0	0.69	0.43	1.30	1	0.1	0.55
Chhattisgarh	16	4.4	0.42	0.31	0.59	15	4.2	0.41
Gujarat	25	2.9	0.60	0.37	0.98	2	0.3	0.53
Haryana	19	4.5	0.99	0.84	0.98	0	0.0	0.00
Himachal Pradesh	12	0.7	0.75	0.63	1.45	6	0.4	0.86
Jammu & Kashmir	17	0.9	0.67	0.29	0.98	4	0.3	0.39
Jharkhand	22	1.7	0.52	0.40	0.67	16	1.3	0.55
Karnataka	27	4.7	0.84	0.47	1.22	6	0.8	0.75
Kerala	14	0.2	0.68	0.41	0.85	10	0.1	0.66
Madhya Pradesh	48	12.2	0.82	0.48	1.27	3	0.5	0.51
Maharashtra	34	4.4	0.50	0.25	0.75	23	2.3	0.45
Odisha	30	4.0	0.45	0.34	0.63	16	1.9	0.43
Punjab	17	6.7	1.16	1.09	1.27	0	0.0	0.00
Rajasthan	32	9.4	0.66	0.16	1.34	4	1.8	0.21
Tamil Nadu	30	2.7	0.96	0.36	2.26	7	0.4	1.51
Uttar Pradesh	70	18.8	0.80	0.60	1.10	1	0.2	0.26
Uttaranchal	13	0.7	0.71	0.48	0.84	9	0.4	0.56
West Bengal	19	5.8	0.75	0.55	1.14	0	0.0	0.00
Total	543	99	0.70	0.16	2.26	154	17.6	0.49

Source: Authors

[a]Rainfed districts are those with net irrigated area less than 25% of the net sown area

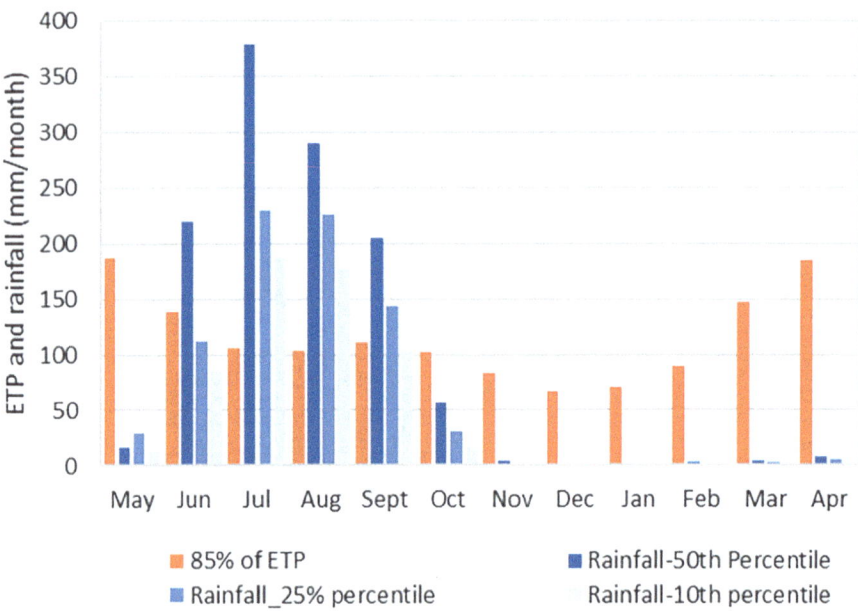

Fig. 3.3 ETP and rainfall in Jharkhand

have some limitations. Nevertheless, the monthly rainfall and ETP patterns show that the State has substantial water deficits in the non-monsoon period (Fig. 3.4). The chronic water deficit, especially during the critical periods of crop growth, affects production in the southern and eastern parts of the states. Many of these water deficit districts need water storage in large to medium reservoirs, minor tanks or in ponds, sub-surface dykes and/or enhanced groundwater recharge and storage to support crop cultivation in non-monsoon periods. Efficient management of harvested water in rainfed areas is most important, together with the best agronomic and agricultural water management practices and the right choice of crops and cultivars to ensure water is used judiciously and productively to enhance water productivity. Prioritizing a location-specific portfolio of smart agricultural water management (AWM) interventions is important in the context of making right investment decisions (Sikka et al. 2022).

Appropriate planning and implementation of smart agricultural water management interventions have a pivotal role in augmenting the availability of surface water, groundwater, and soil moisture. They contribute to enhancing the availability of soil water, minimizing evaporation and seepage, and curbing non-beneficial water losses, all of which collectively increase crop water productivity. Additionally, the augmentation of soil moisture and the implementation of water harvesting strategies, both surface and subsurface storage systems, enhance resilience of rainfed agriculture.

Among the major cereals-producing states, Punjab, Haryana, and Tamil Nadu have significantly higher average PWP than other states (Table 3.3). Punjab and

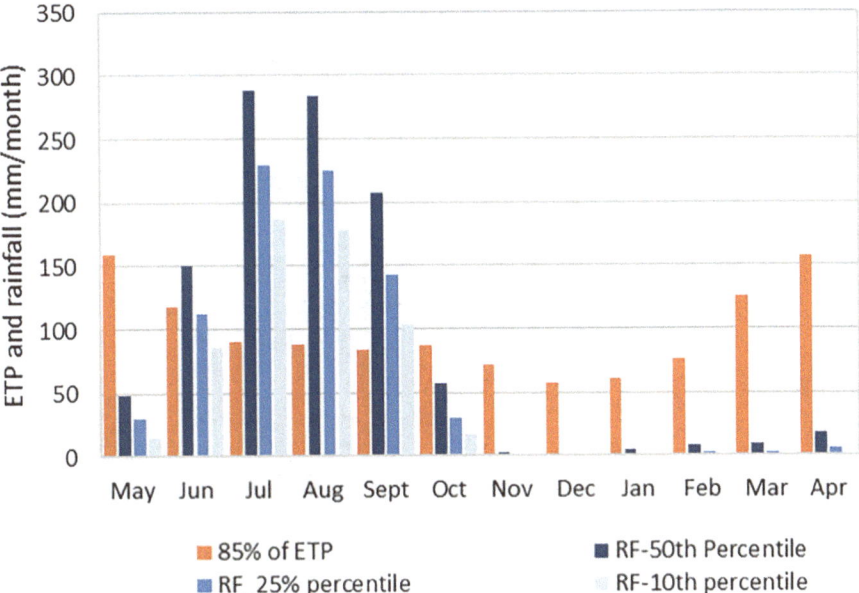

Fig. 3.4 ETP and rainfall in Maharashtra

Haryana have more than 50% cereal area in the Rabi season. With access to groundwater irrigation, these two states have higher Rabi season PWP, and hence higher overall PWP. Tamil Nadu has a significantly higher cereal area in the Kharif Season, and this is the major irrigation season for them, with most rainfall received in the rabi season with north-east monsoon. Moreover, Tamil Nadu has a relatively higher crop diversification, and their reported yields for crops such as Maize are significantly higher than other states. For those reasons, Tamil Nadu has higher cereal yield and PWP in both seasons.

Karnataka, Madhya Pradesh, and Uttar Pradesh have the next set of higher PWPs. Karnataka has about 68% of the crops in the Kharif season; the average cereal yield is substantially higher in Karnataka with a diversified cereal cropping pattern. Because of low water-consuming crops, this State has higher PWP in both Kharif and Rabi. Madhya Pradesh also has a diversified cropping pattern, and with low water consuming crops, it also has relatively higher PWP in both seasons. Uttar Pradesh also has a very high Rabi season cereal crop area and similar cropping patterns to Haryana. But with decreasing access to irrigation from the western to eastern regions, their PWP varies significantly across districts.

The other States have relatively lower access to irrigation. Most of their crops are in the Kharif season and depend on rainfall. Should these States be able to increase the Rabi season crop area or CWU with access to irrigation, they can substantially improve the cereal PWP and total production. This is particularly true in rainfed States such as Odisha and rainfed districts in Bihar and West Bengal, and Rajasthan, where Rabi season's cereal crop area is significantly lower than in other states.

Table 3.3 Cereals production factors in major producing states

State	PWP (kg/m^3)				Production		Share of crop area		
	Kharif	Rabi	Summer	Total	Area (M ha)	Yield (kg/ha)	Kharif	Rabi	Summer
Punjab	0.78	1.83	0.00	1.16	6.7	4.64	48	52	0
Haryana	0.60	1.40	0.00	0.99	4.5	3.99	44	56	0
Tamil Nadu	0.88	1.50	0.73	0.96	2.7	3.71	67	24	9
Karnataka	0.85	0.84	0.70	0.84	4.7	2.31	69	30	1
Madhya Pradesh	0.69	0.90	0.54	0.82	12.2	3.09	36	64	0
Uttar Pradesh	0.57	1.05	0.40	0.80	18.8	3.20	45	55	0
West Bengal	0.63	1.10	0.99	0.75	5.8	2.25	68	27	5
Bihar	0.55	0.92	0.57	0.69	6.0	2.72	48	41	11
Rajasthan	0.41	0.96	0.00	0.66	9.4	2.13	63	37	0
Maharashtra	0.46	0.58	0.21	0.50	4.4	2.12	49	51	0
Odisha	0.45	0.77	0.34	0.45	4.0	2.07	81	7	12

Source: Authors

3.5   WP Improvements in Irrigated Areas

Water scarcity is a major issue, even in many irrigated districts and irrigation projects, marred with low WP. Field-based evidence in irrigated areas has demonstrated that water demand management interventions including laser land levelling, pressurized irrigation systems like drip and sprinklers, piped water distribution networks, alternate wetting and drying (AWD), direct seeded rice (DSR), evapotranspiration (ET)-based irrigation scheduling employing measured actual ET and/or computed ET, and soil moisture-based irrigation scheduling to optimize water application. These measures are pivotal in increasing WP within irrigated command areas. Schmitter et al. (2017) demonstrated the efficacy of establishing local water user group (WUG)-based information platforms to assist farmers in enhancing crop and water productivity in an irrigated area in Ethiopia. Their study indicated that farmers with access to information were able to reduce their water usage similarly to those who had installed soil moisture sensors, such as Wetting Front Detectors and Chameleon sensors. This approach could be replicated in Indian irrigated command areas to disseminate advisories to farmers, enabling them to optimize water utilization for improved water productivity. Recent technological advancements in instrumentation, automation, sensor technology, satellite data, artificial intelligence (AI), machine learning (ML), and big data have paved the way for the development of these technologies for irrigation scheduling and precision farming.

Many major and medium irrigation projects with reservoir storage are categorized as moderate to severe water scarce. Many of the canal command areas are not irrigated fully or do not have access to adequate irrigation. Many irrigation systems reported low irrigation intensity, water use efficiency (WUE), and WP. The general perception of low performance is due to inadequate and/or unreliable access to irrigation supplies. However, studies in Sina and Kukadi Irrigation systems in Maharashtra show that low irrigation intensity and WUE are misperceptions (Amarasinghe et al. 2021a). Farmers in these water-scarce irrigation projects use recharged groundwater from the return flow from canal irrigation and natural rainfall recharge to groundwater in the command area.

However, crops grown in these irrigation systems still have low economic water productivity (EWP), a measure of the value per unit of consumptive water use. Many crops grown are high-water consuming or low-value seasonal crops or annual crops with low PWP (Fig. 3.5). For example, seasonal crops such as sorghum, pulses, and oilseeds, which constitute a major area, have low EWP. Sugarcane and cotton are low EWP crops on account of much higher CWU. Sugarcane has a significantly higher CWU per ha of annual crops. Fruits, on the other hand, require two-thirds of the consumptive water use but generate more than two times the value for the water depleted.

Amarasinghe et al. (2021b), in their study conducted in the Sina irrigation system situated in the semi-arid Ahmednagar district of Maharashtra in India, applied Economic Water Productivity (EWP) alongside the water cost curve as analytical tools. These tools were employed to effectively redistribute consumptive water use for irrigation and identify economically viable cropping patterns and Agricultural

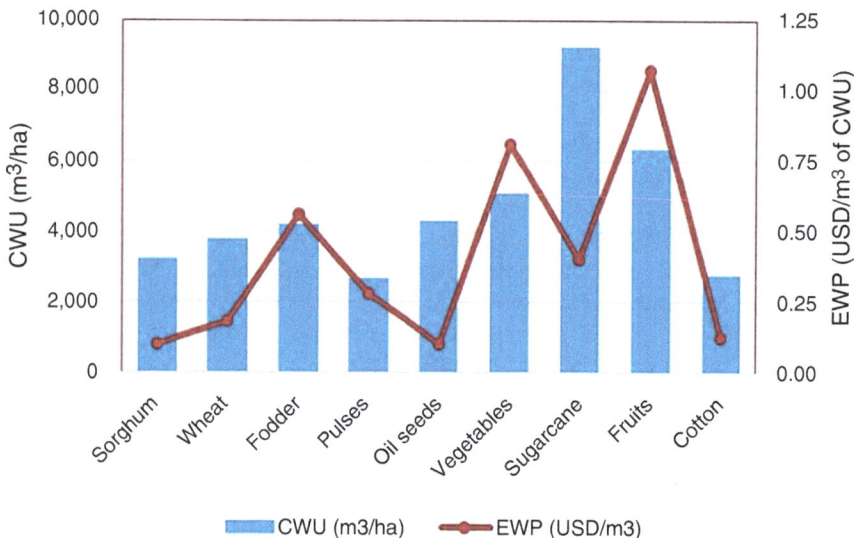

Fig. 3.5 CWU and economic water productivity of crops in the Sina irrigation system. Source: Authors

Water Management (AWM) strategies under varying scenarios, including normal, dry, and wet years. The study recommends that cropping patterns featuring high-value, water stress-tolerant crops such as orchards (e.g., pomegranates) when combined with micro-irrigation techniques can serve as a permanent feature within the cropping scheme. This not only ensures sustainable income even during periods of low rainfall but also enhances income resilience during years with moderate to abundant rainfall by integrating orchards and fodder with key field crops, thereby increasing both EWP and overall output value.

These irrigation systems, however, can increase economic water productivity to improve resilience to droughts (Amarasinghe et al. 2021b). One option is to increase the area under high-value seasonal crops to increase the income under normal rainfall conditions and the area under lower water consuming annual crops that generate more value and withstand moderate to severe droughts. Such an approach can be useful for many non-rainfed districts, let alone canal irrigation systems, with relatively higher access to irrigation.

3.6   Conclusions

This paper discussed why WP improvements should play a major role in improving agri-food systems in India. Many rainfed districts have low access to water. In these districts, inadequate access to irrigation affects agricultural production. However, many of these rainfed districts have the potential for water harvesting and in-situ moisture conservation to improve access to water and improve rainwater productivity. Water harvesting will permit increasing irrigation intensity by expanding crop areas in the dry seasons and providing irrigation at least in critical periods of crop growth to increase/maintain crop yield. These actions will increase the PWP of crops in rainfed districts.

Many other irrigation systems in India also have seasonal water security issues. Climate change can exacerbate the situation. Many of these irrigated districts have excess surface runoff to harvest under normal to moderate drought conditions. The harvested water, whether in ponds, small tanks, or underground, can relieve water scarcity pressures in dry periods.

However, many water-scarce areas, in rainfed or in non-rainfed districts, need to rethink their strategy of cropping patterns and water use management. Water-scarce areas should use the augmented water supply for a diversified cropping pattern to increase economic WP and improve resilience. The utilization of recharged groundwater should become a norm rather than an exception in canal irrigation commands. Furthermore, irrigation systems should be designed to promote the conjunctive use of surface and groundwater by design to proactively mitigate the impact of climate-related risks on agriculture and bolster resilience. It's also important to recognize that water represents only one facet of the agri-food system and therefore thinking beyond water by factoring in the knowledge of other key productivity related factors is important to enhance agricultural water productivity within a specific context.

Moving forward, the process of mainstreaming and expanding context-specific smart agricultural water management interventions at developmental scale demand a strong emphasis on capacity development, provision of explicitly including water management in extension services, and the mobilization of resources through collaborative efforts among institutions, as well as co-financing from development programs.

Acknowledgments We would like to acknowledge funding from CGIAR research initiatives: Nexus Gains (Realizing multiple benefits across water energy food forest biodiversity systems), and ClimBeR (Building Systemic Resilience Against Climate Variability and Extremes). We also acknowledge the Indian Council of Agricultural Research for their support.

References

ADB (Asian Development Bank) (2013) Asian water development outlook 2013: measuring water security in Asia and the Pacific. Asian Development Bank, Mandaluyong City

Ahmed M, Hayat R, Ahmad M et al (2022) Impact of climate change on dryland agricultural systems: a review of current status, potentials, and further work need. Int J Plant Prod 16:341–363. https://doi.org/10.1007/s42106-022-00197-1

AICRPDA (2019) National innovations in climate resilient agriculture managing weather aberrations through real time contingency planning annual report 2018-19. http://krishi.icar.gov.in/PDF/ICAR_Data_Use_Licence.pdf

Amarasinghe UA, Sikka A, Mandave V, Panda RK, Gorantiwar S, Chandrasekharan K, Ambast SK (2021a) A re-look at canal irrigation system performance: a pilot study of the Sina irrigation system in Maharashtra, India. Water Policy 23(1):114–129

Amarasinghe UA, Sikka A, Mandave V, Panda RK, Gorantiwar S, Ambast SK (2021b) Improving economic water productivity to enhance resilience in canal irrigation systems: a pilot study of the Sina irrigation system in Maharashtra, India. Water Policy 23(2):447–465

Brauman KA, Siebert S, Foley JA (2013) Improvements in crop water productivity increase water sustainability and food security—a global analysis. Environ Res Lett 8(2):024030

Fishman R, Devineni N, Raman S (2015) Can improved agricultural water use efficiency save India's groundwater? Environ Res Lett 10:084022. https://doi.org/10.1088/1748-9326/10/8/084022

Foley DJ, Thenkabail PS, Aneece AP, Teluguntla PG, Oliphant AJ (2020) A meta-analysis of global crop water productivity of three leading world crops (wheat, corn, and rice) in the irrigated areas over three decades. Int J Digital Earth 13(8):939–975. https://doi.org/10.1080/17538947.2019.1651912

Giordano M, Turral H, Scheierling SM, Tréguer DO, McCornick PG (2017) Beyond "more crop per drop": evolving thinking on agricultural water productivity. International Water Management Institute (IWMI), Colombo, p 53. https://doi.org/10.5337/2017.202

GOI (Government of India) (2023) Agricultural statistics at a glance 2021. Ministry of Agriculture and farmers welfare. https://eands.dacnet.nic.in/PDF/Agricultural%20Statistics%20at%20a%20Glance%20-%202021%20(English%20version).pdf

Iglesias A, Garrote L (2015) Adaptation strategies for agricultural water management under climate change in Europe. Agric Water Manag 155:113–124. https://doi.org/10.1016/j.agwat.2015.03.014

MOA (2023) Last accessed via https://eands.dacnet.nic.in/

Molden D (2013) Water for food water for life: a comprehensive assessment of water management in agriculture. Routledge, London, p 1736

Molden D, Steduto P, Bindraban P, Hanjra K, J. (2010) Improving agricultural water productivity: between optimism and caution. Agric Water Manag 97(4):528–535. https://doi.org/10.1016/j.agwat.2009.03.023

Rosa L, Chiarelli DD, Rulli MC, Dell'Angelo J, D'Odorico P (2020) Global agricultural economic water scarcity. Sci Adv 6(18):6031

Schmitter P, Haileslassie A, Desalegn Y, Chali A, Langan S, Barron J (2017) Improving on-farm water management by introducing wetting-front detector tools to smallholder farms in Ethiopia. LIVES Working Paper 28. International Livestock Research Institute (ILRI), Nairobi

Sharma BR, Rao KV, Vittal KPR, Ramakrishna VR, Amarasinghe U (2010) Estimating the potential of rainfed agriculture in India: Prospects for water productivity improvements. Agric Water Manag 97(1):23–30. https://doi.org/10.1016/j.agwat.2009.08.002

Sharma BR, Gulati A, Mohan G, Manchanda S, Ray I, Amarasinghe U (2018) Water productivity mapping of major Indian crops. NABARD and ICRIA, New Delhi. http://hdl.handle.net/11540/8480

Sikka AK, Alam MF, Mandave V (2022) Agricultural water management practices to improve the climate resilience of irrigated agriculture in India. Irrig Drain 2022:2696. https://doi.org/10.1002/ird.2696

UNESCO, UN-Water (2020) United Nations world water development report 2020: water and climate change. UNESCO, Paris

Wada Y, Flörke M, Hanasaki N, Eisner S, Fischer G, Tramberend S, Satoh Y, van Vliet MTH, Yillia P, Ringler C, Wiberg D (2016) Modeling global water use for the 21st century: water futures and solutions (WFaS) initiative and its approaches. Geosci Model Dev 9:175–222. https://doi.org/10.5194/gmd-9-175-2016

Zheng H, Bian Q, Yin Y et al (2018) Closing water productivity gaps to achieve food and water security for a global maize supply. Sci Rep 8:14762. https://doi.org/10.1038/s41598-018-32964-4

Transforming Animal Health Sector

4

Habibar Rahman, Jagadish B. Hiremath, Kennady Vijayalakshmy, and K. M. Bujarbaruah

Abstract

The animal health sector within the agriculture sector plays a critical role in fulfilling the food demand for billions of the world's populations by providing safe and nutritional food. The health security of a large human population is also ensured by preventing the spread of zoonotic diseases from animals to humans. If a world must achieve "no hunger" the animal health sector needs a transformation towards enhancing production and productivity. It is also important to meet the increasing global demand for animal-based food products while ensuring sustainability and efficiency. Genetic improvement, nutritional and health management are crucial to harvest the maximum yield from livestock farming. Transforming the animal health sector is essential for ensuring food security, which involves the availability, accessibility, and affordability of safe and nutritious food for all. By focusing on these aspects of the animal health sector transformation, it is possible to enhance food security by producing safe and healthy animal-derived food products. Simultaneously, the promotion of animal health contributes to human health security by reducing the risk of zoonotic

H. Rahman (✉)
ICAR, New Delhi and Regional Representative for South Asia, International Livestock Research Institute (ILRI), South Asia Office, New Delhi, India
e-mail: r.habibar@cgiar.org

J. B. Hiremath
ICAR–National Institute of Veterinary Epidemiology and Disease Informatics, Bengaluru, Karnataka, India

K. Vijayalakshmy
International Livestock Research Institute (ILRI), South Asia Office, New Delhi, India

K. M. Bujarbaruah
Assam Agricultural University, Jorhat, Assam, India

diseases, preserving the effectiveness of antimicrobials, and promoting sustainable and resilient agricultural systems.

Keywords

Animal disease · Animal health · Food security · Health security

4.1 Introduction

Animals play a vital role in the survival and welfare of humans which dates to the prehistoric era. They were a source of meat, draught power, and means of transportation and provided companionship to humans, and significantly impacted the welfare of humans in many ways. The ecosystem is made up of a wide variety of species that are dispersed over the globe and have an impact on the overall health of the planet. Due to high adaptability to changing climate, livestock plays a critical role in the life of hundreds of millions of vulnerable people across the globe and hence are key to food security.

The animal disease burden globally refers to the collective impact of diseases on animal populations worldwide. It encompasses both infectious and non-infectious diseases that affect various species of animals, including livestock, poultry, aquaculture, and wildlife. Although, the global burden of animal diseases is difficult to estimate due to variations in data collection, reporting systems, and regional differences. However, several diseases have significant global implications and contribute to the overall burden. In order to combat the impact of disease burden, the animal health market is expanding significantly, and the value of the animal health products market is around US$24 billion (GADBP-WOAH).

The world human population is increasing rapidly which reached 8.0 billion in mid-November 2020 and is projected to reach 8.5 billion by 2030 and 9.7 billion by 2050 (United Nations). The Asian continent with two most populous countries, China (1.4 billion) and India (1.4 billion) which represent 36% of the world's population. With 40% value of global agriculture, animal production has a bigger role to play in order to achieve the targets of zero hunger (Domestic Animal Diversity Information System-WOAH). The animal health sector can play a crucial role in meeting the food requirements of an increasing human population in several ways such as maintaining animal health, enhancing productivity, promoting sustainable practices, driving research and innovation, fostering collaboration, and ensuring food safety. By focusing on these areas, the sector can help to meet the increasing demand for animal-derived products and support global food security.

4.2   Transforming Animal Health Sector for Food Security

Transforming the animal health sector is essential for ensuring food security, which involves the availability, accessibility, and affordability of safe and nutritious food for all. The two major principles on which transformation can be achieved are prevention or control of economically important livestock diseases and increasing the production and productivity of animals. Following are some key strategies to transform the animal health sector for food security. Animal disease control and prevention encompasses a range of strategies and measures aimed at reducing the incidence, spread, and impact of diseases in animal populations. Here are some key aspects of animal disease control and prevention.

4.2.1   Vaccination and Disease Control Program

Vaccination is a critical tool for animal disease control and prevention. It aids in defending animals against infectious diseases by enhancing their immune systems' capacity to identify and combat particular infections. Vaccination plays a vital role in eradicating certain diseases. By achieving high vaccination coverage and interrupting the transmission cycle of a disease, it is possible to eliminate the disease from a specific region or even globally. Successful examples include the eradication of rinderpest in cattle and PPR is the next potential disease of small ruminants that could be eradicated in the future.

Development and implementation of vaccination programs for key animal diseases are essential to protect animal populations and to enhance their production. Providing access to affordable vaccines and ensuring the availability of veterinary services for disease control and management is critical. The National Animal Disease Control Program of India launched in the year 2019 is one of the world's largest animal disease control programs which aims to vaccinate all the target population of FMD and Brucellosis to control, prevent and eventually eradicate FMD and Brucellosis (DAHD) from India. The National Control Program for Peste des Petits Ruminants (NCP-PPR) targets the control and eradication of Peste des Petits Ruminants (PPR), also known as goat plague. NCP-PPR involves vaccination campaigns, surveillance, diagnostics, and capacity building to prevent the spread of PPR among sheep and goat populations. Similarly, there is a control program for CSF, a major pig disease in India. Under livestock health and Disease Control program of DAHD state specific vaccination programs for regional endemic disease like Anthrax, BQ, HS and ET are carried out (DAHD).

Vaccine development is long process starting from design of vaccine, validation, trail, approval to manufacture, supply and finally vaccination of target population. At each step there are number of challenges of which safety is a major concern. With increasing number of emerging infectious diseases and their pandemic potential it has become much more challenging to develop a vaccine within stipulated time line. Traditionally, vaccines were either live attenuated or inactivated which on inocula-tion produced broad spectrum immune response. The advancement in molecular biology and biotechnology has led to development of vaccine platforms that are tested for safety and employed for development of novel vaccines in record times which was witnessed during COVID-19 pandemics. The Measles Virus Vector, New Castle Disease Virus Vector and Adeno-associated virus vectors are few of the technologies that were studied well, fulfilled safety requirements, quality tested for efficacy and scalability. Due to availability of such vaccine platforms the COVID-19 vaccine was made available in shortest possible time which was otherwise thought impossible. In animal health sector there is need to develop vaccine platforms which are currently lacking and it was always economical to prevent disease by vaccination in animals than any other means of disease prevention.

4.2.2 Biosecurity Measures

Biosecurity is a strategic and integrated approach that encompasses the policy and regulatory frameworks (including instruments and activities) for analyzing and managing relevant risks to human, animal, and plant life and health, and associated risks to the environment (FAO). To stop the spread of harmful infections and poisons, a variety of procedures and practises have been implemented by bioscience laboratories, border patrol, customs officers, and managers of agricultural and natural resource areas. Biosecurity involves national and international stakeholders who play a vital role in biosecurity functions like setting national and international standards, planning large-scale surveillance programs, incursion response activities, and laboratory diagnostic services. It is quite evident that the rate of international human travel is ever-increasing and so is international trade in livestock resulting in the rapid spread of infections from one part of the world to another part.

National biosecurity measures are crucial for preventing and controlling animal diseases that transfer due to international trade. There are some key measures that are to be implemented at different levels to prevent the incursion of transboundary animal diseases at international borders. Biosecurity policy for a country is essential which is comprehensive and addresses the risk of disease transmission across the borders. The creation of biosecurity infrastructure is a primary requirement in order to see the effect of the policy. This includes facilities for quarantine at entry points, a laboratory for disease diagnostics, waste management, and safe disposal of animal carcasses. Capacity building and creating awareness among the stakeholder regarding biosecurity is equally important for the effective implementation of biosecurity policy in the country. The customized training modules about biosecurity practices should target farmers, veterinarians, and other stakeholders. The emergence of transboundary animal diseases (TBAD) is more frequent in the recent past than ever. An important step towards reducing the risk of TBAD is regional cooperation in disease data sharing. Fostering collaboration among government agencies, industry stakeholders, research institutions, and international organizations is a must to coordinate biosecurity efforts. Establishing public-private partnerships to leverage expertise, resources, and information sharing is critical.

4.2.3 Risk Assessment and Surveillance

Risk assessment is a systematic process of evaluating potential risks associated with a specific activity, situation, or hazard. It involves identifying hazards, analyzing their likelihood and potential consequences, and making informed judgments about the level of risk they pose. Risk assessment is commonly used in animal health to identify potential disease threats and prioritize surveillance efforts. Risk assessment may be qualitative or quantitative in nature. The basic steps in risk assessment include hazard identification, risk assessment, risk management, and risk communication. The current practical covers the risk introduction of exotic/TBD into a disease-free country. Such assessment is a critical component of emergency preparedness as it supports the formulation of risk-based biosecurity measures. The process involves listing risk factors and collecting evidence for risk factor categorization into negligible, low, moderate, high, and very high likelihood with low, moderate, or high levels of uncertainty.

Livestock disease surveillance is often organised with the goals of early illness event detection, evaluating the efficacy of intervention strategies, and identifying disease-free or infection-free areas. The information gathered through these surveillance activities will serve as enough support for assessing national disease control and eradication initiatives. The nation's efforts to diagnose and control animal diseases rely heavily on disease diagnostic laboratories at the state and district levels that are under the jurisdiction of the individual State Animal Husbandry and Veterinary Services Departments. The infrastructure and services provided by the government are largely dependent on the livestock health sector, however there are a number of private businesses that engage in contract farming, which also includes health services, in the poultry industry. There are private animal disease diagnostic laboratories throughout the nation in addition to those run by the state and federal governments.

4.2.4 Animal Identification and Traceability

Animal identification and traceability is an important component of biosecurity measures. A system towards this will monitor animal movement and allow quick traceability. Further, it also facilitates rapid response, containment, and control measures. Animal identification in India is an important component of animal health management, traceability, and disease control. The Government of India has implemented various systems and initiatives to facilitate animal identification and traceability. The Government of India has introduced a unique identification system for livestock known as the Pashu Aadhaar (Livestock Aadhaar) program. Under this program, each animal is assigned a unique identification number that is linked to the owner's details, animal breed, age, and other relevant information. Ear tagging is one of the commonly used methods for animal identification in India. Animal identification data is increasingly being integrated into digital platforms, such as mobile applications and online databases. This integration enables real-time tracking of animal movements, health status, and other relevant information, improving overall animal management and disease control efforts. It is important to note that animal identification practices may vary across different states and regions in India. The specific methods and initiatives implemented can depend on factors such as the type of livestock, local regulations, and available resources. Livestock owners and stakeholders are encouraged to comply with the relevant identification requirements and participate in the animal identification programs implemented by the government authorities.

4.2.5 Animal Movement Control

Regulate and monitor the movement of animals, both within the country and across borders, to minimize the risk of disease spread. Implement measures such as licensing, documentation, and veterinary inspections for animal transport. Increased frequency and speed of local and international travel driven by globalization, although benefitting the world economy at large but it poses risk of rapid disease transmission between countries. Hence, international travel and also the trade are an important risk factor that led to transmission of number of TADs. For example, the transportation of poultry and related products contributed to transboundary spread and outbreaks of Avian Influenza worldwide. Similarly, the risk assessment indicates that the risk of introduction of Lumpy Skin Disease in France by the import of vectors in animal trucks is high.

Unlike the trade in livestock products the live animal trade is associated with higher risk of disease transmission between the countries. Globally, Office International des Epizooties (OIE) formulates and regulates the animal health standards which are basis for safe international trade in livestock and livestock products. OIE Terrestrial Animal Health Code developed under the provisions of Sanitary and Phytosanitary agreement of World Trade Organization (WTO) provides the details on animal health and welfare that the 180 member countries should abide by (OIE). The regulations are made based on the principles that the exporting country should be free from listed transboundary animal diseases so as to reduce the risk of disease transmission to importing country.

4.2.6 Response and Contingency Planning

Transboundary animal diseases, which are crucial for the economy and food security, frequently result in animal disease emergencies. Due to their quick spread, these diseases must be controlled in advance; otherwise, they could grow widespread and be very difficult and expensive to eradicate. Emergency preparedness for animal diseases, and in particular contingency planning, should be viewed as a crucial instrument for the management of urgent diseases. Early warning, rapid detection of the introduction of any emergency livestock disease, surveillance, disease reporting, epidemiological analysis, and early reaction, which is to immediately

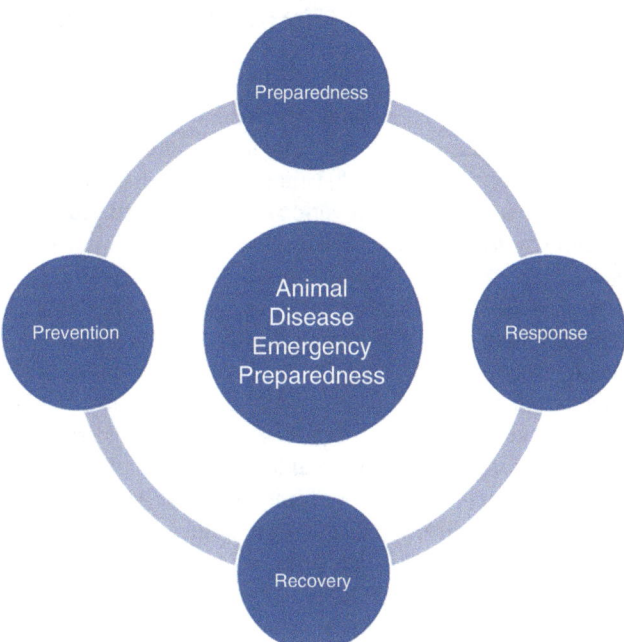

Fig. 4.1 Cycle of animal disease emergency preparedness and contingency planning

implement disease control measures to contain the outbreak and gradually eliminate it, are all part of emergency preparedness planning (Fig. 4.1). Since the government is currently dealing with the advent of new diseases like lumpy skin disease and African swine fever in addition to avian influenza, there is a need for training of professionals involved in the preparation of national emergency contingency plans in order to achieve this goal. The contingency plan is an important document that guides the action to be taken in the event of disease emergency. It will contain all the necessary operational details of response during animal disease emergencies like disease outbreak, methods of detection, list of designated diagnostic laboratories, immediate response, mitigation measures and recovery. Additionally, it includes information on the resources required to handle such an emergency as well as a plan of action for the effective and quick deployment of people and material resources for disease containment and infection elimination. India's government has a plan in place in case serious diseases like FMD, Brucellosis, PPR, CSF, and Anthrax arise. These were created using the contingency plans of other nations (found on the OIE website), but were altered to fit India's requirements and conditions.

4.2.7 Digital Technologies for Effective Animal Health Management

Digital technologies have the potential to significantly transform the animal health sector by improving disease surveillance, diagnostics, treatment, and overall management of animal health. Internet of Things (IoT) and sensors can be used to monitor and collect real-time data on various parameters related to animal health, such as body temperature, heart rate, feeding behaviour, and movement. This data can help detect early signs of illness, monitor animal well-being, and provide valuable insights for disease prevention and management. Digital platforms enable the creation and management of electronic health records for individual animals. Electronic Health Records (EHR) capture comprehensive information on animal health, vaccination history, treatment records, diagnostic test results, and other relevant data. This enables efficient data storage, retrieval, and analysis, facilitating informed decision-making and personalized animal health management.

Telemedicine platforms allow remote communication between veterinarians and animal owners, facilitating virtual consultations, diagnoses, and treatment guidance. Through video conferencing, messaging apps, or dedicated telemedicine platforms, veterinary professionals can provide timely advice and support, reducing the need for in-person visits and improving access to veterinary care, particularly in remote areas. Digital diagnostic tools, such as point-of-care devices and handheld scanners can revolutionize the animal health sector by enabling rapid and accurate diagnosis of animal diseases. These tools can quickly analyze samples and provide immediate results, allowing for prompt treatment and control measures. Digital diagnostic technologies also support remote diagnostic services, where images or test results can be shared with experts for interpretation and guidance. Big Data Analytics and Artificial Intelligence is advancing rapidly with more and more applications in the health sector. The analysis of large datasets through big data analytics and AI algorithms can uncover patterns, trends, and correlations related to animal health. By leveraging these technologies, animal health professionals can gain insights into disease outbreaks, risk factors, treatment efficacy, and other factors affecting animal health. This information can inform preventive measures, optimize treatment protocols, and improve overall health management.

Mobile applications tailored for the animal health sector provide a range of functionalities, including disease surveillance, animal health monitoring, medication reminders, and access to veterinary resources. These apps enable farmers, animal owners, and veterinary professionals to access information, receive alerts, and manage animal health-related tasks conveniently.

Blockchain technology is a decentralized and distributed ledger system that enables secure and transparent recording of transactions and data. This technology has the potential to enhance transparency, traceability, and security in the animal health sector. It can be used to create tamper-proof records of animal health data, vaccine distribution, and supply chain management. Blockchain-based systems ensure the integrity and authenticity of information, improving trust and facilitating data sharing among stakeholders. The adoption above mentioned digital technologies in the animal health sector can transform the sector for improved production and productivity.

4.3   Transforming Animal Health Sector for Health Security

Transforming the animal health sector is crucial for ensuring health security, which encompasses the protection of both human and animal health. Animal health plays a crucial role in human nutrition, as the well-being of animals directly affects the quality and availability of food products that contribute to a balanced and nutritious diet. With ever increasing human population, there are number of challenges and threats to human health and one of the ways to address such challenges is transformation of animal health sector to achieve the health security. The transformation can be achieved through zoonotic disease control and prevention in animals, promoting one health research, mitigating antimicrobial resistance, food safety and environmental sustainability.

4.3.1   Zoonotic Disease Prevention and Control

Prevention and control of zoonotic disease in animals is critical in order to reduce the disease burden in human. The major zoonotic disease reported in India are avian influenza, brucellosis, rabies, leptospirosis, Japanese encephalitis, Kyasanur Forest Disease (KFD), anthrax, and porcine cysticercosis. Antimicrobial resistance is biggest threat which can impact the global health in near future. The major strategies to combat zoonotic disease are surveillance for early detection, cross-sectoral collaboration, public health education, vector control, animal health management and international cooperation. The COVID-19 pandemic in human has reaffirmed the role of animals in emergence of new diseases and also significance of one health approach to understand the zoonotic disease biology. Robust surveillance systems that monitor animal populations, wildlife, and humans are crucial for early detection of zoonotic diseases. This includes active monitoring of disease indicators, such as unusual animal morbidity and mortality patterns, as well as human syndromic surveillance. Early detection allows for rapid response and implementation of control measures to prevent the spread of diseases to humans. Global Influenza Surveillance and Response System (GISRS) of WHO is one such surveillance system which functions with mission of protecting human health from the threat of influenza.

4.3.2   Food Safety

Several technologies can help to address food safety concerns and protect human health. The technologies such as blockchain, IoT, sensors, rapid pathogen detection, food safety mobile apps, food safety analytics, food packaging technologies and genetic testing and authentication have large potential to transform the food industry by addressing the food safety issues.

Blockchain improves traceability and transparency throughout the supply chain and enhances food safety. It allows for the secure recording of transactions and data, enabling the tracking of food products from farm to fork. Further, IoT devices and sensors can monitor various parameters such as temperature, humidity, and contaminants in real-time throughout the food supply chain. These technologies provide continuous data monitoring and can send alerts if any deviations occur, enabling early detection of potential food safety risks.

Number of foods borne pathogens threat human health and cause number of epidemics across globe. Advanced technologies are available for rapid and accurate detection of pathogens in food. These include polymerase chain reaction (PCR), enzyme-linked immunosorbent assay (ELISA), and next-generation sequencing (NGS) techniques. These methods enable quick identification of harmful bacteria, viruses, or parasites in food, helping to prevent foodborne illnesses. Food packaging technologies have evolved significantly. Innovative packaging technologies, such as active and intelligent packaging, can help maintain food safety and extend shelf life. Active packaging contains substances that help prevent microbial growth or control the release of certain gases, while intelligent packaging provides real-time monitoring of food conditions, such as temperature or freshness indicators.

4.3.3 Mitigating Antimicrobial Resistance

Mitigating antimicrobial resistance (AMR) in livestock is crucial for safeguarding human health security. Promoting responsible use of antibiotics in livestock is essential. This includes adhering to appropriate dosage regimens, treatment durations, and withdrawal periods to minimize the risk of antibiotic resistance development. Implementing guidelines and regulations for antibiotic use, training farmers and veterinarians, and promoting antibiotic stewardship programs can help ensure responsible antibiotic use. Strengthening veterinary oversight is crucial to ensure proper diagnosis, prescription, and use of antibiotics in livestock. Regular veterinary inspections, improved access to veterinary services, and capacity-building programs for veterinarians can contribute to better management of animal health, reducing the need for unnecessary antibiotic use.

Implementing effective disease prevention measures and biosecurity practices is key to reducing the reliance on antibiotics in livestock. Vaccination programs, improved hygiene practices, and proper farm management techniques can minimize the occurrence of infectious diseases, reducing the need for antibiotics as a preventive measure. Exploring and promoting alternative approaches to antibiotics in livestock production can help mitigate AMR. This includes the use of probiotics, prebiotics, and phytogenics as feed additives to promote gut health and enhance disease resistance. Additionally, research into novel antimicrobial agents, such as bacteriophages or antimicrobial peptides, can offer alternative solutions for combating bacterial infections in livestock. Establishing surveillance systems to monitor

AMR in livestock is crucial for early detection of emerging resistance patterns. This includes conducting regular surveillance of pathogens and monitoring the levels of antibiotic resistance in livestock populations. Surveillance data can guide policy decisions, support targeted interventions, and enable the tracking of resistance trends over time.

Raising public awareness about the importance of responsible antibiotic use in livestock and the risks associated with AMR is crucial. Educating farmers, veterinarians, and consumers about the proper use of antibiotics, alternatives to antibiotics, and the potential consequences of AMR can drive behavior change and foster responsible practices. Mitigating AMR in livestock requires a multi-faceted and collaborative approach involving governments, veterinary professionals, farmers, researchers, and other stakeholders. By implementing these strategies, it is possible to reduce the emergence and spread of antibiotic resistance in livestock, safeguard human health security, and preserve the effectiveness of antibiotics for both humans and animals.

International Partnership for Transformation of Agri-Food Systems

5

R. B. Singh, B. Mandal, and S. K. Sharma

Abstract

Indian agriculture in the pre-independence era was in highly fragile state resulting in serious food crisis. Initiatives on international cooperation in 1950s and 1960s, brought the first ray of success through international grants, joint research programmes, establishing universities and research institutes and introduction of semi-dwarf high yielding rice and wheat varieties. Subsequently, both bilateral and multilateral cooperation were intensified through Government's effort of establishing international cooperation Divisions in the Ministry of Agriculture and Farmers Welfare and coordinating with the diplomatic channels of the Ministry of External Affairs. The memorandum of understanding, work plan and research proposal templates are well established instruments today for forging international cooperation. Several international organizations, such as CIMMYT, IRRI, ICRISAT, FAO, UNDP, UN-CAPSA, CABI, NACA, APAARI, APCAEM, BMGF, ISTA and ISHS have active research collaborations with the country. Presently, India's international cooperation in agriculture research and education has been broadened significantly connecting more than 65 nations belonging all the continents. During last 75 years, the Indian national agriculture research system has emerged as one of the largest agri-research networks in the world and is rightly playing the critical roles ensuring domestic food security. India's capability in agricultural research and development has made remarkable foot-prints in its neighborhood like Afghanistan, Myanmar and Nepal and now it is ready to serve further in other parts of the world especially in global South. India is implementing National Education

R. B. Singh (✉)
FAO Regional Office for Asia and the Pacific, Bangkok, Thailand

B. Mandal · S. K. Sharma
ICAR, New Delhi, India

© National Academy of Agricultural Sciences, under exclusive license to Springer Nature Singapore Pte Ltd. 2023
K. C. Bansal et al. (eds.), *Transformation of Agri-Food Systems*,
https://doi.org/10.1007/978-981-99-8014-7_5

Policy 2020 for enhancing quality of education. Internationalization of agriculture education system will help developing a good number of Institutes of global excellence. Reorienting agricultural research and education system with more international cooperations will help India achieving many indicators of Sustainable Development Goals.

Keywords

International cooperation · Transformative agriculture · Sustainable development goals · Evergreen economy

5.1   Introduction

Agriculture being the primary source of food and fibre is critical for prosperity of India. Food commodities production is a complex and science driven process primarily involving climate, natural resources, and genetic materials. The sustained success in agriculture depends on policy, planning, funding, and strong research and development programme. The Governmental initiative of agricultural research and development in pre-independent India started way back in 1869 with the proposition of establishing a Department of Agriculture. However, the first agriculture research institute was established in 1905 at Pusa, Bihar. In 1929, the Imperial Council of Agricultural Research came into existence, which later was renamed as Indian Council of Agricultural Research (ICAR), an autonomous apex body responsible for co-ordinating agricultural education and research in India under the Department of Agricultural Research and Education (DARE), Ministry of Agriculture and Farmers Welfare (MAFW).

After independence, India made a spectacular success in agriculture sector. Strong international cooperation, coupled with the comprehensive governmental policy, country wide expansion of agri-research and education system, research and extension programmes, innovation and technology development and adoption transformed India from a food-deficit nation to a food-surplus country. The international cooperation in agriculture began in post independent India with US Educational Foundation, Ford Foundation and Rockefeller Foundation during 1950–1952. Subsequently, the contribution of Dr. N. E. Borlaug by introducing semi-dwarf wheat varieties from Mexico was a turning point in boosting wheat production in India. Since 1973 with the establishment of Department of Agricultural Research and Education (DARE) in the Ministry of Agriculture, the international cooperations were accelerated trough memorandum of understandings with many countries, foreign universities/institutions and international organisations.

Currently, India is emerging as a global agricultural powerhouse. Globally, India ranks second in farm outputs. Agriculture employs more than 50% of the Indian workforce and contributes 17–18% to the country's GDP. The international collaborations during 1950–1980 were pivotal for ushering in the Green Revolution followed by Golden, White, and Blue Revolutions, together known as Rainbow Revolution.

Although there were remarkable successes in taking forward Indian agriculture during last 75 years, yet there are several challenges to ensure food and nutritional security to the population that is largest in the world. In addition, there are other challenges to mitigate climate change, to provide higher income to farmers, to assure quality food for safe health and also to preserve and recycle environmental resources. The national apex bodies like ICAR, DARE, MAFW and other government and nongovernment agencies are engaged in developing vision road map to address the issue of sustained comprehensive food and nutrition security. To reinforce the domestic food requirement and to produce surplus food for serving the cause of core Indian philosophy: "Vashu Dhaiva Kutumbakom", "One Earth One Family", in addition to our internal abilities and capacities, it is also important to extend and strengthen international cooperation.

5.2   Cooperation Facilitation Arrangement

5.2.1   International Relation Division of ICAR and DARE

DARE was established for coordinating and promoting agricultural research and education in the country. DARE provides the necessary government linkages for ICAR, one of the largest national agricultural research systems in the world. DARE is the nodal department of the Government of India for developing foreign cooperation in agriculture research and education. DARE also supports admissions of foreign students in the Indian agriculture universities and ICAR Institutes. ICAR established International Relation Division in 2014 with a mandate to reach beyond borders legitimately for Agri-R&D, to do global technology fore-sighting, to enable research proposals for foreign collaboration and funding, to facilitate SMD/Institute Interface with DARE as a single-window and vice versa and to enable expert visitors from foreign countries to ICAR Institutes.

5.2.2   Department of Agriculture and Farmers Welfare (DAFW)

DAFW facilitates international cooperation through International Cooperation (IC) Division, which has mandate to establish bilateral and multilateral cooperation with other countries. The IC Division is a nodal point of contact for the Food and Agriculture Organization (FAO) and World Food Programme (WFP) of the United Nations. In addition, the Division also coordinates with the other organisations like G-20, BRICS etc. The IC Division has a responsibility to process the proposals for signing Memoranda of Understanding (MoUs) and also for implementing them through Work Plans. The IC Division organises Joint Working Group meetings, exchange of visits of scientists, officials and farmers, facilitates foreign deputations of Ministers and officials of DAFW as well as meetings of visiting foreign dignitaries in India.

5.3   Bilateral and Multilateral Cooperation

Many countries have agro-climates similar to India. Therefore, ICAR technologies can be shared to benefit other countries. The experience, expertise and know-how available with other countries can also be utilized through appropriate customization. Keeping this in view, DARE/ICAR has developed cooperation in research capacity building with Universities and Institutions of global repute. The IR Division of ICAR/DARE and IC Division of MAFW have international cooperation mechanisms under the provisions of Government of India, which is facilitated through MoU, Work Plans, joint working group meeting and research proposal with various Countries, International Organizations, reputed Foreign Universities and Institutions. DARE/ICAR had developed first MoU in 1975 with Peru for scientific and technical cooperation in research on potato improvement and subsequently, developed 69 MoUs with different countries. Similarly, the IC Division of MAFW facilitated signing of MoUs with as many as 61 countries.

MoUs of ICAR were extended to countries across all the continents. ICAR/DARE also extends support for the international cooperations made by other departments/Ministries such as Agriculture and Farmers Welfare, Animal Husbandry and Dairying, Fisheries and Ministry of Science and Technology and Ministry of Environment, Forest and Climate Change. These collaborations strengthen human resources development and exchange of information, technology, and germplasms etc. During 2018–2021, nearly 700 scientists were deputed abroad under these arrangements (Mishra and Arunachalam 2022).

The MoU also has been developed for cooperation with multilateral countries like Association of Southeast Asian Nations (ASEAN), Bay of Bengal Initiative for Multi-Sectoral Technical and Economic Cooperation (BIMSTEC), Brazil, Russia, India, China, and South Africa (BRICS) and South Asian Association for Regional Cooperation (SAARC). The collaborative research programme is also carried out through structured work plans under the 12 MoUs between ICAR and Consultative Group on International Agricultural Research (CGIAR) Centres. The CGIAR plays as a global research platform for a food-secure future by providing technical solutions through exchange of germplasm, technologies, technical cooperation, joint experimentation, joint publications, and capacity building. CGIAR centres work with the ICAR institutions and agricultural universities to address the existing as well as the emerging issues in agriculture. The multilateral cooperation in agriculture with the various international bodies are enumerated below:

5.3.1 Association of Southeast Asian Nations (ASEAN)

ASEAN is an international organization comprising of 10 member countries in Southeast Asia: Brunei, Cambodia, Indonesia, Laos, Malaysia, Myanmar, the Philippines, Singapore, Thailand, and Vietnam. The ASEAN was established in 1967 with the mandate of accelerating cooperative peace and shared prosperity in Southeast Asia, which is achieved through joint research and technical cooperation among ASEAN countries and with the rest of the world. India is a dialogue partner of ASEAN since 1992. Agriculture being an important area of economic development for both India as well as ASEAN Member States, an India-ASEAN Working Group on Agriculture came into action in 2011 to facilitate joint research for increasing production and productivity of crops, livestock and fisheries; to manage natural resources; to develop joint ventures and to exchange technologies and expertise.

The joint working group meeting is held annually and various programmes are proposed in these meetings for enhancing India-ASEAN cooperation in Agriculture, which are grouped under 2 years of Short-Term Plan and five years of Mid-Term Plan. The Short-Term Plan of 2011–2012 was implemented, which includes various activities such as publication of "ASEAN India Newsletters on Agriculture", ASEAN India Fellowships for higher education in Agriculture and allied sciences, conference of Heads of Agricultural Universities, exchange visits of farmers, training programmes, workshop on adaptation/mitigation technologies for climate change, meeting of Agriculture Ministers of ASEAN and India, and ASEAN-India Agri-Expo. The first group of 18 farmers and 9 officials from 9 ASEAN countries toured various ICAR Institutes, farms and industries in 2012. Three training programmes on machineries for precision farming, processing and value addition of soy products and coarse cereals and production and processing technology for value addition of horticultural products were organized for scientists from ASEAN countries in India during 2012.

The 7th ASEAN-India Ministerial Meeting on Agriculture and Forestry was held in 2022, where the Medium-Term action Plan of 2021–2025 was reviewed and the commitment to ASEAN-India cooperation in agriculture and forestry was reaffirmed. India committed to enhance cooperation with ASEAN in food security, nutrition, climate change adaptation, digital farming, nature-friendly agriculture, food processing, value chain, agricultural marketing and capacity building. India urged the ASEAN member countries to support the efforts in increasing the production, processing, value addition and consumption of millets for the health and nutrition of the people.

5.3.2 BRICS: Brazil, Russia, India, China, and South Africa

The BRICS cooperation began in 2001 with the founder member countries, Brazil, Russia, India and China and in 2011, South Africa joined with this multilateral cooperation. The cooperation in the area of agriculture was launched by setting up of Agricultural Expert Working Group (AEWG) in the meeting of Agricultural

Ministers in Moscow, 2010. The first meeting of AEWG was held in China in 2011, and subsequently, a five-year Action Plan (2012–2016) was approved. Five important areas were identified in this Action Plan and each country was assigned to coordinate an action area viz., China: creation of agricultural information exchange system of BRICS Countries; Brazil: development of strategy for ensuring access to food for the most vulnerable population; South Africa: reduction of negative impact of climate change on food security and adaptation of agriculture to climate change; India: enhancing agricultural technology cooperation and innovation, and Russia: trade and investment promotion. A significant outcome of BRICS cooperation is establishment of a virtual BRICS Agricultural Research Platform (BARP) in 2016. The secretariat of BARP was established at New Delhi under the governance of DARE. The objectives of BARP are to facilitate multilateral cooperations among the BRICS member countries for sustainable agricultural development through sharing science and technology.

5.3.3 Bay of Bengal Initiative for Multi Sectoral Technical and Economic Cooperation (BIMSTEC)

The BIMSTEC, an international organisation was established in 1997 for multi-sectoral technical and economic cooperation among seven South and Southeast Asian countries (Bangladesh, Bhutan, India, Myanmar, Nepal, Sri Lanka, and Thailand). Agriculture is one of the 14 sectors of cooperation. In the 2nd Expert Group Meeting on Agriculture Cooperation held in New Delhi in 2008, nine projects were identified of which four projects were coordinated by India: (1) Prevention and control of the trans-boundary animal diseases among the BIMSTEC member countries; (2) Affiliation of Agriculture and Veterinary Universities and Institutions among the BIMSTEC member countries; (3) Development of Agricultural Biotechnology including Biosafety in BIMSTEC member countries, and (4) Development of Seed Sector in BIMSTEC member countries.

Under the BIMSTEC umbrella, so far, meetings of agriculture ministers (BAMM), senior officers, eight agricultural experts, and workshops on good agricultural practices (GAP) and agricultural trade and investment were held. India presented the concept notes on "South-South East Asian Diagnosis Network for Ensuring Bio-security and Bio-safety" and "Human Resources Development in Agriculture" in 2015. As per recommendation of First BAMM, an International Seminar on Climate Smart Farming Systems was organised by ICAR on 11–13 December, 2019 in New Delhi. ICAR organized the expert group meeting in 2021, and identified as many as nine areas for enhancing the capacities by Indian experts/institutions in agriculture and related activities.

5.3.4 Indo Africa Forum Summit (IAFS)

IAFS, a platform to promote the African-Indian relations was first held in 2008 in New Delhi, India. The various areas of cooperations include agriculture, trade, industry and investment, peace and security, promotion of good governance and civil society and information and communication technology. ICAR/DARE was the nodal department for capacity building of African students in the different Agricultural Universities and ICAR-Deemed Universities to pursue M.Sc. And Ph.D. programmes. Under the 1st IAFS, the education programme was launched in 2010 and a total of 49 students were admitted in the various courses. During the second year, a total of 57 students were admitted in the Indian Agricultural Universities. Various training programmes were also conducted for the capacity building of African students in India under the 1st IAFS.

At the 1st IAFS, there was a focus on human resource development. In order to consolidate the outcome of the 1st IAFS, establishment of India-Africa Institutions at pan Africa level was proposed in the 2nd IAFS, which was held at in Ethiopia in 2011. These include Food Processing Cluster to contribute value-addition and to create regional and export markets; Integrated Textiles Cluster to support the cotton industry and its processing and conversion into high value products; Medium Range Weather Forecasting to harness satellite technology for the agriculture, fisheries, disaster preparedness and management of natural resources and Institute of Agriculture and Rural Development. In addition, India is interested to work with African Regional Economic Communities to establish Soil, Water and Tissue Testing Laboratories, Regional Farm Science Centres and Seed Production-cum-Demonstration Centres in the areas of agriculture sector.

The 3rd IAFS was held in New Delhi in 2015. Under the action plan of 3rd IAFS, ICAR/DARE offered India-Africa Fellowship Scheme for post-graduation and doctoral programmes in Indian Agricultural Universities in the academic year 2019–2020. The subject areas covered under this fellowship scheme were Agricultural Economics, Agronomy, Animal Sciences/Husbandry, Plant Breeding, Soil Science, Agricultural Extension, Environmental Science, Biotechnology, Biochemistry, Crop Science, Water Technology etc.

Under the action plan of 3rd IAFS a short-term training programme on "Feed and Fertility Management in Livestock" was organized at the ICAR-National Institute of Animal Nutrition and Physiology, Bengaluru in 2019. Participants from Kenya, Egypt, Sudan and Namibia received training on management of livestock.

5.3.5 South Asian Association for Regional Cooperation (SAARC)

SAARC is the intergovernmental organization of eight South Asian countries that includes Afghanistan, Bangladesh, Bhutan, India, Maldives, Nepal, Pakistan, and Sri Lanka. SAARC was established in 1985 with the objectives of economic development and regional integration. SAARC has created four specialized bodies including South Asian University in New Delhi and eleven Regional Centres including

SAARC Agricultural Centre (SAC) in Dhaka, Bangladesh. The rural people of SAARC countries being heavily dependent on agriculture and livestock, the SAARC leaders recognised importance of cooperation in agriculture sector, which resulted in adoption of the "SAARC Agriculture Vision-2020". The Technical Committee on Agriculture and Rural Development of SAARC identified DARE/ICAR for coordinating various activities in agriculture and rural development. ICAR organises workshops and seminars to support capacity building of participants of SAARC countries on agriculture technologies and good agriculture practices.

5.3.6 G20 Nations

The G20 Meeting of Agricultural Chief Scientists (MACS) is instrumental in promoting joint action to put science-based solutions into practice for achieving sustainable agricultural production systems and food and nutrition security and strengthen cooperation among G20 Nations. Since 2012, G20-MACS launched several concrete activities providing basis for desired progress in the future. The thematic areas of MACS held during 2012–2023, under the Presidency of Member nations are given in Table 5.1.

During the G-20 Presidency of India, MACS-2023 was organised at Varanasi during 17–19 April, 2023. The G-20 Member States, Invited Countries and International Organizations participated in the MACS 2023. In consonance with India's G20 Presidency theme "One Earth, One Family, One Future", the MACS 2023 at Varanasi persuaded dialogues on issues of food security in the context of international year of millets. Four priority areas were discussed for building consensus in the following areas: (a) Food Security and Nutrition- role of frontiers in science and

Table 5.1 The thematic areas of MACS held during 2012–2023-G20 presidencies

Year	Presidency	Theme
2012	Mexico	Research and development innovations
2013	Russia	Food security and nutrition challenges
2014	Australia	Transformative productivity and sustainability lift
2015	Turkey	Agriculture and food security issues-supporting transition towards sustainable agriculture and food systems
2016	China	Agriculture technology innovation and knowledge sharing
2017	Germany	Towards food and water security: fostering sustainability advancing innovations
2018	Argentina	Genome editing, sustainable soil management and climate change impacts
2019	Japan	Global research initiatives-science based decision making
2020	Saudi Arabia	Sustainable agricultural development in drylands-promoting agricultural productivity and sustainability
2021	Italy	Transition towards more sustainable food systems
2022	Indonesia	Sustainable intensification to meet food security and environmental objectives
2023	India	Sustainable agriculture and food systems for healthy people and planet

technology, (b) Building resilience and sustainable agriculture through approaches of climate resilient agriculture and One Health, (c) Digitalization for Agricultural Transformation, d) Public-Private Partnerships for Research and Development (R&D). The meeting also featured Millet-International Initiative for Research and Awareness with emphasis on millets based local food system for agrobiodiversity, food security and nutrition aligning to the International Year of Millets (UN 2021).

The one of the striking outcomes of the G20 MACS 2023 was India's initiative to launch the "Millets and other Ancient grains as International Research Initiative (MAHARISHI)" to facilitate research collaboration on millets and other underutilized grains. Under this programme, the following cost-effective activities are intended to be pursued:

• Establishing a mechanism to establish connection with the researchers for sharing research findings and identifying research gaps.
• Establishing web platforms to exchange data in an open and accessible manner.
• Organising international research workshops and conferences.
• Awarding research and innovation prizes to scientists to support and promote their research interests.

The MAHARISHI secretariat will be established in the Indian Institute of Millets Research (IIMR), Hyderabad with technical support from International Crops Research Institute for Semi-Arid Tropics (ICRISAT), One CGIAR, International Organisations (IOs) and other research institutions. As a follow up of MACS 2023 outcome, India organised two technical G20 workshops: "One Health-Opportunities and Challenges" during 29-31 August 2023 at Bangalore and "Climate Resilient Agriculture" during 4–6 September, 2023 in Hyderabad.

5.3.7 Quad Working Group

The Quadrilateral Security Dialogue among the four member countries, Australia, India, Japan and the United States is commonly known as the Quad. There are working groups on global health, climate, infrastructure, critical and emerging technologies, space, and cybersecurity. Recently, for the cooperation among Quad Countries on Science for Humanity, a working group has been initiated for Advancing Innovations for Empowering NextGen Agriculture (AI-ENGAGE).

5.3.8 Neighbourhood Nations

India has expanded cooperation on agriculture research and education in the neighboring countries such as Afghanistan, Nepal, Myanmar, Sri Lanka. The Afghan National Agricultural Sciences and Technology University (ANASTU) was established by India in Kandahar. ICAR played a critical role in providing technical support. A MoU between ICAR and ANASTU was signed on April 21, 2016. A total

Table 5.2 Enrolment in different levels at ANASTU during 2016–2019

Academic year	Total enrollments	Graduation	Post- graduation	PhD
2016–2017	43	04	39	Nil
2017–2018	88	04	80	04
2018–2019	56	05	48	03
Total	187	13	167	07

187 students were enrolled for UG, PG and PhD programmes during 2016–2019 (Table 5.2).

Similarly, India also extended cooperation with Myanmar through a MoUs signed in 2012 and 2015 that resulted in establishment of Advanced Centre for Agricultural Research and Education (ACARE) in Myanmar. In 2018, Hon'ble President of India dedicated ACARE to the people of Myanmar. DARE/ICAR provided technical support through capacity building programme on crop breeding, post-harvest technology and agriculture knowledge management. ICAR facilitated in starting new PG programmes in various subjects during 2017–2018.

Joint working group meeting between India and Nepal on agriculture focused on study visits and trainings, collaborative research and agri-business. Nepal-Aid Fund provided fellowships to the master students in Indian Agricultural Universities. There is a longstanding cooperation with ICAR and Sri Lanka Council for Agricultural Research and Policy Cooperation. India supported training of Sri Lankan scientists and students in India. India's support in the establishment of university, R&D center, laboratories, and development of faculty and course curricula for agricultural sciences has made a significant foot-print in the neighborhood. These achievements in handholding for institution building have proven India's capability in human resource development in other countries.

5.4 Cooperation with International Organisations

5.4.1 Consultative Group on International Agricultural Research (CGIAR)

CGIAR was established in 1971. It includes 15 International Centres that collaborate globally with governments, civil societies and private businesses. The major objectives of CGIAR are to diminish hunger and enhance human health through international cooperation in agricultural research and partnership. India provides strong support to CGIAR centres and ICAR/DARE has MoUs and 112 research projects with 12 CGIAR Centres/Institutes for strengthening national agriculture research system (ICAR 2017). Besides the historic introduction of wheat and rice HYVs in the 1960s leading to Green Revolution, the global partnership of India through the CG Centres helped India to harness the global germplasm, exchanging over 2 lakhs of germplasm (Fig. 5.1) for developing landmark varieties in all major commodities in different regions of the country. For example, the Borlaug Institute

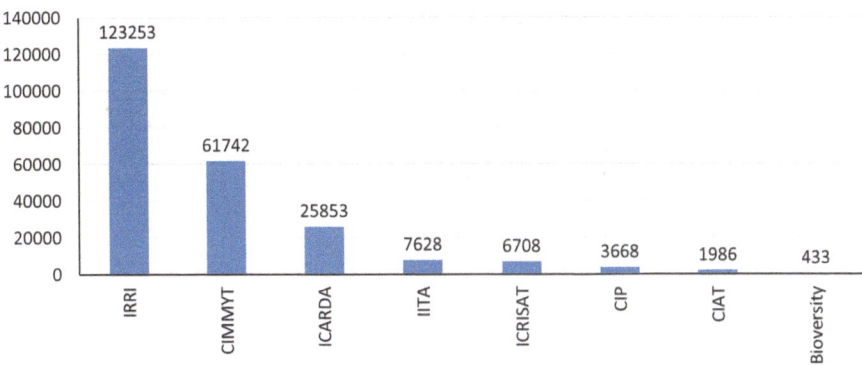

Fig. 5.1 Germplasm exchanged with CGIAR centres

for South Asia (BISA) established in 2014 at PAU, Ludhiana is strengthening the research on wheat and maize in India and South Asia. The potato germplasm from CIP has helped India develop 8 potato varieties by CPRI, including the two processing varieties which fetch 20-25% higher price in wholesale over HYVs, and occupy over 1,00,000 ha area in India. CIP has also trained a large number of potato scientists in modern technologies.

5.4.2 Food and Agriculture Organisation (FAO) of the United Nations

India is the founding member of FAO. Ground level operations of FAO in India began in 1948 and the partnership has grown steadily in the areas of crops, livestock, fisheries, food security, management of natural resources, and climate smart agriculture. Sustainable agriculture productivity, increasing farm income, natural resource management, transboundary cooperation and employment opportunity in the agriculture sector are the priority areas of collaboration.

In January 2021, FAO India, in collaboration with NITI Aayog and the MoAFW, organized a National Dialogue on Indian Agriculture Towards 2030 highlighting the pathways for enhancing farmers' income, nutritional security and sustainable food and farm systems. The proceedings of the Conference were brought out as an open access book presenting an array of policy options for a transformative agriculture. This forward-looking book is an invaluable source of science-based information for the stakeholders locally and globally. Recognizing the global leadership of India in implementing the Farmers' Rights, one of the main outcomes of the ITPGRFA, FAO had organized a Global symposium on Farmer's Rights, 12–15 September 2023, in collaboration with India's PPVFRA, New Delhi.

5.4.3   World Food Programme (WFP)

WFP is a part of United Nations that works to provide food assistance at global level. WFP was established in 1961 and is having its presence in as many as in 80 countries. WFP is supporting food and nutrition security programmes in India since 1963. The Indian programme of WFP is focused on food distribution, fortification and food insecurity mapping.

5.4.4   Asia-Pacific Association of Agriculture Research Institutions (APAARI)

ICAR being founder member since 1990 contributes US$ 10,000 annually to APAARI. Since 2018, the Director General of ICAR is the Vice-chair of APAARI. APAARI supports capacity building of ICAR scientists on regular basis.

5.4.5   Network of Aquaculture Centres in Asia-Pacific (NACA)

DARE/ICAR is the founder member and donor of NACA since 1992. The Central Institute of Freshwater Aquaculture, Bhubaneswar is an important lead center from India. ICAR led several activities e.g., production of fish spawn, infrastructure development, exposure visits of scientists, international meetings and workshops. In 2019, National Bureau of Fish Genetic Resources, Lucknow conducted a regional consultation on 'genetically responsible aquaculture: sustainability of genetically fit bloodstock and seed of certified origin in Asia aquaculture.'

5.4.6   Centre for Agriculture and Bioscience International (CABI)

In 2017, DARE/ICAR signed a MoA with CABI to establish cooperation in information management, pest management, sanitary and phytosanitary standards and capacity building. In addition, ICAR has collaboration and activities with several other internationally global organizations such as, Asian and Pacific Centre for Agricultural Engineering and Machinery, International Society for Horticulture Sciences and International Seed Testing Association.

5.5   Other Foreign Aided/Collaboration Research

Besides participating in project funded by UN Agencies, India is currently participating in nearly 50 other foreign/collaborative research projects funded by USAID, DIFD, BMGF, JIRCAS, CSIBO etc. ICAR-CIFT, Kochi collaborated with several universities in UK in the field of Anti-microbial Resistance. The collaboration with Bill and Melinda Gates Foundation has further been intensified and diversified, including work on molecular markers for improving reproduction of cattle and buffaloes, and a mega project on "Application of next-generation breeding, genotyping and digitalization approaches for improving genetic gain in Indian staple crops."

Rice Genome Project is a brilliant success story of International Collaborative Research. For India, rice is the most important food crop. With an annual production of 124 million tons of milled rice, India is the second largest producer and largest exporter of rice in the world. The production is, however, heavily constrained by various biotic and abiotic factors. Indian scientists have strived to breed improved rice varieties with higher yield potential and ability to withstand adverse climatic conditions and disease and pest attacks. The promise of creating such varieties has received a boost after the unveiling of complete genome sequence of rice by an international group of scientists including India.

The International Rice Genome Sequencing Project (IRGSP) was launched in 1998, under the leadership of Japan. It is the first crop plant whose complete genetic blue print has been unraveled. The $200 million project involved over 250 scientists from ten countries, namely, Japan, USA, Taiwan, China, France, India, South Korea, Thailand, Brazil and U. K. (each contributing varying amounts to this project) who deciphered the code of a '*japonica*' rice variety "Nipponbare".

India joined the IRGSP in 2000 with a joint ICAR-DBT project costing Rs. 48 crore (nearly US$ 10 million). In India, this work was accomplished by the ICAR-National Institute for Plant Biotechnology (NIPB) and University of Delhi, South Campus (UDSC) both at New Delhi. The 28-member team of Indian scientists was led by Prof. N. K. Singh at the ICAR-NIPB and Prof. A. K. Tyagi at UDSC.

Each grain of rice has 12 chromosomes, which bear the genes that determine all the morpho-physiological traits of a rice variety. India chose the sequencing of rice chromosome number 11 which is a rich repository of disease-resistance genes. The IRGSP scientists identified virtually all of the 389 million base pairs of the rice genome to which Indian contribution was 15 million base pairs. Total 37,544 protein-coding genes were identified. India achieved its sequencing target of phase 2 in November, 2002 and phase 3 in October, 2004 both ahead of the international deadlines and ICAR-NIPB contributed 6.7 million base pairs of the sequence. Decoding of rice genome is a major milestone in the Genetic research in India and also the world.

The rice genome provides complete information on all the rice genes. The physical map of rice chromosomes can be compared to a city map where individual genes are like its inhabitant with specific address, identity and functions. This reference rice genome information is accessed by scientists worldwide in their genetic studies and had received more than 5000 citations till 2005 (IRGSP Nature 8 August 2005).

The majority of world's poor rely on rice for their daily meals and rice genome information is helping accelerated breeding of improved rice varieties for the food and nutritional security to the human population which is anticipated to touch 10 billion by 2050. Knowledge of the rice genome will help breeders develop new rice varieties with high yield potential and resistance to diseases, pests, drought and salinity as well as having better nutritional quality.

The rice genome also provides a reference for comparing genetic differences among rice varieties. Once a gene is identified for a trait of interest, alternative forms of this gene can be searched in other varieties by a process called allele mining. These novel alleles of known genes and their fruitful assortments provide the basis of future super rice varieties.

Other Research Initiatives

NITI Aayog has emphasized that climate change management should be tackled through Multi-stakeholder Partnerships: Promoting SDG17 to Combat Climate Change. The SDGs aim to strengthen the three pillars of sustainable development: economic, social and environmental. SDG17 promotes the "right way" of collaboration between all different actions of society through the formation of multi-stakeholder partnerships to enhance synergy among SDG 1–16. The Biological Diversity Bill 2023 and the National Research Foundation Bill 2023 provide new opportunity to reengineer science-informed policies and fortify the synergy of Science Social Responsibility with Corporate Social Responsibility as well as intensify international collaboration to meet the SDGs.

All the above-mentioned international initiatives and activities are productive and register a win-win situation to the collaborative countries. India is making further efforts to enhance its global outreach systematically. The international partnerships would establish ICAR's brand value beyond India.

5.6   Aligning Agriculture Education with National Education Policy 2020

The National Education Policy-2020 (NEP-2020) of India was declared on 29 July 2020. NEP-2020 visualizes reshaping Indian agricultural education for developing professionals those have the ability to solve the problem of declining profitability and/or productivity ensuring better livelihood and food security. ICAR on 28 September 2021 released a roadmap and implementation strategy for NEP-2020 in Agricultural Education System (ICAR 2021). Before this, the National Academy of Agricultural Sciences had brought out a National Policy Document (Policy Paper No. 99) on realigning Agricultural Education with NEP-2020 (NAAS 2021). Some of the major highlights of NEP 2020 are: increasing the gross enrolment ratio (GER), defining Minimum Standards of Quality of Higher Agricultural Education, improvement in research output and outcome, making dynamic and relevant changes in curriculum and infrastructure to stay relevant and providing placement along with right skills and functional competencies. The multiple exit and entry options or flexibility in the higher education (UG, PG and Ph.D.) programmes are to be adopted also in the higher agriculture education system in the country. The ICAR Deemed Universities are in a process to be transformed into Multidisciplinary Education and Research Universities.

Agriculture is the focal point for the country's journey towards *Atmanirbhar Bharat* (self-reliance) with farmers as main stakeholder at its core. The closing years of the second decade of twenty-first century brought one of the most remarkable transitions in the international cooperation for strengthening the agriculture research and education in the history of agriculture by transitioning India from a resource-based to a science-based system of agriculture. The Green Revolution has to be transformed into Evergreen Revolution and Green Economy (Singh 2011, 2013).

The NEP 2020 focuses on reorienting India's school and higher education systems, as well as instilling research-based studies and innovations in national educational system. The Indian Council for Agricultural Research (ICAR), on the other hand, has been implementing several measures of NEP-2020 on the ground for years and is thus in line with the NEP's goals but many reforms and new measures are to be taken up for more scale and speed in holistic manner. As part of its focus on innovation and research-based learning, the ICAR-Agricultural University (AU) system, with its network of 76 universities, offers degree courses at the undergraduate level in 12 disciplines, with postgraduate programmes in 96 disciplines and PhD programmes in 73 subjects. AUs have created virtual classrooms and e-courses for UG programmes supported by a centralised Academic Management System and now being assisted for online classes by the recently released "KrishiMegh." In all fields of higher agricultural education, post-graduate courses are being converted into e-courses and blended learning is increasingly promoted. As a recommended choice, the NEP recommends a four-year Bachelor's degree programme that includes multidisciplinary education with different multi-entry and exit options to students giving emphasis on making students job providers rather than job seekers.

New curricula are being implemented for Undergraduate program of different disciplines as per guidelines of NEP-2020 with more weightage on skill development courses in first 2 years giving flexibility and choice to the students in selection of skill development courses from a basket of 'Skill Employment and Entrepreneurship Development (SEED) Modules'. Also, provisions have been incorporated for Exit and Entry options for students to get "Certificate" by end of first year and "Diploma" at the end of second year along with necessary completion of Student READY/Industry placement/Industry exposure/Hands on training in related domain of skill acquired to get first-hand experience to become eligible for the award of Certificate on exit. Similarly, internship is necessary for those exiting after diploma course. These students are expected to acquire competency and confidence to start their own enterprise, as well as will have adequate competency for getting jobs at national and international level.

Implementation of NEP-2020 in present higher agricultural education system of India have number of challenges with regards to institutional and infrastructural capacities in light of global enhancement in technologies and tools in higher education system. Low and declining budget in agricultural education and research, mushrooming of new institutions without allotting adequate resources, splitting and bi-furcation of existing agricultural universities, disconnect of education with employability and lack of adequate skills and competence are the important issues to be addressed urgently (Singh 2014).

Whole higher education system in the country requires reorientation of system of evaluation, monitoring, impact assessment, accountability and its harmony with digitalization and good governance. India can take the advantage of adopting appropriate models for strengthening of higher education from the institutes of global excellence at abroad and addition of newer courses encompassing global initiatives for making higher agricultural education system more focused, competitive and visible. Shortage of faculty in the SAUs is a major constraint to gear up the system in its right spirit and full potential. We can learn from the policy of Netherlands for reforming in higher education and research institutes and emerging as best Institutes or Universities at the global level. Netherlands stopped the proliferation of agricultural research institutes during the second half of the last century with overlapping mandates and in 1997 established the world class Wageningen University and Research Centre (WUR) by merging various institutions.

In India, as mentioned by NEP-2020, Agricultural Universities comprise approximately 9% of all universities but the enrolment in these universities is less than 1% of all enrolments in higher education which shows that GER is very poor in higher agricultural education system in the country. Both capacity in terms of intake of number and quality of students must be improved to increase the performance and competitiveness. The design of agricultural education is required for making multitasking professionals with high problem solving and critical thinking ability by providing more courses and trainings under real working conditions. Institutions offering agricultural education must benefit the local community directly; and for this developing Agricultural Technology Parks to promote technology incubation and business orientation need to be promoted. The NARES should assess manpower needs in terms of demand and supply of skilled human resource and reorient the intake and academic programmes accordingly to make students as Future Ready to deal with the national and international issues.

ICAR is expanding the scale and speed of agricultural education in the country through its IARI-Mega University Hub concept in year 2023 and opening undergraduate programmes in agriculture and allied sciences at its 15 different national institutes and centres expanding academic programs throughout India with hub partners' cooperation. In future, a strategy is required by IARI to touch a number of 3000 students as per guidelines of University Grant Commission and excel itself as a university of global excellence. ICT, digitalization, biotechnology, nanotechnology, agro-processing, Artificial Intelligence (AI), precision agriculture, carbon positive farming and systems simulation should be integrated as a core part of learning soft power to improve system efficiency over time and space.

References

ICAR (2017) ICAR–CGIAR agricultural cooperation. New Delhi, Indian Council of Agricultural Research, p 97

ICAR (2021) Implementation strategy for national education policy-2020 in agricultural education system. New Delhi, Indian Council of Agricultural Research, p 114

Mishra JP, Arunachalam A (2022) International and national collaboration in agricultural research and development. In: Pathak H, Mishra JP, Mohapatra T (eds) Indian agriculture after independence. ICAR, New Delhi, pp 358–370

NAAS (2021) Policy Paper No. 99-new agricultural education policy for reshaping India. National Academy of Agricultural Sciences, New Delhi, p 23

Singh RB (2011) Towards an evergreen revolution—the road map. New Delhi, National Academy of Agricultural Sciences, p 61

Singh RB (2013) Climate smart agriculture towards an evergreen economy. New Delhi, National Academy of Agricultural Sciences, p 32

Singh RB (2014) Transforming agricultural education for reshaping India's future. In: R B Singh (ed) Proceedings NAAS 11th agricultural science congress, Bhubaneswar, p 17–39

UN (2021) A/Res/75/263 International Year of Millets, 2023 - UN Resolution adopted by the General Assembly on 3 March 2021

Managing Plant Genetic Resources for Food and Agriculture as Global Commonwealth

6

Chikelu Mba, Ndeye Ndack Diop, Stefano Diulgheroff, Bonnie Furman, Wilson Hugo, Shawn McGuire, Arshiya Noorani, Lucio Olivero, and Shoba Sivasankar

Abstract

The conservation and sustainable use of plant genetic resources for food and agriculture are imperative for attaining and sustaining global food security and nutrition. The international community has invested significant efforts over the past several decades in developing instruments and mechanisms that underscore the interdependence of countries in safeguarding these resources, using them sustainably and in the fair and equitable sharing of the benefits that arise from their use. The Food and Agriculture Organization of the United Nations is a major driver of these endeavors and continues to both convene its members and stakeholders to achieve consensus on relevant themes and support developing countries to implement required activities. The support, which the Organization provided to its Member States in the implementation of the Second Global Plan of Action for Plant Genetic Resources for Food and Agriculture, the internationally agreed framework for this theme, over the past two years, 2021 and 2022, is described in this paper. A call is made for continued partnerships at various scales in the conservation and sustainable use of these resources going forward.

C. Mba (✉) · N. N. Diop · S. Diulgheroff · B. Furman · W. Hugo · S. McGuire · A. Noorani · L. Olivero
Plant Production and Protection Division, Food and Agriculture Organization of the United Nations, Rome, Italy
e-mail: Chikelu.Mba@fao.org

S. Sivasankar
Plant Breeding and Genetics Section, Joint FAO/IAEA Centre of Nuclear Techniques in Food and Agriculture, Department of Nuclear Sciences and Applications, International Atomic Energy Agency, Vienna, Austria

63

Keywords

Plant genetic resources for food and agriculture · Conservation · Sustainable use ·
Ex situ · In situ · On-farm · Plant breeding · Seed systems

6.1   Introduction

Plant genetic resources for food and agriculture (PGRFA), refer to "any material of
plant origin, including reproductive and vegetative propagating material, containing
functional units of heredity" .... "of actual or potential value for food and agriculture" (FAO 2009). PGRFA therefore encompass current and obsolete crop varieties;
farmers' varieties/landraces; near relatives of cultivated varieties, i.e. crop wild
relatives (CWR); wild food plants; and special genetic stocks (such as elite and
current breeders' lines and mutants) (FAO 1983). While conceivably including the
deoxynucleic acids and other hereditary materials of these plants, PGRFA usually
refer to whole plants that are used, or could potentially be used, for food and
agriculture and their propagules. PGRFA are therefore found on-farm; safeguarded
in genebanks, i.e. ex situ or in their natural habitats, i.e. in situ.

   Food insecurity and malnutrition have been worsening progressively over the
past several years with the efforts to reverse the trend being confounded by the
exacerbating effects of an ever-increasing global population, climate change, scarce
agricultural water resources and arable lands, strife, pandemics and many socioeconomic drivers (FAO, IFAD, UNICEF, WFP and WHO 2018, 2019, 2020, 2021,
2022). The situation is so dire that the attainment of the United Nations Sustainable
Development Goals (SDGs), in particular SDG 2 on the eradication of hunger and
malnutrition (UN General Assembly 2015), by 2030 now seems improbable. Nonetheless, the Food and Agriculture Organization of the United Nations (FAO),
through its Strategic Framework—with the aspirational pillars of better production,
better nutrition, a better environment and a better life—outlined an innovation-
driven blueprint for reversing this trend (FAO 2021).

   As approximately 80% of foods are plant-based, the conservation and sustainable
use of PGRFA should constitute core elements of efforts to attain food security and
nutrition. In the following sections, the critical importance of PGRFA is underscored
by a review of the intergovernmental endeavors and the ensuing multilateral processes, mechanisms and instruments for safeguarding and using them as a common
global heritage. The support which the Food and Agriculture Organization of the
United Nations (FAO) provided to its Member States for the conservation and
sustainable of PGRFA over the past two years is also provided.

6.2 International Cooperation on the Conservation and Use of PGRFA

Over the past five decades, there has been a continuing series of intergovernmental collaborations on the conservation and sustainable use of PGRFA (Sonnino 2017) underscoring therefore both the universal recognition of the critical importance of PGRFA to food security and nutrition and the interdependence of countries in managing them. This recognition is reflected in the enduring concerted and significant efforts and resources that have been invested in various normative processes and instruments aimed at making PGRFA freely available, especially for research and development, and the fair and equitable sharing of the benefits that accrue from their use.

FAO commenced the publication of a newsletter on plant genetic resources in 1957 and in 1959, at the 10th Session of its Conference, called for immediate action for the collecting and conservation of landraces and CWR (FAO 1997). Shortly afterwards, major technical meetings on the theme were convened. The Technical Meeting on Plant Exploration and Introduction was held in 1961 and led to the subsequent establishment of an FAO Panel of Experts on Plant Exploration in 1963 to advise FAO on, and set international guidelines for, the collecting, conservation and exchange of plant germplasm. Later, the International Technical Conference was organized jointly by FAO and the International Biological Programme (IBP) in 1967. The initiatives resulted in streamlined procedures for plant germplasm conservation and distribution. Importantly, the Panel of Experts articulated the guidelines and plan of action for the establishment of a global network for ex situ conservation which formed the bases for the establishment of exploration centers in regions of greatest genetic diversity (Frankel and Hawkes 1975; Scarascia-Mugnozza and Perrino 2002). The International Board on Plant Genetic Resources (IBPGR) was consequently set up in FAO as a coordinating centre for the global efforts to conserve and use PGRFA sustainably. The CGIAR center, the International Plant Genetic Resources Institute, which was later renamed Bioversity International and currently the Alliance of Bioversity International and CIAT, arose from the IBPGR.

6.2.1 The Global System on Plant Genetic Resources for Food and Agriculture

The Global System on Plant Genetic Resources for Food and Agriculture (Global System), which was created under the auspices of FAO's Commission on Genetic Resources for Food and Agriculture (Commission), consists of a set of international policy instruments and mechanisms that are aimed at the promotion of the safeguarding, availability and sustainable use of PGRFA (FAO 2010; Frison et al. 2011). FAO has continued to leverage the Global System to provide the normative platform for international cooperation on the conservation and use of PGRFA.

1. The three principal agreements under the Global System are the following:
   (a) The *International Undertaking on Plant Genetic Resources for Food and Agriculture* was adopted by FAO's Conference in 1983 based on the underlying principle that PGRFA being a heritage of humankind should be available without restriction. Its objectives are "'.... to ensure that plant genetic resources of economic and/or social interest, particularly for agriculture, will be explored, preserved, evaluated and made available for plant breeding and scientific purposes'" (FAO 1983).
   (b) The *Convention on Biological Diversity* (CBD), which is the international agreement for "'the conservation of biological diversity, the sustainable use of its components and the fair and equitable sharing of the benefits arising out of the utilization of genetic resources'" (United Nations 1993).
   (c) The *International Treaty on Plant Genetic Resources for Food and Agriculture* (International Treaty) resulted from a revision to the International Undertaking, in response to the CBD. The International Treaty was adopted by the 31st session of the FAO Conference on 3 November 2001 and entered into force on 29 June 2004, with the remit of PGRFA only and is in harmony with the CBD.
2. In June 1992, the United Nations Conference on Environment and Development (UNCED) proposed certain actions to strengthen the Global System. These included the preparations of periodic reports on the state of the world's PGRFA and a rolling global cooperative Plan of Action on PGRFA (FAO 1997). The periodic reports and rolling global plans that arose from these recommendations are the following:
   (a) **The first Report on the State of the World's Plant Genetic Resources for Food and Agriculture (First Report)** (FAO 1997). Information for compiling the First Report was obtained primarily from 154 Country Reports, which had been prepared based on guidelines developed by FAO. Through these, countries produced status updates on "indigenous plant genetic resources, national conservation activities (ex situ and in situ), in-country uses of plant genetic resources, national goals, policies, programmes and legislation, and international collaboration". Additional information was obtained from the FAO-managed database, the World Information and Early Warning System on Plant Genetic Resources for Food and Agriculture, which contained countries' responses to FAO's two questionnaires on PGRFA and forest genetic resources, respectively. Information provided by CGIAR centres; obtained from the then recently conducted external review of the CGIAR genebanks; and FAO's electronic conferences on plant breeding and genetic diversity—in which about 200 individual scientists participated, was also incorporated in the First Report. The First Report was the first comprehensive worldwide assessment of the state of the conservation and use of plant genetic resources and was based mostly on reporting by countries and CGIAR centres. It detailed the diversity, genetic vulnerability and genetic erosion of PGRFA and on the dedicated institutional and human capacities for managing them.

(b) **The Global Plan of Action for the Conservation and Sustainable Utilization of Plant Genetic Resources for Food and Agriculture (GPA)** was adopted along with the "Leipzig Declaration" in 1996 (FAO 1996). The GPA, envisaged as a rolling plan that would be subject to continuing review, was a costed plan for operationalizing the Global System. The GPA consisted of 20 priority activity areas, covering four main themes: In situ Conservation and Development; ex situ Conservation; Utilization of Plant Genetic Resources; and Institutions and Capacity Building. It was adopted as the internationally agreed framework for the conservation, exploration, collecting, characterization, evaluation and documentation of crop genetic resources.

(c) **Second Report on the State of the World's Plant Genetic Resources for Food and Agriculture (Second Report)**. The Commission, at its Twelfth Regular Session in 2009, endorsed the Second Report as the authoritative global assessment of the status of the conservation and sustainable use of PGRFA (FAO 2010). The preparatory process for the Second Report was again a country-driven and was integrated with the monitoring of the implementation of the GPA based on a set of agreed indicators. In addition to the country reports, additional information was provided by the CGIAR and other regional and international genebanks. The Second Report reflected the changes that had occurred since the adoption of the First Report while also identifying the gaps and needs for the sector. Additionally, some thematic background studies were published along with the Second Report in order to provide context.

The draft Third Report on the State of Plant Genetic Resources for Food and Agriculture (Third Report) was considered by the Commission at its 19th Regular Session in July 2023; it is expected to be endorsed in 2025. The thematic background studies on climate change; nutrition; genotyping and phenotyping of PGRFA; novel biotechnologies; and germplasm exchange will provide context for the Third Report.

(d) **Second Global Plan of Action for Plant Genetic Resources for Food and Agriculture (Second GPA)**. The GPA remained the internationally agreed framework for the conservation and sustainable use of PGRFA and for the fair and equitable sharing of the benefits that accrue from their use for 15 years, i.e. from 1996 to 2011, when the Second GPA was adopted by the FAO Council and replaced it. The Second GPA is the result of the updating of the GPA, to take into account the findings of the Second Report, especially the identified gaps and needs. The Second GPA consists of 18 Priority Activities, grouped into four main themes: In Situ Conservation and Management; Ex Situ Conservation; Sustainable Use; and Building Sustainable Institutional and Human Capacities (FAO 2012). Countries provide periodic updates to FAO on the implementation of these 18 Priority Activities.

6.3   FAO's Activities in Support of the Implementation of the Second GPA

FAO strengthens institutional and human capacities in its Member Countries for the conservation and sustainable use of PGRFA—and thereby implement the Second GPA. In this regard,

- The Organization works with countries to safeguard PGRFA in situ (i.e. in their natural habitats), ex situ (i.e. in genebanks) and to enhance on-farm diversity, in particular farmers' varieties/landraces. This involves the conservation, characterization and documentation and publication of the associated data according to internationally agreed standards, especially with the aim of making available the full range of heritable variations that could be used in plant breeding and directly by farmers and other end-users.
- FAO also supports countries' efforts to breed a diverse suite of progressively superior crop varieties that are nutritious and adapted to the farmers' changing milieu and end-user preferences, release them officially, disseminate widely the associated information—to enhance adoption by farmers; and produce early generation seeds.
- FAO's work on seed sector development also enhances farmers' timely access to sufficient quantities of the quality seeds and planting materials of the most suitable crop varieties, especially through the development and implementations of seed policies, laws and regulations; seed production, processing, storage, quality assurance and marketing; and the strengthening of the entrepreneurial skills of small- and medium-scale seed enterprises. A significant portfolio of FAO's work on seed delivery systems involves the provision of quality seeds to farmers to re-start crop production after disruptions that are caused by natural disasters and strife.

These are achieved through both normative work and operational activities. The normative work is implemented primarily through the Commission, which is the intergovernmental platform for reaching international consensus on policies for the conservation and sustainable use of genetic resources for food and agriculture and the fair and equitable sharing of the benefits that accrue from their use. The Intergovernmental Technical Working Group on Plant Genetic Resources for Food and Agriculture (Working Group) is the sector-specific subsidiary body of the Commission, with mandate for PGRFA.

The operational activities are implemented through projects in countries, which are funded through a variety of sources. FAO also convenes stakeholders to share experiences and foster collaborations on relevant themes. Another aspect of FAO's work in support of the implementation of the Second GPA is the development of tools. The Organization reports on these activities, which it implements to support its Members' efforts to conserve and use PGRFA sustainably—and thereby implement the Second GPA—to the sessions of both the Commission and the Working Group. The highlights of these support activities—in 2021 and 2022, which were reported to the 19th Regular Session of the Commission in July 2023 (FAO 2023a), are summarized below.

6.3.1 In Situ Conservation and On-Farm Management of PGRFA

The *Proceedings of the First International Multi-stakeholder Symposium on Plant Genetic Resources for Food and Agriculture*, which was held virtually in March 2021, was published (FAO 2022a) and the outcomes presented at the Ninth Session of the International Treaty's Governing Body (FAO 2022b). Additionally, a webinar on the role of the conservation and sustainable use of crop wild relatives and wild food plants was organized, in collaboration with the Treaty on 28 February 2023 (FAO 2023b).

FAO, in collaboration with international and local partners, supported several activities on in situ conservation and on-farm management of PGRFA, in particular through projects, including those funded by the Global Environment Facility (GEF) in Cuba, Ecuador, Mauritania and Peru. The conservation and use of local crops and varieties in Senegal (maize and horticultural species) and Algeria, (medicinal and aromatic plants) were also supported by FAO.

6.3.2 Ex Situ Conservation of PGRFA

6.3.2.1 Application of the Genebank Standards for Plant Genetic Resources for Food and Agriculture

Three Practical Guides for the Application of the Genebank Standards for Plant Genetic Resources for Food and Agriculture covering: (1) conservation of orthodox seeds in seed genebanks (FAO 2022c); (2) conservation in field genebanks (FAO 2022d); and (3) conservation via in vitro culture (FAO 2022e) were published.

FAO also supported various ex situ conservation activities in several countries, including Armenia, Azerbaijan, Malawi, Mongolia, the Philippines, Samoa and Venezuela (Bolivarian Republic of). As an example, 124 germplasm accessions of local crops were collected, characterized and multiplied for conservation in the genebank and distribution for use in the appropriate agroecological zones of Malawi.

The operation of community seed banks in Peru and in Southern Africa (Angola, Botswana, Malawi, Namibia, the United Republic of Tanzania and Zimbabwe) was strengthened, the latter under the auspices of GEF's Dryland Sustainable Landscapes Impact Program in Southern Africa.

6.3.3 Sustainable Use of PGRFA

6.3.3.1 Global Conference on Green Development of Seed Industries

FAO organized the Global Conference on Green Development of Seed Industries as a virtual event in November 2021 (FAO 2023c) with over 2200 participants from

126 countries. The themes of the conference were: advanced technologies; conservation of PGRFA; crop varietal development and adoption; and seed systems. The proceedings of the conference, which contain ten recommendations identified by the steering committee of the event, were published (Ruane et al. 2022). The priority activities for implementing these recommendations were discussed at the First FAO Roundtable Forum on Sustainable Seed Systems Management held in November 2022. These activities focus on adopting innovations; strengthening institutional and human capacities; safeguarding crop genetic resources, including in their natural habitats; breeding a diverse portfolio of well-adapted, progressively superior crop varieties; and developing capacities along the seed value chain (FAO 2023d).

6.3.3.2 Strengthening Seed Systems

FAO supported its Members in the development of responsive seed systems that would enable farmers, in particular small-scale farmers, to have sustained access to affordable quality seeds and planting materials of well-adapted, productive, nutritious crop varieties that are resistant to biotic and abiotic stresses. The typical interventions included the strengthening of capacities for the enhanced adoption of crop varieties, including biofortified ones; community-level seed production and delivery systems; prebasic and basic seed production and supply; capacity development for seed testing laboratories and international accreditation; training and provision of seed processing equipment; and strengthening seed certification systems. Through a variety of projects, relevant activities were implemented in Afghanistan, Armenia, Azerbaijan, Cambodia, Cote d'Ivoire, El Salvador, Egypt, Ethiopia, Georgia, Haiti, Kyrgyzstan, Mozambique, North Macedonia, Sierra Leone, Sri Lanka, Tajikistan and Venezuela.

Other interventions included the support to farmers in Haiti to obtain clean taro planting materials from the Centre for Pacific Crops and Trees (CePaCT); the establishment of demonstration plots and for the procurement and distribution of 85 tons of potato seeds and 27 tons of early-generation seeds of two elite varieties in Tajikistan. Additionally, around 200 farmers in Armenia, Kyrgyzstan, North Macedonia and Tajikistan were trained in seed multiplication. Also, Azerbaijan received support in the evaluation of European potato varieties and in the production and storage of pest and disease-free seed potatoes in vitro and in greenhouses and fields. In Niger, the production of in vitro potato seeds was also supported. In Georgia, FAO's interventions resulted in the establishment of a National Seed Producers Association while nurseries were supported with the production and export of fruit trees. Other countries where the seed delivery systems were strengthened through FAO's work in the previous two years included Cambodia, Egypt, Ethiopia and Sri Lanka.

FAO assisted 12 countries across various regions, Armenia, Azerbaijan, Gambia, Georgia, Kyrgyzstan, Mali, Mozambique, Nicaragua, North Macedonia, Rwanda, Sudan and Tajikistan, in the development of national seed policies, legislation and regulations. For Georgia, for instance, the support resulted in the development of a legal framework for the certification of fruit tree nursery materials and the establishment of a repository of pest-free propagating materials. The national seed certification agency of Sierra Leone received assistance in the review of the country's seed policy and regulations while FAO continued its support to Mozambique and Nicaragua for the development of the respective nation's seed law.

6.3.3.3 Strengthening Plant Breeding

FAO's interventions resulted in strengthened institutional and human capacities for the development and adoption of improved crop varieties in 11 countries, Afghanistan, Cote d'Ivoire, Dominica, Georgia, Niger, North Macedonia, Republic of Moldova, Suriname, Trinidad and Tobago, Uzbekistan and Venezuela (Bolivarian Republic of). In Georgia, this resulted in the verification of the genetic identity of grapevine cultivars using molecular marker systems; the genetic improvement and the strengthening of value chains, including improved access to markets, for berries in the Republic of Moldova. The production of soy, through enhanced access to early generation seeds and certified seed production, was strengthened in Afghanistan. Seed value chains were strengthened for enhanced production and distribution of certified rice seeds in Côte d'Ivoire with the concomitant facilitation of linkages between seed producers and research centres as means to enhance the production of early generation seeds. Further, FAO facilitated enhanced cooperation between the Kingdom of Saudi Arabia and the United Arab Emirates in molecular crop breeding for improved abiotic stress tolerance. Farmers in Dominica, Suriname and Trinidad and Tobago also leveraged FAO's interventions to access three new cassava varieties each, which were introduced as disease-free plantlets.

Through a GEF-funded project, Sri Lanka's capacity for the implementation of its National Biosafety Framework in accordance with the Cartagena Protocol on Biosafety to the CBD was strengthened. The outputs included the drafts of the Biosafety Regulation of Sri Lanka for LMOs/GMOs and the National Biosafety Master Plan. Also, various guidelines, manuals and strategies, including for the assessment, management and communication of risks, were developed. Training programmes in the management of the biosafety workflow were conducted for staff of the National and Sectoral Competent Authorities while the facilities of four laboratories were upgraded and staff trained for the detection of living modified organisms.

The Joint Centre of FAO and the International Atomic Energy Agency (IAEA) for Nuclear Techniques in Food and Agriculture (CJN) implemented 79 crop improvement-related national and regional Technical Cooperation Projects (TCPs) in over 100 countries. The interventions included human capacity building, technology transfer, infrastructure upgrade and technical advice related to the efficient use of mutation breeding in crop improvement. Overall, 72 new crop varieties, which were released in several countries in 2021 and 2022 were results of prior and/or ongoing support provided by FAO through CJN. Also, CJN facilitated collaborations among researchers from more than 50 institutions across 39 different countries through five crop improvement-themed collaborative projects implemented under the auspices of IAEA's Coordinated Research Projects mechanism. As of December 2022, data on 3400 mutant crop varieties derived from 228 crop species and which had been released for cultivation in 72 countries were accessible through the FAO/IAEA Mutant Variety Database.

Other crop improvement-related work of CJN included the hosting of the Global Research Symposium on the Management of Banana Fusarium Wilt TR4, which was held in Ecuador in March 2022 and the inter-regional TCP to combat TR4. Additionally, 53 requests for irradiation services for 489 crop accessions/varieties, which were received from 28 Members were actioned.

Rehabilitation of Seed Systems

FAO supports countries to restart crop production systems after disasters and strife—such as drought, floods, tropical storms and the COVID-19 pandemic, civil unrest and wars—in particular through the procurement and distribution of quality seeds and planting materials of well-adapted crop varieties to needy vulnerable farmers. Seeds and planting materials worth USD 50 million and USD 83 million were procured and distributed in the last 2 years, i.e. 2021 and 2022, respectively. These interventions are informed by the results of seed security assessments, as were recently conducted in five countries, Afghanistan, Burkina Faso, Somalia, Sudan and the Syrian Arab Republic. In the past 2 years (i.e. 2021 and 2022), vulnerable smallholder farmers affected by diverse crises in over 70 Member Nations were able to access quality seeds and planting materials of well-adapted crop varieties through FAO's emergency seed relief work.

Several countries affected by the Russia–Ukraine conflict received assistance with the supply of quality seeds and planting materials. These included Armenia and Lebanon, where vulnerable farming households received quality seeds of improved varieties of winter wheat, and the Republic of Moldova and Ukraine, where farmers received the seeds of improved varieties of sundry cereals and vegetables.

Farmers whose production systems were affected by a combination of drought and desert locust invasion, such as Afghanistan; the Horn of Africa, i.e. Djibouti, Eritrea, Ethiopia, Kenya, Somalia, South Sudan and Uganda; and Nigeria, received quality seeds and planting materials through FAO's emergency relief interventions. Also, displaced people and their host communities in Cameroon, Mozambique, Papua New Guinea and Uganda were also provided with quality seeds so as to carry on crop production activities.

These seed relief interventions typically include development components. For instance, the strengthening of seed quality assurance systems and the decentralization of farmer-led seed production entities were part of the seed relief interventions in crisis-affected countries, such as the Democratic Republic of the Congo, Haiti, Madagascar and South Sudan. In Venezuela (Bolivarian Republic of), the setting up of demonstration plots to enhance the adoption of improved crop varieties coupled with seed multiplication constituted components of the efforts to rehabilitate the national seed system.

6.3.4 Building Sustainable Institutions and Human Capacities

Human and institutional capacities for the conservation and sustainable use of PGRFA were strengthened in various countries through interventions that included enhanced partnerships and linkages and through collaboration with various partners, such as within the United Nations system-especially the World Food Programme, the International Fund for Agricultural Development and the World Meteorological Organization—and the CGIAR Centres, the Global Crop Diversity Trust, the West and Central African Council for Agricultural Research and Development, the International Seed Federation and the International Seed Testing Association (ISTA).

6.3.4.1 Capacity-Building Activities

Institutional and human capacities were strengthened through the training of agricultural extension officers and the implementation of farmer field schools (FFS) to promote ecosystem-based practices that enable the conservation of biodiversity in the United Republic of Tanzania and Zimbabwe, to mainstream biodiversity considerations into the agricultural sector. In Haiti, capacities for in vitro propagation of disease-free taro germplasm were strengthened. In 2022, following a pause due to the COVID-19 pandemic, CJN implemented 33 training courses for 704 researchers, of which 317 were women and 387 were men. Moreover, five fellows interned at its Plant Breeding and Genetics laboratory for various lengths of time in the course of the same year.

FAO assisted Azerbaijan and Pakistan to join the OECD seed certification scheme while the Organization's interventions also supported the ISTA accreditation of seed laboratories in Azerbaijan and Mozambique. In Tajikistan, national experts and 385 farmers were trained on crop variety maintenance, evaluation and registration and on the techniques for producing quality potato seeds, integrated pest management and promoting horticulture. Experts, trainers and producers, including women producers, were trained on sustainable rice production in Mauritania in a bid to strengthen the rice sector.

Pilot demonstrations and training activities were implemented in Georgia, North Macedonia, the Republic of Moldova and Uzbekistan as means to enhance the capacity of national experts and farmers in the use of improved varieties. Other capacity strengthening interventions included the support provided to Venezuela (Bolivarian Republic of) for varietal maintenance and seed production and in Niger, for in vitro potato production at the national laboratory. FAO's work also resulted in strengthened capacities for various institutions and experts in Mali, Mauritania and Niger in quality control, seed testing and seed certification; and additionally for Niger, crop varietal adoption.

In collaboration with the Secretariats of the CBD and the Treaty, the Secretariat of the Southern African Development Community (SADC) was also supported in the review of its Regional Biodiversity Strategy and to prepare for the Fifteenth Meeting of the Conference of the Parties to the CBD.

6.4   Conclusion

A description of the continuing evolution of the Global System has been used to underscore the importance that the global community rightly places on the imperatives of safeguarding PGRFA, harnessing their potentials to address developmental goals and the fair and equitable sharing of the benefits that emanate from their use. The underlying principle of the interdependence of countries in the management of PGRFA is also emphasized. The activities, which were implemented by FAO over the past two years to support countries in the conservation and sustainable use of PGRFA—using the Second GPA as a template—have also been highlighted. Nonetheless, FAO, though a critical stakeholder in the global management of PGRFA, is by no means alone in these efforts. The centres of the CGIAR, the Global Crop Diversity Trust, and numerous regional and national institutions and agricultural research and extension systems play critical roles in the management of PGRFA for the benefit of current and future generations. It is evident, however, that needs of countries remain unmet. The Third Report, which is expected to be published in 2025 will, inter alia, document the gaps and needs that remain. Addressing these needs will require continued collaborations at global, regional and local levels to apply the best practices, scientific and technological methodologies in both conserving and using PGRFA sustainably. Partnerships will remain key just as the convening roles of FAO—to enable consensus building and the harmonisation of policies and regulations—will be critically needed. In like manner, funding will be required to both strengthen institutional and human capacities and to implement required activities.

Acknowledgement   The efficient secretarial support provided by Elena Rotondo and Sara Tripodi is gratefully acknowledged.

References

FAO (1983) Resolution 8/83. International undertaking on plant genetic resources. Food and Agriculture Organization of the United Nations, Rome

FAO (1996) Global plan of action for the conservation and sustainable utilization of plant genetic resources for food and agriculture and the Leipzig declaration. Food and Agriculture Organization of the United Nations, Rome, p 63

FAO (1997) The state of the world's plant genetic resources for food and agriculture. Food and Agriculture Organization of the United Nations, Rome, p 511

FAO (2009) International treaty on plant genetic resources for food and agriculture. Food and Agriculture Organization of the United Nations, Rome. Available at https://www.fao.org/3/i0510e/i0510e00.htm. Accessed 11 September 2023

FAO (2010) The second report on the state of the world's plant genetic resources for food and agriculture. Food and Agriculture Organization of the United Nations, Rome, p 370

FAO (2012) Second global plan of action for plant genetic resources for food and agriculture. Food and Agriculture Organization of the United Nations, Rome, p 91

FAO (2021) FAO's strategic framework 2022-31. Food and Agriculture Organization of the United Nations, Rome, p 38

FAO (2022a) Proceedings of the first international multi-stakeholder symposium on plant genetic resources for food and agriculture: technical consultation on in situ conservation and on-farm management of plant genetic resources for food and agriculture. Food and Agriculture Organization of the United Nations, Rome. https://doi.org/10.4060/cc3716en

FAO (2022b) Cooperation with the commission on genetic resources for food and agriculture. In: Ninth session of the governing body. International treaty on plant genetic resources for food and agriculture. Food and Agriculture Organization of the United Nations, Rome, p 11. https://www.fao.org/3/ni842en/ni842en.pdf. Accessed 11 September 2023

FAO (2022c) Practical guide for the application of the Genebank Standards for Plant Genetic Resources for Food and Agriculture: conservation of orthodox seeds in seed genebanks. Commission on Genetic Resources for Food and Agriculture, Rome. https://doi.org/10.4060/cc0021en. Accessed 11 September 2023

FAO (2022d) Practical guide for the application of the Genebank Standards for Plant Genetic Resources for Food and Agriculture: conservation in field genebanks. Commission on Genetic Resources for Food and Agriculture, Rome. https://doi.org/10.4060/cc0023en. Accessed 11 September 2023

FAO (2022e) Practical guide for the application of the Genebank Standards for Plant Genetic Resources for Food and Agriculture: conservation via in vitro culture. Commission on Genetic Resources for Food and Agriculture, Rome. https://doi.org/10.4060/cc0025en. Accessed 11 September 2023

FAO (2023a) Implementation and review of the second global plan of action for plant genetic resources for food and agriculture. In: Nineteenth regular session FAO's commission on genetic resources for food and agriculture. Food and Agriculture Organization of the United Nations, Rome, p 16. https://www.fao.org/3/nm435en/nm435en.pdf. Accessed 11 September 2023

FAO (2023b). https://www.fao.org/cgrfa/resources/news/detail-events/en/c/1629970/. Accessed 11 September 2023

FAO (2023c). https://www.fao.org/events/detail/global-conference-on-green-development-of-seed-industries/en. Accessed 11 September 2023

FAO (2023d). https://www.fao.org/director-general/news/news-article/en/c/1626124/. Accessed 11 September 2023

FAO, IFAD, UNICEF, WFP and WHO (2018) The state of food security and nutrition in the World 2018. Building climate resilience for food security and nutrition. Food and Agriculture Organization of the United Nations, Rome, p 181

FAO, IFAD, UNICEF, WFP and WHO (2019) The state of food security and nutrition in the world 2019. Safeguarding against economic slowdowns and downturns. Food and Agriculture Organization of the United Nations, Rome, p 212

FAO, IFAD, UNICEF, WFP and WHO (2020) The state of food security and nutrition in the world: transforming food systems for affordable healthy diets. Food and Agriculture Organization of the United Nations, Rome, p 287

FAO, IFAD, UNICEF, WFP and WHO (2021) The state of food security and nutrition in the world 2021. Transforming food systems for food security, improved nutrition and affordable healthy diets for all. Food and Agriculture Organization of the United Nations, Rome, p 211

FAO, IFAD, UNICEF, WFP and WHO (2022) The state of food security and nutrition in the world 2022. Repurposing food and agricultural policies to make healthy diets more affordable. Food and Agriculture Organization of the United Nations, Rome, p 231

Frankel OH, Hawkes J (1975) Crop genetic resources for today tomorrow. IBS series, vol 2. Cambridge University Press, Cambridge

Frison C, Lopez F, Esquinas-Alcazar JT (eds) (2011) Plant Genetic resources and: stakeholder perspectives on international treaty on plant genetic resources agriculture. Food and Agriculture Organization of the United Nations, Rome, p 321

Ruane J, Mba C, Xia J (eds) (2022) Proceedings of the global conference on green development of seed industries. FAO, Rome. https://doi.org/10.4060/cc1220en. Accessed 11 September 2023

Scarascia-Mugnozza GT, Perrino P (2002) The history of ex situ conservation and use of plant genetic resources. In: Engels JMM, Rao VR, Brown AHD, Jackson M (eds) Managing plant genetic diversity. CABI, Wallingford, pp 1–22

Sonnino A (2017) International instruments for conservation and sustainable use of plant genetic resources for food and agriculture: an historical appraisal. Diversity 9(4):50. https://doi.org/10.3390/d9040050. Accessed 11 September 2023

UN General Assembly (2015) Transforming our world: the 2030 agenda for sustainable development, 21 October 2015, A/RES/70/1. https://www.refworld.org/docid/57b6e3e44.html. Accessed 11 September 2023

United Nations (1993) Convention on biological diversity (with annexes). Concluded at Rio de Janeiro on 5 June 1992; entered into force 29 December 1993. Treaty Ser 1760(30619):79–307

Mainstreaming Millets for Food and Nutritional Security

C. Tara Satyavathi and B. Venkatesh Bhat

Abstract

The need to provide food security for the burgeoning impoverished populations of India culminated in adoption of new high yielding varieties of rice and wheat and intensive farming technologies which sidelined the production of hitherto main staple foods of the masses, i.e., millets, to marginal situations where no other food crops can be grown. The decline in the cultivation of nutritionally rich millets was further facilitated by the financial incentives and subsidies provided to rice and wheat cultivation and policy bias against millets. However, with the recent awareness about nutritional and health benefits ushered by millet cultivation and consumption, and the imperatives of promoting millets cultivation to mitigate climate change effects on food production, millets are becoming important again. In this direction, to strengthen the millets value chain, it is required to enhance supply and demand of millets through promotion of awareness and policy initiatives. The present status, strategies for bringing millets to the mainstream of food and nutritional security, through enhancing supply and consumption by increasing consumer awareness, facilitating manufacture and marketing of millet products as well as providing policy push for the complete millets value chain are discussed.

Keywords

Millets · Production · Consumption · Promotion · Nutrition · Policy interventions · Production incentives · Start-ups · Farmer producer organizations · International year of millets

C. T. Satyavathi (✉) · B. V. Bhat
ICAR-Indian Institute of Millets Research, Hyderabad, India
e-mail: director.millets@icar.gov.in

K. C. Bansal et al. (eds.), *Transformation of Agri-Food Systems*,
https://doi.org/10.1007/978-981-99-8014-7_7

7.1   Introduction

Millets are the warm season food crops adapted to dryland agriculture ecosystems of the arid and semi-arid tropics. In India, millets are grown in regions characterized by low to moderate precipitation (200–800 mm rainfall) across most of the states. Bajra or pearl millet (*Pennisetum typhoides*), jowar or sorghum (*Sorghum bicolor*) and ragi or finger millet (*Eleusine coracana*) are considered major millets as they are grown in larger area. Other millets which are grown in less area are called small millets (Seetharam 2015). Small millets consist of kodo millet (*Paspalum scrobiculatum*), sawan or barnyard millet (*Echinochloa frumentacea*), kutki or little millet (*Panicum miliare*), cheena or proso millet (*Panicum miliaceum*), kangni or foxtail millet (*setaria italica*), and korale or brown top millet (*Brachiaria ramosa*). Millet crops provide food, nutritional and fodder security to the communities and livestock in dry lands areas. India is the leading producer and consumer of millet crops and their products (Rao et al. 2018).

Globally, millet grains production is about 91.48 million tons, produced from 74.15 million hectares during 2019, with sorghum and pearl millet comprise more than 90% of the area and production (FAOSTAT 2023). The rest of the production includes finger millet and small millets. Millets are also being recognized as food grains for nutrition and health and nearly organic in cultivation.

7.1.1   Present Status of Production and Consumption of Millets

India is the largest producer of millets for food purposes. Most of the states of India grow one or more millet crop species. During 2022, 15.3 million tons of millet grains were produced in India from 12.7 million ha area. This constituted about 6% of the national food grain basket (DES 2023). During this period, pearl millet was grown in 6.77 million hectares that produced 8.9 million tons grain, sorghum was grown in 4.48 million ha yielding 4.38 million tons, finger millet was grown in 0.97 million ha yielding 1.68 million tons, and small millets cultivated in 0.46 m ha area that produced 0.34 million tons. Thus, major millets, i.e., pearl millet, sorghum and finger millet made for more than 95% of the area, while small millets constituted less than 5% of the area under millet grain crops. Rajasthan state has maximum area under millets (38% of the area in the country), followed by Maharashtra (19% area) and Karnataka (13% area). Maharashtra and Karnataka have the maximum area under sorghum while Rajasthan, Gujarat, Uttar Pradesh and Maharashtra have more area under pearl millet. Ragi has the maximum area in Karnataka, Tamil Nadu and Uttarakhand, besides Maharashtra. Small millets area is maximum in Madhya Pradesh, Uttarakhand and Chhattisgarh.

Though millets surrendered 56% area to other crops, millet production increased from 11.3 to 15.3 million tons during the last five decades due to enhanced productivity, which went up by more than two times overall while it's more than tripling in pearl millet. Adoption of improved varieties and hybrids has played an important role in the higher yield per unit area in these crops. The demand for millets is dominantly been for the household consumption for decades, but it's gradually declined due to the lack of awareness about health benefits, and easy availability of fine cereals (Rao et al. 2018).

7.1.2 Inclusion of Millets in Contemporary Consumption Channels

Due to the continuous research efforts aimed at diversifying the food uses of millets. In the last 10–15 years by several R&D institutes, SAUs and private agencies, a wide range of ready-to-eat and ready-to-cook products based on millets are available now, similar to those made from rice and wheat. Several state governments have piloted the inclusion of millet foods in PDS, ICDS, etc. programs. Some of the notable interventions are pilot inclusion of millet foods by the governments of Odisha, Madhya Pradesh, Karnataka, Telangana, Andhra Pradesh and Tamil Nadu.

The private food companies such as ITC, Britannia, MTR, Parle Agro, Soulful and 24 Mantra have launched many products either as fully or partially millet based formulations. Grocery e-commerce stores such as *big basket* are offering a large number of millet-based products and experiencing millions of consumer searches for millet products. Besides, the efforts by the R&D institutes like ICAR-IIMR, CFTRI, NIFTEM, IIFPT and SAUs with support from DST and RKVY RAFTAAR have built the traction of millet-based startups which has manifested as 400+ millet startups in a 5-year span.

Strengthening the millet market and transforming this niche market into a main segment requires a collaborative framework of various stakeholders—R&D, academia, processors, regulators, doctors, nutritionists, chefs, house wives, etc. However, there are several challenges on the way in scaling up the millets market to mainstream category. As millets are gaining importance in recent years, these challenges including the disruptions in the supply chain including procurement issues, shelf-life concerns, absence of standards and grades for minor millets, lack of sufficient technical know-how and research gaps, lack of awareness among the majority of people, branding, etc. to be addressed by the governments, and thus to tap the potential of this segment.

Mainstreaming millets for food and nutritional security basically involves promotion of awareness about the goodness of millets among the public, access to millet products in easily consumable form, incentives for enhancing production, among others. There is a greater need for the involvement of enterprises that scale up the availability of millet products in desirable forms. Further, ease of production, availability of quality millet grains, processing and marketing are key issues for facilitating manufacture of millet products. The present and the ongoing efforts in this regard are discussed hereunder.

7.2   Promotion of Awareness About Millets

There is a popular belief that millets are the foods consumed mainly consumed by the poor. This is despite that the millets are nutrient dense and they were cultivated and consumed for centuries. Due to the decades of efforts and occurrence of unprecedented situations like the Covid-19 pandemic, there is a shift in perceptions of the communities and millets are increasingly perceived as healthy and nutritious alternatives to fine cereals (Chapke et al. 2022). As the lifestyle diseases are surging in both urban and rural populations, millets are now more accepted in the diets of many segments of the consumers. At this juncture, strategic promotion of millets with nutritional know-how would place the millets as better choice foods in terms of nutrition and health (Gopalan et al. 2007; Longvah et al. 2017).

India as a country will be able to showcase the millets/traditional cereals as future alternative healthy options that offer nutritional security to all the needy consumers across the world, besides, offering a climate resilient option to the world. In the last decade, research efforts on diversifying uses of millets has been attempted by several R&D institutes (IIMR 2021), SAUs and private agencies. Today, a wider range of ready-to-eat and ready-to-cook products made are available, on par with rice and wheat.

7.2.1   Promotion Campaigns

Government has been very frequently promoting the use of millets through show-casing in national and international events throughout this "international year of millets" which is reaching far and wide audiences. Devising and implementing a campaign with ambassadors of public standing (celebrities, social media influences) to spread the message in social media may have quicker and widespread reach. Fairs, exhibitions, festivals and campaigns will be organized during entire period. National Level 'Eat Millets Campaigns' or similar ones may be organized to promote awareness about the usefulness of millets to reach out to the larger audience for creating demand for millets which in turn would enhance the production of millets. The campaign on social media and TV need to be further reinforced by advertisements in print media, short documentary films, publicity through vans, unique schemes of popularizing millet kitchen carts through empowering unemployed youth, recipe books in local languages, exhibitions, seminars and sponsorships of sport and cultural events.

7.2.2   India-State and Central Governments

Several state governments have piloted the inclusion of millet foods in PDS, ICDS, etc. programmes that include Odisha, Karnataka, Tamil Nadu, Madhya Pradesh and Andhra Pradesh. The Ministry of Food Processing Industries (MoFPI), Government of India has initiated establishing Common Incubation Facilities (CIFs) to facilitate food manufacture (including millets) under One district- One Product (ODOP) programme (https://mofpi.gov.in/pmfme/one-district-one-product). MoFPI has approved 5 millet-based incubation centres (CIFs).

7.2.3   Indian Council of Agricultural Research

ICAR-Indian Institute of Millets Research (IIMR), Hyderabad (https://www.millets.res.in) has pioneered in processing diversification through retrofitted forty machineries and developed more than 50 diversified millets value added processing technologies that involve milling, baking, popping, flaking, puffing, cold and hot extrusion, convenience food formulations (IIMR 2021). Through these technologies the protein and starch digestibility are improved. To make up for the nutrients lost during processing, the products such as millet cookies, millet vermicelli, millet pasta, *khichdi* mix and millet bread are fortified with natural nutrient rich ingredients like garden cress or spinach (for iron) and gingelly seed (for zinc).

Millets normally pounding in to flour and used for roti making that is also difficult to prepare due to absence of gluten that gives elasticity, to avoid the inconvenience ICAR-IIMR has done an intervention by modifying the starch content with hydro thermal treatment for easy sheeting Millet based ready to Eat—puffs, flakes, muesli, extruded snacks, cookies, Ready to cook—millet vermicelli, millet pasta, millet semolina (medium, fine and coarse) instant millet mixes and milk-millet based beverages. IIMR has also developed millet- mushroom based products to provide the balanced nutrition of carbohydrates and proteins.

Through Centre of Excellence (CoE), IIMR has put up a platform to help in R&D efforts in developing new millet food products or fine-tuning the existing ones and also on shelf life. To minimize the loss of nutrients and increase the physiological and chemical accessibility of micronutrients, processes such as soaking, malting, thermal processing, fermentation are deployed. These processes also decrease anti-nutrients such as phytates in the foods and often enhance the components responsible for increasing bioavailability (Rao et al. 2017). IIMR has been making efforts to develop novel foods including nutraceutical and functional foods, plant (millets)-based vegan proteins, 'Express foods', analogue rice, to reach new audiences managing the life style diseases.

ICAR-IIMR has launched a mobile friendly Android app called *Millet First* for providing all the information about growing millets, value addition technologies, markets, etc. at finger tips. IIMR has made millets processing machines in six clusters for the tribal communities across five states of North-East India.

Millets Technology Business incubator (TBI) of IIMR, branded as "Nutrihub" and has put forth proactive measures to build a network of millets value chain entrepreneurs and has so far done handholding of more than 400 millet-based start-ups (https://www.millets.res.in/pdf/success_stories/Compendium_Success_story_30-09-21_for_web.pdf) and their brands, and more success stories are coming up. These start-ups have been connected with both niche markets and public-funded captive markets. Further, ICAR has enabled these enterprises to connect with

Government e-market place (GeM) to cater to the demands of central and state government procurement.

7.2.4 UN: Commemoration of International Year of Millets (IYM) 2023

Developing Strategy by UN for working with Global South countries for capacity building, technology transfer and support, value chain development and incubation have been crucial for the success of IYM. All these efforts are to make sure that millets diversify the food plate of all for better nutrition, health and address SDG goals of mitigating malnutrition and zero hunger, and also help in diversifying agriculture. These also strengthen local food systems.

7.3 Strategies for Enhancing Consumption of Millets

Due to the showcasing of millets by the governments, civil societies and enterprises, the awareness about goodness of millets for nutrition and health is spreading to the educated classes. More and more of diverse millets are back in grocery shops, retailing and online portfolios, and make up significant commodities part of organic food portfolios. However, backward integration of the demand still remains a challenge. Further, the profit due to the demand generated through value addition is yet to percolate down to the millet farmers, which may however happen in a couple of years. Targeted production and marketing of millets products are briefed hereunder.

7.3.1 Millet Products for Targeted Consumer Segments

1. *For mid-day meals:* Programmes by central and state governments are essential for ensuring the nutrition to women, children and common public, and procurement security for farmers. Localized procurement and distribution of millet products that possess lower shelf life and supplying the excess grain to other locations can be a more economically viable model.
2. *For diabetic people:* Millet grains contain more dietary fibre and higher amylase-inhibitory activity which makes millets-based foods lower in glycemic index. Thus, millet foods can play an important role in delaying the onset diabetes (hyperglacemia) and its diet-based management. These properties minimize the incidence of diabetes mellitus and gastro-intestinal tract related disorders among the people using millets as staple food (Rao et al. 2017).
3. *Organic and natural food:*Millets are considered eco-friendly crops due to their lower requirement of water, chemicals and management interventions. As millets are extremely resistant to pests and diseases, these crops are easily amenable for a mixed cropping system using non-chemical pesticide management techniques.

Planting a few rows of millets in between the rows of more susceptible legumi-
nous crops are a common practice in dryland farming.

4. *Adoption of millets to manage diseases*: Studies have revealed that incidence of
life style diseases such as hyperglycemia, obesity, cardiovascular diseases, incidence of colon cancer, etc. care reduced or well managed by adopting millet-
based diet (Rao et al. 2017). The study by Anitha et al. (2021) demonstrated that
regular consumption of millets can help lower the risk of diabetes and obesity,
contributing to lowered risk of cardiovascular diseases. The research studies have
also demonstrated the effectiveness of millets in combating anemia and
deficiencies of calcium.

7.3.2 Connecting with FPOs for Better Bargaining in Pricing

Farmer Producers Organizations (FPOs) may be strengthened in such a way that
pooling, cleaning, farm-gate processing, storage and supply to market will enable
them to reap the economic benefits for their enhanced incomes, which will have a
huge motivation for increasing the millets cultivation area. Farmers in India buy all
the inputs from retailers and sell their produce to wholesalers, also accounting for the
transportation costs and losses on them. This unorganized approach where farmers
are lacking both Backward and Forward linkages is making their cultivation a
non-remunerative business. There are 100 Millets based FPOs supported by
NFSM under the National Mission on Nutricereals of which 6 millet FPOs from
Telangana, Andhra Pradesh, Karnataka and Madhya Pradesh states are being hand-
hold by ICAR-IIMR to model them.

7.3.3 Mission Millets

Under a series of programmes beginning 2011, Government of India has taken
several initiatives to promote millet cultivation and processing. In 2018, Indian
decided to include millets in NFSM (National Food Security Mission) and in the
public distribution system (PDS). Further, Indian government had approved 2018 as
National Year of Millets to boost production and consumption of millets (Press
Information Bureau 2018). A sub-mission on "nutricereals" was initiated for 5 years
beginning 2019, under National Food Security Mission, exclusively for millets.
Plans are also being formulated millets in mid-day meals in schools. Several state
governments have millet promotional activities in place—Karnataka, Tamil Nadu,
Andhra Pradesh, Odisha and Rajasthan, to name the pioneers. These steps are
anticipated to mainstream millets to benefit farmers and consumers.

Owing to their nutritional, economic and climatic characteristics, Indian Government has started efforts towards showcasing the millets in the country and across the
world. The interventions covered under NFSM-Nutri Cereals (https://www.
nutricereals.dac.gov.in/) include Cluster Front Line Demonstrations (FLDs), production and distribution of certified seeds of millets, bio-fertilizers and micro

nutrients, and cropping system-based training. These involved formation of FPOs in crop clusters, processing units for FPOs, Seed Hubs, and distribution of seed minikits. Three National Centres of Excellence (CoEs) were established at CCS Haryana Agricultural University, Hisar for pearl millet, IIMR, Hyderabad for millets and University of Agricultural Sciences, Bengaluru for small millets. Being the leader in millets production, the Government of India is exercising its leadership to position millets for the global markets. In response to the UNGA's declaration of 2023 as the 'International Year of Millets', the efforts for mainstreaming the millets both nationally and globally are on.

7.3.4   Creating Additional Demand

1. *Using grain as poultry and feed*: In USA and Japan, millets, especially, sorghum grain is being fed to poultry and swine. The grains in the form of ground, dry rolled, soaked, pelleted, steam rolled, extruded, flaked, popped, etc. are being used as poultry feed. Replacing the maize with sorghum or other millets in broiler ration will improve the body weight gaining because of high feed the conversion ratio. In India, it is used as an alternative to maize grain when the prices are 80% of maize.
2. *Use in bioethanol production:* Sweet sorghum varieties have a potential uses as complementary feedstock to sugarcane; sugars and starch in sweet sorghum could be fermented to ethanol and lignocellulose could be gasified to methanol. The high energy sorghums are capable of 75–100% more energy than sweet sorghum also. The distilled ethanol finds purposes for various applications such as cooking, lighting, gas engines, boilers, etc. The bioethanol can be sourced from stalks of sweet sorghum and is now being propagated through technology partners IIMR and ICRISAT with some industries who are involved in piloting use of 'sweet sorghum stalks' as feedstock for ethanol manufacture to supplement sugarcane molasses-based ethanol for bioethanol to be used for blending bioethanol with petrol up to 20%.
3. *Potable alcohol:* Ethanol is the most important fermentation product of grain sorghum, and making it potable as a beverage alcohol is possible with process modifications. There are number of distillation units in India that are into manufacturing potable alcohol from sorghum and bajra grain especially when it is mould-infested. States like Maharashtra have given policy support grain-based distilleries. Other fermented products such as citric acid and lactic acid can also be made from sorghum grains, besides riboflavin, antibiotics and microbial polysaccharides.
4. *Policy support for regional foods thru PDS*: The current procurement and distribution system of the country which is dominated by fine cereals like rice and wheat has caused drastic reduction in the production and consumption of nutritious millets. Regional procurement of millet grains at designated spots and use of the grains in State PDS, Mid-Day meals, etc. would increase production and consumption of millet grains and the nutritional security of masses. Some

state governments such as Maharashtra, Madhya Pradesh, Odisha, Karnataka have initiated distribution of millets through PDS.

5. *Exploring export potential:* India traditionally export millets in commodity form and not significantly. World over sorghum and other millets are used for feed purposes except in India, SE Asia, and Africa. However, in the changed context, the millets exports are emerging in terms of diversified food uses. India is the world's leader in the production of millets with 15% share of world millet production. India is the second-largest exporter of millets. India has exported millets of worth 28.5 million USD during 2019–2020, with top 10 country destinations being Nepal, Saudi Arabia, Pakistan, UAE, Tunisia, Sri Lanka, Yemen, Libya, Namibia and Morocco (FAOSTAT 2023).

7.4 Strategies for Enhancing Supply of Millets

7.4.1 Enhance Production and Productivity of Millets

1. *New cultivars with high production potential:* enormous progress has been made in post- independent India to improve the productivity, adaptability and product diversification in millets by developing high-yielding varieties and hybrids suitable for regional and national adoption. The first hybrids developed in sorghum and pearl millet established the seed industry in India. Hybrids could be successful in sorghum (often cross-pollinated crop) and pearl millet (cross-pollinated crop) due to the availability of the male sterility system for hybrid seed production. Private sed industry has significantly contributed to the development of hybrids in pearl millet, forage sorghum and grain sorghum. IIMR, along with All India Research network on Improvement of Millets, has developed a large number of cultivars in all millets, many of them being rich in micronutrients such as iron, zinc and calcium (Yadava et al. 2020).

2. *Enhancing adoption high yielding cultivars:* a major input for enhancing farm productivity in terms of production of millets is the adoption of high yielding and climate change resilient varieties to both replace older varieties/land races and area expansion. Quality seed of improved varieties is sparsely available in all millets except in bajra where private hybrids form bulk of the seeds planted. Therefore, to improve the availability of quality seeds of millets across the states where they are grown, 24 seed hubs have been sanctioned under NFSM Sub-Mission on Nutricereals, to cater to 10% increase in availability of quality seeds each year. These seed hubs would source the quality breeder seed from breeder seed production centres which are being supported under this mission.

Millet Crop Cultivars Notified in India During 2010–2023

Millet crop	Number of cultivars notified	Number of national releases
Pearl millet	94	65
Sorghum	70	28

(continued)

Millet crop	Number of cultivars notified	Number of national releases
Finger millet	44	2
Foxtail millet	8	3
Barnyard millet	7	2
Little millet	17	5
Kodo millet	12	5
Proso millet	8	3
Total	260	113

Source: ICAR-Indian Institute of Millets Research, Hyderabad

3. *Strengthening seed supply chain:* Breeder seed and certified seed production of latest high yielding varieties of millets were initiated under seed hubs to enhance the seed replacement ratio. Due to the sanction and functioning of 18 EBSP centres under the DA&FW sponsored NFSM Sub-mission on nutricereals, additional 751, 852 and 1140, 575 and 851 quintals of breeder seed were produced during 2018–2019, 2019–2020, 2020–2021, 2021–2022 and 2022–2023 respectively, in millet crops. This additional seed can cover an additional 10% area. These quality seed made available to farmers support increasing farm productivity. Under seed hubs, a total of 4887, 6231 and 4637 quintals of quality seed were produced during 2020–2021, 2021–2022 and 2022–2023.

7.4.2   Expand Area Under Millets Cultivation

1. *Rice fallows*: Growing sorghum and pearl millet after rice crop has been practiced by farmers in many states. In the past decade there has been gradual increase in area under sorghum in rice fallows of Guntur district (Andhra Pradesh) due to difficulties in going for second paddy crop. This is spreading to other locations in states such as Orissa and Karnataka. Millets' production can be further augmented in better endowed non-traditional 'wheat-rice cropping systems' regions of Punjab, UP, Bihar, etc. to diversify agriculture by adopting short duration varieties of millets.
2. *Increasing the cropping intensity in dryland agriculture*: The field crop production in dryland regions of the country is limited to single crop (during kharif or rainy season) in a year due to non-availability of irrigation during other seasons. This has limited the national cropping intensity which stands to be 142% (DAFW 2023). To enhance farmers' income and food production in the country, proper crop planning to utilize all three seasons of kharif, rabi and summer in the dryland conditions is important. In many parts of the country, sorghum is grown in both kharif and rabi seasons. Sorghum and pearl millet are most adapted to use in all-year cropping system in the dryland agriculture in warmer seasons. Most of the small millets are of shorter duration (65–90 days) and can be easily grown in warm seasons in the post-kharif fallows. Thus, inclusion of millets in rotation can

significantly increase the cropping intensity in dryland agriculture and result in additional income generation for the farmers (Rao et al. 2018).

7.5 Processing Technologies and Infrastructure-Existing Challenges

Despite the plethora of advantages of using millets as a food crop, there are challenges associated with the processing of millet grains and shelf life. Millet grains are different from rice and wheat in terms of their kernel size. Also, different millet grains have different sizes, and a single machine can't be used to mill all millets. Further, due to the smaller size of millet grains, it is difficult to remove the oil-rich millet germ from the endosperm for making flour. The presence of lipid along with moisture in the flour affects the shelf-life leading to rancidity.

There is a lack of specialized equipment decortication and polishing of millets, with the existing equipment being transferred without much modification from rice and wheat processing. As a result, primary processing equipment for millets often tends to be inefficient, resulting in poor quality. Additionally, few systematic studies have considered the optimization of storage technologies and conditions for millets, leading to high post-harvest losses and limited-period supply chains. Development of shelf-life extension approaches for millets and millet products also requires more scientific and industrial effort.

7.6 Commercialization of Millet Products

Following years of constant engagement, the food industry has recognized the unique nutritional qualities and health benefits of millets. Many companies now have built a sizeable portfolio of millet products. Presently, there are products ranging from millet flour to multi-grain mixes, instant mixes, beverages, breakfast cereals and snacks. Millet product technologies to manufacture Ready-to-Cook items such as Vermicelli, Pasta, Instant Dosa, Upma, Kheer mixes, etc., and Ready-to-Eat types including cookies, muffins, flakes, energy bars and extruded snacks are available for commercialization. Consumers are increasingly buying these products which is indicative of increasing demand.

Nutrihub, the millets technology business incubator of IIMR has successfully materialized more than 400 millet start-ups during the past 5 years (https://www.nutrihubiimr.com). The cumulative investments so far in the niche market is estimated to be about Rs. 2000 crore, and is expected to swell exponentially in the coming years with increase in market size. More than 500 startups are presently working in millet value chain in the country.

7.7   Training and Capacity Building

Training and capacity building is a very important part of mainstreaming millets to achieve a critical mass for taking off. This calls for empowerment of stakeholders involved at different levels across the value chain such as farming millets, farm-gate processing, millets value addition, commercialization, marketing etc. Currently there are limited training opportunities for the public on various millet production, processing and value addition technologies and recipes with millets. ICAR-IIMR, Hyderabad has been organizing regular training programmes on value addition, cooking with millets, entrepreneurship opportunities, etc. The states of Odisha, Karnataka, Madhya Pradesh, etc., have been engaged with IIMR for training women, tribal communities, etc., on millets value addition for public distribution. Nevertheless, there is a need for a central framework for collating the innovative technologies across the source institutions and organizing training to impart the innovations to potential users or user-trainers.

7.8   Policy Interventions

Six decades post green revolution, India has become self-sufficient in food grains production and achieved higher productivity in most of the food crops. The then need of increasing the quantity of food production, however, ignored quality of food produced in terms of nutrition if offered. This scenario led to the incidence of 'hidden hunger' or 'micronutrient malnutrition'. Today, India is grappling with the triple burden of under-nutrition, over-malnutrition and hidden hunger. In retrospect, decrease in consumption of nutritious grains such as millets is another major cause for the hidden hunger. The diversion of resources in terms of inputs subsidy, price support, procurement, inclusion in PDS and other incentives for attaining food security through highly productive crops—rice and wheat-have resulted in decreased area under millets cultivation and thus deprived the consumers of the benefits of millets, the 'nutricereals'.

7.8.1   Seven *Sutras*

The Government of India has launched a set of seven sutras in during IYM 2023 (Verma 2023). These seven sutras outline areas are: enhancement of production and productivity of millet crops, critical studies on nutrition and health benefits of millets, Value addition technologies for consumer benefits, new initiatives in processing, and recipes development, entrepreneurship/startup/collective development, awareness creation and promotion including international outreach, besides strategic policy interventions for mainstreaming of millets. These programmes have been assigned to respective ministries and other relevant government bodies specializing in the niche areas.

7.8.2 Promoting Household Consumption by Supporting PDS

The household consumption of millets is an important component of overall consumption and this can be strengthened if millets are included in the Public Distribution System (PDS). For implementing this at the grassroot level decentralized local food system may be more efficient as they can potentially neutralize regional and seasonal variation. The PDS initiative will potentially boost the rural consumption of millets.

7.8.3 Inclusion of Millets in Welfare Schemes of the Government

The second component consist of the government programmes that are looking for including nutritious millets foods and fortified foods to supplement current dietary ration for addressing public health challenges of nutrition for pregnant and lactating mothers' health, child food, anemia in adolescent girls and women, etc. In this regard, the inclusion of millet foods in government welfare schemes such as ICDS, Mid-day meal scheme, etc. is a promising proposition, which is also a means of mainstreaming the millets.

7.9 Outlook

The importance and significance of millets as nutricereals is gradually entering our socioeconomic consciousness which can potentially result in beneficial nutritional outcomes for the consuming population. As the awareness about nutrition and goodness of millets spreads in the society and the technologies to make versatile food products out of millets becomes available, the means to scale up manufacture to address the increasing demand becomes crucial. Additionally, the cost of the millet foods need to be competitive to sustain long term interest of consumers. The traditional food use of millets and regional recipes also need to be emphasized and inclusion of millets in PDS is an important step in this direction. The challenges of availability of good quality and adequate quantity of raw millets produced needs to be addressed through assured procurement at MSPs. Commissioning more clusters for processing of millets will facilitate bringing more area under millets production and farmers will be able to realize more value for their produce. Thus, the consumers as well as the business and farmers stand to benefit in the process of mainstreaming millets.

India being the largest producer and consumer of millet for food, needs to achieve seamless integration of backward linkages, sizable production of all millets, a strong millets value chain, manufacturing technologies for a range of value-added products, consumer awareness, policy support, branding and enterprise, for catering to domestic as well as export market potential. This would enable sharing the heritage crops and commodities of India with the world communities as future cereal foods that are in tandem with conferring nutritional and functional health benefits.

References

Anitha S, Joanna KP, Takuji WT, Rosemary B, Ananthan R, Ian GD, Devraj PJ, Kowsalya S, Prasad KDV, Mani V, Bhandari RK (2021) A systematic review and meta-analysis of the potential of millets for managing and reducing the risk of developing diabetes mellitus. Front Nutr 8:687428. https://doi.org/10.3389/fnut.2021.687428

Chapke RR, Bhat BV, Shivaramane BGK, Tonapi VA, Rao CS (2022) Emerging potential of millets for food, fodder, nutrition and income security. Policy paper. ICAR-Indian Institute of Millets Research & ICAR-National Academy of Agricultural Research Management, Hyderabad

DAFW (2023) Annual report 2022-23. Department of Agriculture and Farmers Welfare. Ministry of Agriculture & Farmers Welfare, Government of India. https://agricoop.gov.in/Documents/annual_report_english_2022_23.pdf

DES (2023) Food grains area, production and yield. Directorate of Economics and Statistics, Department of Agriculture and Farmers Welfare. Ministry of Agriculture & Farmers Welfare, Government of India. https://eands.dacnet.nic.in/

FAOSTAT (2023) Statistical database. Food and Agriculture Organization of the United Nations, Rome

Gopalan C, Sastri BVR, Balasubramanian SC (2007) Nutritive value of Indian foods national institute of nutrition. Indian Council of Medical Research, Hyderabad

IIMR (2021) Technologies at a glance. ICAR-Indian Institute of Millets Research, Hyderabad. https://millets.res.in/technologies/Technologies_at_glance.pdf

Longvah T, Ananthan R, Bhaskarachary K, Venkaiah K (2017) Indian food composition tables. National Institute of Nutrition, Indian Council of Medical Research, Hyderabad

Press Information Bureau (2018). Government decides to declare Year 2018 as "National Year of Millets" production of millets will definitely help in providing nutritional security & preventing malnutrition, especially to the poor. https://pib.gov.in/newsite/PrintRelease.aspx?relid=177889

Rao BD, Bhaskarachary K, Christina GDA, Devi SG, Tonapi VA (2017) Nutritional and health benefits of millets. ICAR_Indian Institute of Millets Research (IIMR), Hyderabad, p 112. http://millets.res.in/m_recipes/Nutritional_health_benefits_millets.pdf

Rao BD, Bhat BV, Tonapi VA (2018) Nutricereals for nutritional security. Hyderabad, Indian Institute of Millets Research, p 148

Seetharam A (2015) Genetic improvement in small millets. In: Tonapi VA, Patil JV (eds) Millets: ensuring climate resilience and nutritional security. Daya Publishing House, New Delhi, pp 233–275

Verma M (2023) India's wealth: millet for health. National Institute for Communication, Government of India, New Delhi. https://www.publicationsdivision.nic.in/journals/Journalarchives/Yojana/Yojana-English/2023/January/Yojana_2023_January_pdf.pdf

Yadava DK, Choudhury PR, Hossain F, Kumar D, Mohapatra T (2020) Biofortified varieties: sustainable way to alleviate malnutrition, 3rd edn. New Delhi, Indian Council of Agricultural Research, p 86

Natural Resource Management for Nutritional Security

Suresh Kumar Chaudhari

Abstract

Natural Resource Management plays a significant role in achieving food and nutritional security amidst declining soil health, increasing input cost/cost of cultivation, land degradation and climate change. Food insecurity, non-availability of food and nutrient-deficient food grains lead to chronic malnutrition in human. Therefore, it is essential to prioritize and implement sustainable soil management practices across different landscapes to safeguard food and nutritional security in the country. In India, widespread occurrence of nutrient deficiency have direct impact on animal and human health through food chain. For achieving nutritional security, several NRM strategies namely agronomic manipulations/crop diversification, bio-fortification, balanced nutrient application (including micronutrients), conservation agriculture, regenerative agriculture, organic farming, sustainable resource management have been discussed in this chapter. Similarly, developing and implementing soil governance and legal frameworks and policies at global, regional and national level is the need of the hour.

Keywords

Nutritional security · Natural resource management · Soil health · Sustainable agriculture practices

S. K. Chaudhari (✉)
Krishi Anushandan Bhawan-II, Pusa, New Delhi, India
e-mail: ddg.nrm@icar.gov.in

© National Academy of Agricultural Sciences, under exclusive license to Springer
Nature Singapore Pte Ltd. 2023
K. C. Bansal et al. (eds.), *Transformation of Agri-Food Systems*,
https://doi.org/10.1007/978-981-99-8014-7_8

8.1   Introduction

All soil functions are crucial for our well-being, but it's hard to rate their relative importance. Nevertheless, providing food and agriculture to the world population is fundamental for the advancement of human beings (FAO)

The world is falling behind in its efforts to put an end to hunger, food insecurity, and malnutrition in all its forms by 2030. Around 10.5% of the world's population, equivalent to 828 million people globally, were affected by hunger in 2021. This indicates an alarming increase of 46 million and 150 million people since the end of 2020 and the start of the COVID-19 pandemic a year earlier, respectively. Approximately 21.3% or 144.0 million children under the age of 5 are stunted due to chronic undernutrition. An estimated 45 million children are too thin for their height, a condition known as Cachexia wasting, and 5.6% or 38.3 million children are overweight. The costs of malnutrition on the global economy are high, with loss of productivity and health care spending amounting to as much as 5% of the global GDP, corresponding to USD 3.5 trillion annually. The immediate causes of malnutrition are multidimensional, including inadequate availability and access to safe, diverse, and nutritious food, lack of access to clean water, poor sanitation, and inadequate health care, as well as inappropriate child feeding practices. The root causes of malnutrition are far-reaching and encompass a range of economic, social, political, cultural, and environmental influences.

India ranks 107th among 121 countries and scores 29.1 in the Global Hunger Index. The majority of children in India are either undernourished or malnutrition. Achieving nutritional security is a challenging task amidst declining soil health, depletion of soil organic carbon, increasing input cost/cost of cultivation, accelerated soil erosion, land degradation-desertification and climate change (Oliver and Gregory 2015; Rattan et al. 2009; Shukla et al. 2014). In this context, food and nutrition security became of paramount significance and has to be looked into its four dimensions, viz. *availability*, *accessibility*, *utilization*, and *stability*. Several Natural resource management strategies, such as agronomic manipulations, crop diversification, bio-fortification, balanced nutrient application, resource conservation technologies, etc. play a significant role in attaining nutritional security and fulfilling sustainable development goals (SDGs). In this chapter, an attempt has been made to address Nutritional Security through different strategies and approaches vis-à-vis natural resource management.

8.2   Need and Importance of Nutritional Security

Food is the main output of agricultural activities and in turn this is the main dietary input for nutrition. There is little food or nutrition without agriculture. But, the mere availability of food is not enough to ensure adequate nutrition. Each year, World Food Day (WFD) is observed on 16 October in order to highlight the need for nutritious food and its access to millions of people, particularly those who cannot

Table 8.1 Nutrients and primary functions

Minerals	Primary functions
Sodium (Na)	Balance of fluids, nerve signalling, and contractions of muscles
Chloride (Cl)	Balance of fluids, production of acid in stomach
Potassium (K)	Balance of fluids, nerve signalling, and contractions of muscles
Calcium (Ca)	Keeping bone and teeth healthy, nerve transmission, contractions of muscles, helps in clotting of blood
Phosphorus (P)	Keeping bone and teeth healthy, keeping balance between acid and base
Magnesium (Mg)	Helps in synthesis of protein, nerve signalling, contractions of muscles
Sulfur (S)	Helps in synthesis of protein
Iron (Fe)	Act as the carrier of oxygen, act as trigger point for energy production
Zinc (Zn)	Helps in DNA and protein synthesis, helps in growth development and improvement in immune system however it deficiency cause dwarfness, reduce growth and immunosuppression
Iodine (I)	Helps in thyroid related hormone and helps in metabolism, deficiency of iodine cause Goiter and brain damage
Selenium (Se)	Act in antioxidant mechanism and its deficiency cause muscular dystrophy
Copper (Cu)	Act as cofactors in many enzymes (SOD), helps in Fe metabolism, helps in bone growth and development, deficiency cause anemia and degeneration of nervous system
Manganese (Mn)	Co factor in many enzymes
Fluoride (F)	Keeping bone and teeth healthy
Chromium (Cr)	Helps in glucose metabolism through insulin metabolism
Molybdenum (Mo)	Co factor in many enzymes
Vitamin A	Helps in vision health, immune system, skin health and other health issue

Source: Collected

afford a healthy diet. The theme for WFD for 2022 was "**Leave NO ONE Behind**". Nutrient elements serve numerous crucial functions within the organism (Table 8.1).

It is evident that food insecurity and non-availability of food, as well as nutrient-deficient food grains, leads to chronic malnutrition in humans such as Kwashiorkor (acute protein-energy malnutrition), kerato malacia /loss of vision (Vit A deficiency), Rickets (Vit D deficiency), Beriberi (Vit B1 deficiency), scurvy (Vit C) and pellagra (nicotinic acid deficiency). The food insecurity, coupled with a non-nutritious diet, imposes problems of obesity, stunting, wasting, underweight, anaemia, and severe nutrient deficiencies (Narwal et al. 2017). These problems are not only afflicting the poor but are mostly prevalent in women and children. The nutritional status in developing countries like India is related to the level of social status, education, and living standard. While under-nutrition is generally related to lower socio-economic status (SES), over-nutrition is linked with higher SES (Shukla et al. 2002; Subramanian and Davey 2006; Griffiths and Bentley 2005; Osmani and Sen 2003).

Thus, the problem of malnutrition is complex and its prevention requires deliberate efforts of awareness, education, health, nutrition of women's and the decision to empower them for nation nutritional security. Further, it is absolutely clear that there is a need to go for out-of-the-box policy decisions and innovative ideas. The role of different stakeholders like community groups, agencies, NGOs, private and public sectors, policymakers, and liberal collaborations between nations is very essential. Need to focus on agricultural research and ideas to increase efficiency of water, nutrients and energy, carbon sequestration, nutrient bio-fortification, nutrition-sensitive agriculture etc. must be made with dedicated groups and enhanced budgetary allocation and partners.

8.3   Strategies and Approaches in Achieving Nutritional Security through NRM

Achieving nutritional security is critical to the nation's prosperity. It refers to a condition where everyone can access sufficient, safe, and nutritious food to meet their dietary needs and preferences for active and healthy life (Engler-Stringer 2014). The fundamental principles and practices of food production *aim to stimulate and improve biological cycles within the farming system, maintain and enrich soil fertility, minimize various forms of pollution, avoid the use of synthetic fertilizers and pesticides, preserve genetic diversity in food, take into account the extensive socio-ecological effects of food production, and generate ample amounts of high-quality food.*

Managing soil resources is a formidable challenge in achieving food, nutritional, environmental, and livelihood security while also conserving natural resources for future generations without any further degradation. The significance of caring for soil cannot be overstated, as it plays an essential role in sustaining healthy life on Earth. As per the Vedas, *"upon this handful of soil our survival depends care for it, and it will care for your food, the fuel that you need, that will shelter you and surround you with beauty. Abuse it and the soil will collapse and die, taking us all with it"*. The root cause of deteriorating human/animal health can be attributed to poor soil health. A healthy soil can raise vigorous plants and provide nutritious produce to keep us healthy. From an agricultural perspective, *"soil health refers to the capacity of soil to function within ecosystem boundaries to sustain biological productivity, maintain environmental quality, and promote plant, animal, and human health. A healthy soil would ensure proper retention and release of water and nutrients, promote and sustain root growth, maintain soil biotic habitat, respond to management, resist degradation, and act as a buffer for environmental pollution"* (Brevik and Sauer 2015). The United Nations Millennium Development Task Force on Hunger has recommended Soil Health Improvement as one of the five measures to enhance agricultural productivity and fight hunger.

Soil organic matter (SOM) is inevitable in improving soil health and also play a significant role in global carbon cycles [i.e. from improving soil microbial diversity (micro-scale) to mitigating climate change effects (global scale)]. Therefore, enhancing soil organic carbon (SOC) through different approaches such as conservation agriculture *or minimum tillage*, organic farming and integrated nutrient management (INM) (balanced and replenishing nutrients to offset nutrient mining), in-situ decomposition of crop residue, etc. to sustain soil health and enhance productivity.

The SOC is a unique indicator that has a significant impact on various physical, chemical, and biological attributes of soil. Studies conducted in India have demonstrated that SOC is the most crucial parameter for the formulation of the soil quality index (SQI), indicating that soils with higher organic matter content are better at performing essential functions for food and nutritional security, as well as environmental conservation. The soil physical properties, such as bulk density, aggregate stability, and moisture retention, are largely affected by SOC. The higher SOC results in lower soil bulk density, improved aggregation, and stability of water-stable macro aggregates (≥ 250 µm), which positively affects soil water retention and transmission properties (Benbi 2016). The production of organics produced by soil microbes primarily leads to an increase in aggregate formation and stability. The SOC significantly influences the chemical properties of soil, including nutrient availability, exchange capacity, and its capacity to act as a proton buffer. Depending on the nature and content of SOC, it contributes 25–90% of the CEC of surface layers of mineral soils. The SOC also influences soil biological properties or processes, such as mineralization, microbial biomass, and enzyme activities.

Soil biological health is equally important as soil chemical and physical health in order to maintain higher crop productivity. Soil's biological processes affect soil's physical and chemical properties, making these three interdependent and equally important. The concept of soil health goes beyond the quality and ability of the soil to produce a certain crop. It is based on the ecological characteristics of the soil, mainly those related to the soil biota, such as its diversity, food web structure, activity, and range of functions (Sharma and Adhya 2016). In general, when we refer to soil health, biological health is often overlooked; nevertheless, soil physical and chemical health greatly rely on soil biodiversity and biological processes. Soil and its living organisms are integral components of agricultural ecosystems, playing a critical role in maintaining soil health, ecosystem functions, and productivity. These services not only aid in the functioning of natural ecosystems but also constitute an important resource for the sustainable management of agricultural systems. The value of "ecosystem services" produced by soil biota worldwide may exceed US$1542 billion/year (FAO 2002). Approximately 40–48 million tons of nitrogen are biologically fixed by agricultural crops on a yearly basis. In contrast, 83 million tons of nitrogen are industrially fixed each year for the production of fertilizer. These figures underscore the critical role that nitrogen fixation plays in supporting agricultural productivity and the global food supply. Despite the fundamental role of soil biota in maintaining sustainable and efficient agricultural systems, it is still largely ignored in the majority of agricultural development initiatives. Neglecting or abusing soil life inexorably weakens soil functions, resulting in a significant loss of fertile lands and an over-reliance on chemical fertilizers for maintaining agricultural production.

Natural resource management (NRM) plays a significant role in fulfilling sustainable development goals (SDGs) through a sustainable soil management approach (Lal 2021). Moreover, natural resources (land/soil, water, and vegetation) management have closely inter-linked with more than 7 out of 15 SDGs in alignment with the tagline of G20 i.e. 'Vasudhaiva Kutumbakam'-One Earth-One Family-One Future). In fact, more than seven SDGs directly linked with soil resources, namely SDG:1 (End poverty), SDG:2 (Zero hunger) SDG: 3 (Good health and well-being), SDG:6 (Glean water and sanitation), SDG:9 (industry, innovation, and infrastructure), SDG:13 (Climate Action), SDG: 15 (Life on Land), etc. Therefore, protecting soil resources is vital for providing/advancing food security and nutrition.

Healthy soil plays a crucial role in ensuring food security and nutrition as it produces almost 98% of our food requirements (FAO 2015). *Healthy soils* provide essential macro- and micro-nutrients, water, oxygen, and support for plants. However, soil health is threatened by various factors, including loss of soil nutrients, declining organic carbon, biodiversity, pollution, salinization, erosion and land degradation (Hou 2023; Naorem et al. 2023). Consequently, there has been a decline in the quality and quantity of food production, which can ultimately impact human health and safety (Gashu et al. 2021). To overcome these challenges, it is essential to enhance our understanding of balanced and INM, the advantages and disadvantages of conservation/regenerative agriculture, and climate change adaptation and mitigation strategies and incorporate advanced technologies such as sensors, AI, precision agriculture, crop modelling, digital farming, and big data into practice (Hou 2023).

Soil is one of the most vital natural resources for supporting life on Earth, alongside air and water. It stores plenty of nutrients that are necessary for the growth, development and survival of plants, livestock and humans. Use of soil nutrients, especially by the intensively cultivated crop regions, removes substantial amounts of nutrients around the year. Thus, maintenance of native soil fertility in a prerequisite for improving crop yield and maintaining it. It is essential to replenish the soil nutrients every year that were taken away from the field. This is especially important when the measures for proper replenishment of depleted nutrient pools are not in place due to the removal of crop residues from agricultural fields.

An effective strategy to enhance crop yields and maintain them at a high level should involve an integrated approach towards managing soil nutrients in conjunction with other complementary measures. This approach recognizes that soils store most of the essential plant nutrients needed for plant growth, and how we manage these nutrients will significantly impact plant growth, soil fertility, and agricultural sustainability. Soil nutrient management involves two main aspects: inputs, which add nutrients to the soil, and outputs, which remove nutrients from the soil, primarily through crop harvesting.

Meanwhile, the depletion of nutrients takes place when the removal is done continuously, but replenishment has not been done properly. The main reasons for the depletion include erosion, removal from vegetation cover, monocropping system, high nutrient-demanding crops, use of excess fertilizers, weed flora, oxidation, mineralization, leaching, removal of residues etc. Thus, their replenishment is quite necessary.

To ensure food and nutrition security, it is essential to prioritize and implement sustainable soil management (SSM) practices across all levels to enhance soil quality. Some key NRM interventions (Chaudhari et al. 2015) for achieving nutritional security are (1) integrated nutrient management (INM) encompassing the use of both mineral fertilizers and organic manures such as cattle manures, crop residues, urban/rural wastes, composts, green manures and bio-fertilizers, (2) soil test based site-specific nutrient application using 4 R's approach, consists of using the Right source, applying the Right rate, at the right time and in the right way, (3) crop rotation with legumes, (4) in-situ and ex-situ crop residue management and (5) controlling soil erosion/land degradation adopting appropriate soil and water conservation measures/amelioration technique, (6) agronomic manipulations/crop diversification, (7) resource conservation technologies, (8) conservation agriculture, (9) precision farming and (10) regenerative agriculture and (11) agronomic bio-fortification, and (12) climate change adaptation and mitigation strategies.

Some of the policy interventions required to ensure sustainable higher nutritive produce are; (1) strengthening soil governance and legal frameworks at a global level, (2) developing and implementing legal frameworks and policies at regional and national levels and (3) increased investment in soils and adaption of proven and sound management practices (Rojas et al. 2016)

The Indian government has initiated a range of schemes that encourage the optimal use of soil resources. The aim of these initiatives is to increase agricultural productivity and improve the profitability of farmers. This is achieved through the technology support provided by the Indian Council of Agricultural Research (ICAR) and State/Central Agricultural Universities. Some of these initiatives (Chaudhari et al. 2016) are National Mission for Sustainable Agriculture (NMSA), National Mission on Soil Health Card, Biogas and manure management schemes, Paramparagat Krishi Vikas Yojana, Watershed Management component of Pradhan Mantri Krishi Sinchayee Yojana and Nutrient Based Subsidy scheme. In India, majority of children are either undernourished or malnutrition, thus Indian Government has made a stupendous effort in convincing United Nations to declare this year as 'International Year of Millets 2023'. The International Year of Millets (IYM) provides an opportunity to raise awareness and policy attention towards the nutritional and health benefits of millets. Millets have been found to thrive in harsh and changing climate conditions, making them an ideal crop for cultivation. Furthermore, encouraging the sustainable production of millets can guarantee access to wholesome food for people (Fig. 8.1).

8.4 Conclusions

Soil is a pillar to support the idea of "One Health (Oliver and Gregory 2015; Keith et al. 2016). The goal of the One Health approach (http://www.onehealthinitiative. com) is to enhance health by unifying and promoting association among various

Fig. 8.1 Addressing
nutritional security through
including millets in diet/
international year of millets
(IYM 2023) [Photo Source:
Soma Jayaraman)

disciplines related to human, animal, and ecosystem health. These disciplines
include ecology, veterinary, agriculture, public health, microbiology, and economics
(Keith et al. 2016). It encapsulates the idea that the health of an individual, population, and ecosystem are interconnected (Gibbs 2014).

The key principles of the One Health approach include:

1. *Interdisciplinary Collaboration:* Collaboration and cooperation among
   professionals from various fields, such as human health, animal health, and
   environmental health, are promoted by the One Health approach. By fostering
   interdisciplinary partnerships, it aims to enhance communication, information
   sharing, and joint efforts to address health threats.
2. *Prevention and Preparedness:* One Health promotes a proactive disease prevention and preparedness approach. Monitoring and identifying potential health risks
   seeks to prevent disease emergence and spread through early detection, surveillance, and risk assessment.
3. *Holistic Perspective:* This approach recognizes the interrelationships of human,
   animal, and environmental health. It acknowledges that addressing health
   challenges requires considering the interactions and interdependencies among
   these three components and that interventions in one sector can affect the others.
4. *Environmental Stewardship:* One Health emphasizes the importance of
   protecting and preserving the environment for the health of all living beings.
   Climate variabilities and climate change can affect the emergence and spread of
   diseases, so safeguarding ecosystems and promoting sustainable practices are
   integral to the One Health approach.

5. *Land Degradation Neutrality*: The government of India has made a commitment to restore ~26 million ha of degraded lands by 2030 (COP 14). The LDN approach helps in restoring degraded lands also improving existing good lands by adopting a landscape approach for sustaining ecosystem functions and services in perpetuity.
6. *Public Awareness and Education:* One Health increases public awareness and understanding of the interconnectedness of human, animal, and environmental health. It advocates for education and outreach initiatives to promote responsible practices, such as proper hygiene, responsible use of antibiotics, and sustainable agriculture.

The One Health concept has gained recognition and support from various organizations worldwide, including World Health Organization (WHO), Food and Agriculture Organization (FAO), and World Organization for Animal Health (WOAH). By embracing the One Health approach, it is possible to improve disease surveillance, response, and prevention strategies, leading to better human, animals, and environmental health.

References

Benbi DK (2016) Organic matter turnover in soil. In: Pathak H, Sanyal S, Takkar PN (eds) State of the Indian agriculture – soil. National Academy of Agricultural Sciences, New Delhi, pp 36–50

Brevik EC, Sauer TJ (2015) The past, present, and future of soils and human health studies. Soil 1: 35–46

Chaudhari SK, Adlul I, Biswas PP, Sikka AK (2015) Integrated soil, water and nutrient management for sustainable agriculture in India. Indian J Fertil 11(10):51–62

Chaudhari SK, Biswas PP, Abrol IP, Acharya CL (2016) Soil and nutrient management policies. In: Pathak H, Sanyal S, Takkar PN (eds) State of the Indian agriculture – soil. National Academy of Agricultural Sciences, New Delhi, pp 332–342

Engler-Stringer R (2014) Food security. Encyclopedia of quality of life and well-being research. Springer, Dordrecht, pp 2326–2327

FAO (2002) Soil biodiversity and sustainable agriculture. In: Background paper for the Ninth Regular Session of the Commission on Genetic Resources for Food and Agriculture (CGRFA). FAO, Rome, pp 14–18

FAO (2015) Healthy soils are the basis for healthy food production. Fact sheet. FAO, Rome

Gashu D, Nalivata PC, Amede T, Ander EL, Bailey EH, Botoman L, Chagumaira C et al (2021) The nutritional quality of cereals varies geospatially in Ethiopia and Malawi. Nature 594(7861): 71–76

Gibbs EPJ (2014) The evolution of one health: a decade of progress and challenges for the future. Vet Rec 174:85–90

Griffiths P, Bentley M (2005) Women of higher socio-economic status are more likely to be overweight in Karnataka, India. Eur J Clin Nutr 59(10):1217–1220

Hou D (2023) Sustainable soil management for food security. Soil Use Manag 39(1):1–7

Keith AM, Schmidt O, McMahon BJ (2016) Soil stewardship as a nexus between ecosystem services and one health. Ecosyst Serv 17:40–42

Lal R (2021) Feeding the world and returning half of the agricultural land back to nature. J Soil Water Conserv 76(4):75A–78A

Naorem A, Jayaraman S, Sinha NK, Mohanty M, Chaudhary RS, Hati KM, Mandal A, Thakur JK, Patra AK, Srinivasarao C, Chaudhari SK, Dalal RC, Lal R (2023) Eight-year impacts of conservation agriculture on soil quality, carbon storage, and carbon emission footprint. Soil Tillage Res 232:105748

Narwal RP, Malik RS, Malhotra SK, Singh BR (2017) Micronutrients and human health. Encycl Soil Sci 2017:1443–1448

Oliver MA, Gregory PJ (2015) Soil, food security and human health: a review. Eur J Soil Sci 66(2): 257–276

Osmani S, Sen A (2003) The hidden penalties of gender inequality: fetal origins of ill-health. Econ Hum Biol 1(1):105–121

Rattan RK, Patel KP, Manjaiah KM, Datta SP (2009) Micronutrients in soil, plant, animal and human health. J Indian Soc Soil Sci 57(4):546–558

Rojas RV, Achouri M, Maroulis J, Caon L (2016) Healthy soils: a prerequisite for sustainable food security. Environ Earth Sci 75:1–10

Sharma MP, Adhya TK (2016) Management of soil biological quality. In: Pathak H, Sanyal S, Takkar PN (eds) State of the Indian agriculture – soil. National Academy of Agricultural Sciences, New Delhi, pp 85–99

Shukla HC, Gupta PC, Mehta HC, Hebert JR (2002) Descriptive epidemiology of body mass index of an urban adult population in western India. J Epidemiol Community Health 56(11):876–880

Shukla AK, Tiwari PK, Prakash C (2014) Micronutrients deficiencies vis-a-vis food and nutritional security of India. Indian J Fertil 10(12):94–112

Subramanian SV, Davey SG (2006) Patterns, distribution, and determinants of under- and over-nutrition: a population-based study of women in India. Am J Clin Nutr 84(3):633–640

Combating Micronutrient Deficiencies: Pharmaceuticals and Food Fortification

9

K. Madhavan Nair

Abstract

Vitamin and mineral deficiencies is a significant risk factor in the global burden of disease. Widely practiced strategies to address these deficiencies are supplementation and food fortification. The focus of this article is on global and national evidence base for the success of pharmaceutical supplementation and food fortification as strategies in preventing and controlling micronutrient deficiencies. Both these strategies have been found to impact in preventing and controlling dietary deficiencies of micronutrients. Food fortification demands country specific regulation based on standards and extent of spread of deficit in specific micronutrients, evidence base for impact, demand and availability, introduction of mandatory or voluntary and needs careful contextual consideration. Continuous monitoring using biomarkers and taking contextual corrective steps such as redesigning and withdrawing low impact interventions should form integral part of the policy system. The present evidences on certain micronutrients appear to be inadequate to formulate guidelines either for or against the use of single or multi vitamins and minerals supplements to prevent chronic diseases. Consideration for other public health strategies such as point of use fortification with multiple micronutrient powder (MNP) sachet for children and multiple micronutrient supplement (MMS) for pregnant women merits consideration for large scale trials. An attempt has been made to understand the challenges, knowledge gaps and research priorities in the area.

Keywords

Micronutrients · Fortification · Supplementation · Multiple micronutrient supplement · Multiple micronutrient powder · India

K. M. Nair (✉)
Former Scientist-F, ICMR-National Institute of Nutrition, Hyderabad, Telangana, India

9.1 Introduction

9.1.1 Vitamins and Minerals or Micronutrients

Micronutrients constitute vitamins and minerals which are essential nutrients needed in very small amounts. They are not synthesized in our body and therefore need to obtain from daily diets. Micronutrients carry out diverse roles such as biosynthesis of enzymes, hormones and other bio molecules needed for growth and development. A dietary inadequacy leads to deficiencies that can cause visible clinical and even alarming health conditions (pregnancy out comes), and also manifest in non-clinical forms such as reduced work capacity, lethargy etc. This can accompany compromised scholastic achievements, low labour productivity and liability from infectious diseases. The aim of this article is to review contemporary information on inadequacies of micronutrients and public health strategies practiced across the globe. It focuses on global and national standards setting, evidence base for the successful pharmaceutical supplementation (single vs multiple micronutrient) and food fortification as strategies in reducing micronutrient deficiencies and in health and disease conditions. Other evidence based public health strategies such as point of use fortification (home fortification) with multiple micronutrient powder (MNP) and multiple micronutrient supplement (MMS) for pregnant mothers as a health supplement is also considered. The challenges, knowledge gaps and research opportunities in the area are also highlighted.

9.1.2 Micronutrient Deficiencies

9.1.2.1 Global Prevalence

Vitamin A deficiency (VAD) mainly affects children and about 250 million children are at risk. On the other hand, anaemia and iron deficiency affect over 2 billion people, especially among vulnerable sections of the population-infants, children, adolescent girls and women of reproductive age. Iodine deficiency disorder (IDD) is estimated to affect more than 1.5 billion people and over 200 million people suffer from goitre and 20 million with mental impairment (Stevens et al. 2013). Low- and middle-income (LAMI) countries including India bear higher burden of the above three micronutrient deficiencies. Vulnerable segments of the population in particular pregnant women, infants, children, adolescents, and the elderly pose a higher risk of deficiency than others.

9.1.2.2 Micronutrient Deficiency in India

Periodic surveys from the ICMR-National Nutrition Bureau (NNMB) show that the habitual diets of Indians are micronutrients deficient (NNMB 2012, 2016). Assessment of clinical prevalence of IDD and VAD in children and serum vitamin A inadequacy in children and anaemia in the general population corroborates with the dietary inadequacies of these micronutrients (NFHS-3 2006; NFHS-4 2016; NFHS-5 2021; Laxmaiah et al. 2012; Nair et al. 2016; MoHFW 2019; Sarna et al. 2020).

Table 9.1 Percentage of preschool age children, school age children and adolescent aged 10–19 years classified as having deficiency of iron, zinc, vitamin A, vitamin D, vitamin B12, folate and iodine (MoHFW 2019)

Micronutrient	Percentage		
	Pre-school age children	School age children	Adolescent
Iron deficiency-ferritin <15 ng/mL	40	24	28 (female 40/male 18)
Zinc <70 µg/dL	19	17	32
Vitamin A <20 ng/mL	18	22	16
Vitamin D <12 ng/mL	14	18	24
Vitamin B12 <203 pg/mL	14	17	31
Folate erythrocyte <151 ng/mL)	23	28	37
Iodine (urine <50 µg/L)	5.4	–	

Table 9.1 provides the most recent comprehensive source of information on micronutrient deficiency in the country.

9.2 Strategies to Control and Prevent Micronutrient Deficiencies

9.2.1 Global Strategies

It is well known that people who have access to a balanced diet can meet the daily requirements of all the nutrients from normal diet. Since foods contain many phytochemicals that nurture health, it is recommended that individuals should be encouraged to acquire a balanced diet before introduction of any nutrient supplement. Many of the deficiencies of vitamins and minerals are preventable through nutrition education and by ensuring a food environment conducive for practicing consumption of locally available diverse food (WHO, FAO 2006). Public health interventions are required when daily dietary intakes are insufficient to meet an individual's daily requirements of vitamins and minerals. The following are the strategies of public health relevance to address micronutrient deficiencies.

9.2.1.1 Dietary Improvement

Enhanced agriculture production of foods containing vitamins and minerals, such as fruits, vegetables, nuts, seeds, and animal products their availability and consumption are the hallmark of diet based strategy. Though dietary diversity is considered as a long term sustainable strategy, it is difficult to put in place in LAMI countries for people to practice. Nutrition education and behavioural change communication forms an integral part of this strategy and should be a national priority.

9.2.1.2 Food Fortification

Food fortification strategy involves deliberate addition of vitamins and minerals to staple foods. This strategy is implemented for correcting a prevailing deficiency or deficiencies of single or multiple micronutrients in a population or target population groups (FAO 1995). Generally, vitamins and minerals are added to staple food products, such as wheat flours, rice, salt, milk and oil. The amount added to a food vehicle is context specific and decided based on the inadequacy of specific micronutrients at population level. Food fortification is also considered as the best global welfare investment, since it is a cost effective option to ameliorate lives and advance development of a country in a reasonable period of time.

9.2.1.3 Health Supplements/Dietary Supplement

Dietary supplements are generally taken by individuals as part of their daily diet when they perceive dietary inadequacy and diet requires supplementation of extra amounts of vitamin and minerals. They are intended for the purpose of supplementing the normal diet and not intended to treat or cure any deficiency. These are regulated in India under the Food safety and Standards Act (FSSA) and recommend not more than one RDA of vitamins and minerals in such supplements (FSSAI 2022).

9.2.1.4 Pharmaceutical Approach (Prophylactic and Therapeutic Supplements)

According to Codex the purpose of a supplement is to enable individuals to consume additional intake of vitamins and/or minerals from the normal diet. They are considered to be good sources of vitamins and minerals as it provides concentrated form of one or more micronutrients which are marketed in various formulations such as capsules, tablets, powders, solutions. These formulations are to be consumed in prescribed amounts and not in any conventional food form. Most widely used supplements include supplements of vitamin A, vitamin D, vitamin E, vitamin C, vitamin B12, folic acid; minerals like calcium, iron and zinc. There exists confusion with regards to this category; whether it is to be regulated under food or drug. In India, these formulations are to be given under the guidance of a health professional.

9.2.1.5 Public Health and Disease Control Measures

These strategies include policies and programs that prevent and treat infections, parasites, and other diseases that affect micronutrient absorption and utilization. These strategies have been shown to prevent vitamin and mineral deficiencies and also help in prevention of general under nutrition (FAO 1997).

9.2.2 National Strategies

In India, national strategies have been designed considering deficiencies of certain key macronutrients (protein and energy) and micronutrients such as vitamin A, iron and iodine. Although some of the strategies have been implemented since 1970, the national policy recommendation is to tackle them synergistically, through a multi-pronged approach.

The primary source of data for the national policy on nutrition has been the periodic surveys done by NNMB (National Nutrition Monitoring Bureau) since 1975. These surveys have highlighted that cereal-based Indian diets are acutely inadequate in iron, zinc, and vitamins A, B2, B12, D and folic acid (NNMB 2012, 2016). The intake of micronutrient containing foods such as animal and plant foods especially pulses, nuts and oil seeds, vegetables and fruits have not improved over the years (MoHFW 2019).

The current National Health Policy of India addresses micronutrient malnutrition under the heading "Interventions to address malnutrition and micronutrient deficiencies" (NHP 2017). The national policy stresses the need for addressing the deficiency through a well planned strategy on vitamin and mineral interventions. The key strategy to reduce micronutrient deficiency continues to be supplementation and food fortification, screening and treating anaemia and public awareness. It envisages continuation of supplementation of IFA (iron folic acid) among the vulnerable segments of the population, calcium supplementation during pregnancy, vitamin A supplementation among children under the age of 5 years and mandatory use of iodine fortified salt. Further, the policy recommends exploring fortified staple foods and point of use/home fortification using sachet of micronutrient for addressing deficiencies through various nutrition-sensitive interventions platform.

9.3 Nutrition Metrics

9.3.1 Nutrient Requirements and Recommended Dietary Allowances

There are global efforts to harmonize methodologies for computation of metrics for standards for nutrient referred to as dietary reference values. It is important to understand the significance of these metrics in determining adequacy/inadequacy of nutrients at population level. They can also influence the food production and availability of income elastic foods especially micronutrient containing pulses, nuts and oil seeds, fruits, vegetables and flesh foods. ICMR-NIN in 2020 revised the nutrition metrics for Indians and provided three metrics.

1. The estimated average requirement (EAR) which is the average (mean) nutrient requirement of a healthy population.
2. The recommended dietary allowance (RDA) is computed from EAR by adding 2 times the standards deviation of the mean nutrient requirement value (EAR +2SD). Thus the RDA of micronutrients displays the 98th percentile of the requirement distribution of each micronutrient and constitutes individual nutrient level 98 or INL 98 value.
3. The third metrics widely applied at population level for monitoring purpose is the tolerable upper limits (TUL) which represents the safe intake level of a given

micronutrient from all sources (diet + fortified food + supplement etc.) in a day. Although TUL for certain vitamins such as vitamin B12 have not been determined due to lack of appropriate outcome measure of toxicity, the margin of safety of micronutrients such as vitamins A, D and iron and zinc are narrow. National food regulatory agencies have adopted TUL for advocating availability of food products supplements/fortified foods based on post market surveillances and constant monitoring in a given context.

9.3.2 Rationale for Vitamin and Mineral Supplements and Fortification

LAMI countries including India are lagging behind in guaranteeing food and nutrition security leading to greater risk of micronutrient malnutrition among vulnerable segments of the populations. Dietary inadequacy of multiple micronutrients such as iron, zinc, iodine, folate, vitamin B12, and other B vitamins and vitamins C and D have been documented. However, the severity of deficiencies based on clinical or sub clinical forms of these micronutrients in populations have been documented only for vitamin A in children and folic acid and vitamin B12 in mothers and iodine and iron in general population. Further, on a background of triple burden of diseases (under nutrition, micronutrient deficiency and overweight and obesity) emerging in India, approaches to tackle them needs contextual assessment and evidence based strategies.

The two interim measures globally implemented and defined are supplementation and food fortification. Accordingly a pharmaceutical approach or supplementation is suggested only when the severity and spread of micronutrient deficiency across the population is severe-to-moderate deficiency, while staple food fortification is recommended when the deficiency is not severe but widespread.

9.3.3 Standards of Pharmaceutical Supplement

Pharmaceutical approach for correcting micronutrient deficiency considers formulations of vitamins and minerals relatively at a higher concentration well above RDA in tablet or capsule formats. These preparations may contain single or multiple micronutrients indented for parenteral use at individual levels for a given period of time. In case of National Programmes to treat a clinical deficiency condition such as anaemia or goiter or night blindness or rickets doses and duration that have been clinically tested are introduced. With respect to iron, vitamins A and D the doses administered for correcting severe forms of deficiency conditions usually go beyond TUL; for example TUL for iron 45 mg/day and supplement 60–180 mg (Nair et al. 2018).

9.3.4 Standards of Micronutrients in Food Fortification at Population Level

9.3.4.1 WHO Guidelines
WHO recommends use of the three metrics, EAR, RDA and TUL to compute and describes a method of establishing specific food fortification level of a micronutrient (WHO, FAO 2006). WHO suggests setting the target median intake level for fortification of a micronutrient with the intake of target population at the 97.5th percentile of their nutrient requirement distribution (RDA) and the probability of inadequacy (PIA) at 2.5% level. However, this process might shift (1) the target median intake above the RDA and (2) the intake among a large segment of the population may shift beyond the TUL.

9.3.4.2 National Guidelines
Food fortification of iodine, iron, folic acid, vitamin B12, vitamins A and D has been implemented in India by the FSSAI since 2018 in the country (FSSAI 2018). The standards of vitamin and mineral for food fortification implemented were set based on the single reference RDA values reported in ICMR-NIN (2010), Nair and Augustine (2016), and Nair (2019). However, these standards need updating with the introduction of new metrics in line with the recently released Nutrient requirements and RDA for Indians by the ICMR-NIN (2020), which adopted an internationally harmonized methodology. It recommends setting the 'target median intake level' that overlaps the requirement distribution in a population for setting standards of micronutrient fortification (Ghosh et al. 2023). The scientific principles and the method described are as follows:

- The risk of inadequacy/adequacy of micronutrient intake is assessed by computing the relative place of the normal micronutrient intake on its requirement scale. That is, if the dietary intake is more to the right on the requirement distribution scale then lower is the risk of inadequacy.
- The probability of inadequacy (PIA) is an indicator that estimates the percent of a given population that is at risk of inadequate/adequate intake for a specific micronutrient.
- The threshold of probability of inadequacy for a population with adequate intake of micronutrient should be $\leq 50\%$ and their intake and requirement distribution overlaps when they have nutrient adequacy.
- Intervention strategies such as diet diversification, fortification, or supplementation aimed to improve vitamin and mineral intake should ideally target a probability of inadequacy of 50%.

The two important indicators that decide assessment of an optimal level of micronutrients for staple food fortification are (1) fixing the highest acceptable proportion of the population at risk of inadequacy in vitamin and mineral intake post introduction and consumption of the fortified food, and (2) fixing the lowest acceptable proportion of population that would cross the TUL post consumption of the fortified food. Taking these two into consideration, a method for easy computing of an optimum level of micronutrient was developed which can help monitor the long-term changes in intake from multiple sources. To achieve the above objectives the following five steps have to be established (NAAS 2022; Ghosh et al. 2023).

Step 1: Observe the usual distribution of micronutrient intakes in specific population subgroups. Identify which population subgroups (based on age and sex) have the highest probability of inadequacy (PIA) of micronutrient.

Step 2: If the PIA of any population sub-group is more than 50% calculate the minimum additional micronutrient intake required to bring the PIA of all such groups close to 50% to overlap the requirement by an iterative process.

Step 3: Estimate the distribution of the usual consumption of the chosen food vehicle for fortification (cereals, salt, milk and oil) by this group.

Step 4: By simulation select an appropriate level of fortification in the chosen vehicle so that PIA is nearer to 50%, and proportion at risk of having intake more than TUL is less than 1%, so that the fortification level of selected micronutrient is safe to the extent of 99%.

Step 5: Calculate the reduction in the PIA that would be expected to occur in all the 39 subgroups of the population at this level of fortification

9.4   Outcome Measures of Vitamin and Mineral Nutrition

9.4.1   Clinical/Sub-clinical and Biomarkers

Dietary inadequacies of vitamins and minerals can manifest in clinical and subclinical forms of deficiencies. Clinical forms of their deficiency are symptomatic and oblivious while subclinical forms are diagnosed based on blood or urinary levels of biomarkers of micronutrients. The public health approaches for their mitigation is also based on the above criteria of diagnosis; whether it is clinical or sub clinical. Many countries have adopted micronutrient supplementation at prophylactic or pharmacological level or/and micronutrient fortified staple foods as strategies. These two approaches are generally introduced concurrently. The former approach is expected to prevent and the latter is to control the recurrence of micronutrient deficiencies from the population. Outcome measures selected for evaluation also depends on various factors including target group.

Table 9.2 summarizes impact of interventions with iron, folic acid, zinc, vitamin A, and multiple micronutrients on biomarkers and/or on health outcomes. Majority of the studies have been carried out in the vulnerable segments of the population viz. pediatric populations and pregnant women for the following reasons. These target groups (1) suffer from various grades (severe-moderate-mild) of micronutrient deficiency and (2) the impact of intervention is quantifiable due to their beneficial effects on growth (in terms of stunting, underweight, and wasting), development, and morbidity and in certain cases even mortality and (3) anaemia negatively impact the health and development indicators among pregnant mothers, infants and children.

Table 9.2 Summary of global public health strategies to address micronutrient deficiencies

Micronutrient	Populations at greatest risk	Strategies/mandatory/voluntary	Health impact/clinical symptom/biomarker
Iodine	General population	Mandatory/fortification of common salt	Effective in controlling goiter/increased urinary excretion of iodine in children
Iron	Vulnerable segments/less in males	Fortification/supplementation/dietary diversification/public health measures. Short term–Pharmaceutical preparation of iron and folic acid is widely used among vulnerable segments. test and treats method. Pregnant women: IFA/multiple micronutrient supplementation[a]. Medium term–wheat/rice fortification. Long term strategy–dietary diversification. A combination of iron and folic acid (and vitamin B12) either as supplement/fortification of staple food (fortified rice in India)	Iron deficiency anaemia/anaemia/biomarkers of iron (haemoglobin, ferritin, soluble transferrin receptor)
Vitamin A	Children <5 years and pregnant women	Voluntary fortification of milk and oil. Supplementation. Increased dietary diversity	Lower immunity, higher morbidity due to infections, blindness, mortality and lower serum RBP and retinol
Vitamin D	Geographically defined and General population	Mandatory fortification of oils, fats, milk fortification. Supplementation. Exposure to UV radiation	Rickets, disrupted Ca and bone metabolism/serum vitamin D <50 µmol/L (<20 ng/mL)
Folate	Pregnancy mother-fetus, adolescents, elderly	Mandatory fortification of flour. Supplementation	NTDs, anaemia, risk of chronic disease / RBC folate, serum folate and homocysteine
Vitamin B12	Elderly, vegetarian populations	Voluntary fortification supplementation. Dietary diversity. Intra muscular administration	Pernicious anaemia, neurological involvement/Serum B12
Vitamins B1, B2, B3, B5, B6, biotin, etc.)	Neurological disturbances in young and older infants and children. Alcoholic	Mandatory for some B vitamins in white flour and breakfast cereals, supplementation. Minimally processed cereals	Mortality due to infantile and dry beriberi, Wernicke-Korskakoff syndrome, pellagra/plasma thiamine mono/di phosphate, erythrocyte transketolase activity coefficient, etc

[a] 15 micronutrients–iron, iodine, zinc, copper, selenium, vitamins A, D, E, B1, B2, B3, B6, B12, C, and folic acid

9.5   Pharmaceutical Supplementation

9.5.1   Efficacy and Effectiveness

There are attempts in literature to assess impact of both efficacy and effectiveness of supplements of micronutrients on various outcome measures. Tables 9.3 and 9.4 summarizes global evidence based on impact of single (Table 9.3) and multiple micronutrient supplementation (Table 9.4) among adults, children under 5 years and pregnant women (Roy and Pitkin 2007; Fairfield and Stamfer 2007; Traber 2007;

Table 9.3 Summary of the review of evidence on the use of single micronutrient supplement in chronic disease prevention in the general adult population

Micronutrient supplement	Conclusions
• β-Carotene: 2 large trials on lung cancer prevention in smokers and male asbestos workers • In healthy American men • Four trials on CVD • Vitamin A • Vitamin A +β-carotene • Vitamin A + zinc • Vitamin E: four trials	Increase in lung cancer incidence and deaths and no impact in checking other types of cancer No effect on cancer, higher risk for thyroid and bladder cancer No benefits No trials on vitamin A Increase in Lung cancer and CVD deaths No effect on esophageal or gastric cardia cancer, reduced noncardia stomach cancer
	Decreased cardiovascular deaths, no effect on CVD Decreased risk for prostate cancer in smokers Inconclusive results on development of age related cataract, No effect on lens opacity
Vitamin B2 and niacin (Chinese trial)	Lower risk for nuclear cataracts, no effects cortical cataracts, mortality rates, stroke, upper GI dysplasia, or cancer
Vitamin B6	No effect in preventing cognitive decline in elderly
Folic acid and vitamin B12	Effective in women of childbearing age to prevent NTDs No effect in short-duration studies to prevent cognitive decline in elderly
Calcium ± vitamin D	BMD, fracture risk in postmenopausal women Ca alone increases BMD, no reduction in fracture risk Vitamin D alone: no increase BMD or reduction in fracture risk Ca + vitamin D: decrease the risk for hip and non-vertebral fractures/may increase the risk for kidney stones/no effect on colorectal cancer risk
Niacin, folic acid, vitamins B2, B6 and B12	No positive effect on chronic disease occurrence in general population

Roy and Pitkin (2007), Fairfield and Stamfer (2007), Traber (2007), and Heaney (2007)

Table 9.4 Summary of efficacy of multiple micronutrient supplement in chronic disease prevention in the general adult population

Multiple vitamin and mineral	Conclusions
Vitamin E, A, β-carotene, Se +Zn (China) Vitamin E, C, β-carotene, Se + Zn (France)	3–7 micronutrients. There were some limitations Reduction in incidence and mortality rates for esophageal and gastric and noncardia gastric cancer/lower incidence in younger persons/ higher incidence in elderly Lower incidence in overall cancer and mortality in men/No change in individual cancer Lower incidence of prostate cancer in men having normal PSA and higher in men with increased levels of PSA/no benefits or harm on CVD/only modest and inconsistent effects on cataract Lower development of intermediate-stage age related macular degeneration
Vitamins C, E, β-carotene + Zn	There was a lower risk of cancer mortality that was not significant. Daily intake associated with a reduction in total cancer in men with a history of cancer but not significant from men without any history or free of cancer
Multivitamins in the prevention of cancer in men: The Physicians' Health Study II Randomized Controlled Trial Vitamin E, C and ß-carotene 11.2 years follow up Vitamin and mineral supplements in the primary prevention of CVD and cancer: an updated systematic evidence review for the U.S. Preventive Services Task Force	No evidence of an effect of doses of MVM on CVD, cancer, or mortality in healthy individuals without nutritional deficiencies Two large trials reported a small, borderline-significant impact from intake of multivitamin for >10 years on cancer and no effect on CVD in men No effect in women
Vitamin D and risk of cause specific death: systematic review and meta-analysis of observational cohort and randomized intervention studies	Inverse associations of circulating 25-hydroxyvitamin D with risks of death due to CVD, cancer, and other causes. Significantly reduced overall mortality among older adults. Recommended further investigations to establish the optimal dose and duration

Prentice (2007) and Greenwald et al. (2007)

Heaney 2007; Prentice 2007; Greenwald et al. 2007). The need for vitamin and mineral supplementation in these segments of population vary considerably; in children and pregnant mothers the impact is more related to primary functional out comes and which may lead to extra nutritional-functional outcomes and among the older adults it is for secondary health benefits such as on cardiovascular diseases (CVDs), diabetes and cancer.

It is interesting to link the health benefits of supplementation to outcome measures of primary prevention of CVD, diabetes and cancer which are life style related health problems among adult population. In contrast in vulnerable segments of the population the primary reasons for supplementation is to correct prevailing deficiency due to increased demand of micronutrients. By correcting micronutrient deficiencies existing in children and pregnant mothers could impact on outcome measures of child growth (nutritional status) and development (mental and motor skill development) and mortality and on indicators of morbidities due to lower respiratory tract infections and diarrhea. All the above health outcome measures are beyond specific nutrient-functions and can be considered as extra-nutritional functional outcomes.

The results of all the studies are on the expected lines; both single/multiple micronutrients showed positive impact on blood biomarkers, while impacts were not evident on health outcomes such as child development and growth (Black et al. 2021) or pregnancy outcomes (Nair et al. 2004). However, there is evidence that multiple micronutrients formulations may be more impactful in enhancing micronutrient status and in reducing anaemia in school children (Table 9.4). Though mortality data was not part of such studies, all-cause mortality may show significant difference among zinc or vitamin A supplemented children compared to control. This has been attributed to the heterogeneity in pooled data and doubtful pooled effect estimate with extremely wide confidence intervals. These findings on extra nutritional functional benefits have been questioned in recent times and policy changes have been recommended (large dose administration of vitamin A for children under 5 years) in India. The routine use of multiple micronutrient supplements to reduce infections in elderly people is not convincing and contradictory (El-Kadiki and Sutton 2005). Further like in the case of vitamin A and mortality in children, the study results are heterogeneous and are parallel confounded by outcome measure.

9.5.2   Multiple Micronutrients Supplement (MMS) in Pregnancy

Based on the recent review of literature on the safety and efficacy of supplementation of nutrients in pregnancy it was recommended that supplementation is a safe and frugal method to minimize risk of preeclampsia, gestational diabetes mellitus and small for gestational age amongst others (Brown and Wright 2020). The UNICEF/WHO/UNU have designed a multi micronutrient health supplement containing 15 micronutrients (vitamin A, D, E, B1, B2, B6, B12, C, Niacin, Folic Acid, Fe, Zn, Cu, I, Se) in amounts of one RDA for pregnant women of US, except folic acid at 400 μg and iron at 30 mg. This formulation is known as UMINAP (UN International Multiple Micronutrient Antenatal Preparation) (UNICEF/WHO/UNU 1999). The composition of MMS, the nutrient reference values set by the Codex (2005) and India (ICMR-NIN 2020) and the TUL for micronutrients are given in Table 9.5. Over the years several clinical trials and systematic reviews and meta-analyses comparing MMS have been carried out. The global evidence on the relative effectiveness compared with the standard of care of supplementation of iron (with/without folic acid) documented (Tuncalp et al. 2020). The following are the summary of the findings.

Table 9.5 Composition of UNICEF/WHO/UNU multiple micronutrient supplement (MMS), US Recommended Dietary Allowances (RDA), codex nutrient reference values (NRV) and Indian estimated average requirement (EAR), RDA and tolerable upper limits (TUL) for pregnant women

Micronutrient	MMS[a]	RDA[a]	NRV[b]	EAR[c]	RDA[c]	TUL[c]
Vitamin A (µg)	800	800	800	406	900	3000
Vitamin C (mg)	70	70	60	65	80	2000
Vitamin D3 (µg)	5	5	5	10	15	4000
Vitamin E (mg)	10	10	10	–	10 (AI)	ND
Vitamin B1 (mg)	1.4	1.4	1.4	1.6	2.0	ND
Vitamin B2 (mg)	1.4	1.4	1.6	2.3	2.7	ND
Vitamin B3 (mg)	18	18	18	17	20	ND
Vitamin B6 (mg)	1.9	1.9	2	1.9	2.3	ND
Folic acid (µg)	400	600	200	290	342	1000
Vitamin B12 (µg)	2.6	2.6	1	2.2	2.45	ND
Iron (mg)	30	30	14	21	27	45
Iodine (µg)	150	175	150	160	220	1100
Zinc (mg)	15	15	15	12	14.5	40
Se (µg)	65	65	NP	–	40 (AI)	ND
Cu (mg)	2	1.5–3	NP	–	1.7 (AI)	ND

ND not determined, *AI* adequate intake
[a] UNICEF/WHO/UNU (1999)
[b] Codex (2005)
[c] ICMR-NIN (2020)

It reduced the risk of LBW (low birth weight) and SGA (small for gestational age); resulted in comparable (60 mg iron) lowering of maternal anaemia prevalence (5.8% and 8.3%) and changes in Hb levels (0.2 + 1.59 g/dL and 0.1 + 1.55 g/dL); lower improvements in serum ferritin in UNIMAP (30 mg iron) than with 60 mg iron (15.3 ± 32.5 ng/mL vs 20.9 ± 32.6 ng/mL). Thus there is a clear benefit and cost-effectiveness of UNIMAP-MMS which may offset the limitations of IFA supplementation only (Haider and Bhutta 2015; Gomes et al. 2022a; Ranjith et al. 2022). Currently, it is being debated whether to adopt the UNIMAMAP-MMS or formulate a India specific MMS according to the Indian EAR/RDA of micronutrients for pregnant women or to consider personalized recipe of MMS during pregnancy based on type and extent of micronutrient deficiencies for clinical trials (Gomes et al. 2022b; Kurpad and Sachdev 2022).

9.6   Fortification

9.6.1   Historical Perspective (Success Stories)

A summary of the success stories of vitamin and mineral fortification of food vehicles across the globe is listed in Table 9.6. Voluntary addition of vitamin A added to margarine was the first fortified product, which became a mandatory

Table 9.6 Global impact of vitamin and mineral fortification as a public health strategy

Year	Micronutrient	Impact
1923	Iodine	Mandatory Iodized salt in India in 1998; countries legislated mandatory salt iodization-124; voluntary iodization-21; use of iodized salt—88% of the global population (Zimmermann and Andersson 2021)
1933	Vitamin B1	Canada, mandatory fortification of flour and elimination of Beriberi (WHO, FAO 2006)
1941	Vitamin B3	USA, Mandatory fortification of flour and elimination of pellagra (WHO, FAO 2006)
Early 1940s	Vitamins B1, B2 and B3	Many countries, cereal products became common practice (WHO, FAO 2006)
1954	B vitamins + iron	Chile, Flour products, country with very low prevalence for anaemia
1992	Vitamin A	Venezuela, wheat and maize flour country with vitamin A sufficiency in general population and reduction in anaemia in children
1998	Folic acid	USA, mandated folic acid fortification and now implemented in 60 countries (Pfeiffer et al. 2005)
2000	Vitamin D	USA and Canada, vitamin D fortification of milk and dairy products (WHO, FAO 2006)
2015	Iron	Systematic review on evidence of the effectiveness of flour fortification for reducing the prevalence of anaemia is limited (Pachón et al. 2015)
2017	Iron	Mandatory in 87 countries and legislation to fortify wheat flour, maize flour, and/or rice (Marks et al. 2018)

practice. Regular use of this eliminated clinical prevalence of vitamin A, xerophthalmia within a year after the introduction in UK and Denmark. On similar lines mandatory fortification of vitamin D to milk prevented the widespread problem of childhood rickets in Canada and other countries. The practice of enrichment of white flour (wheat flour) for bread making and other grain products with niacin was implemented in the US which subsequently became a mandatory practice of fortification and has resulted in elimination of pellagra by the year 1950 (WHO, FAO 2006).

9.6.2 Effectiveness of Mandatory Fortification

Robust evidence on mandatory introduction of micronutrient fortified foods on prevention and control of deficiencies of certain micronutrient is available in literature (Das et al. 2013). Reduction in the incidence of clinical prevalence of IDD (goitre by 19–64%, neonatal mortality by 65.7% and infant mortality by 56.5%) after introduction of salt with iodine; NTDs upon introduction of folic acid fortified wheat flour and an overall improvement in weighted mean difference (WMD) in Hb in the population (among women of child bearing age 5.7 g/L, pregnant women 6.9 g/L and in infants and children under 5 years 7.4 g/L) (Bhutta et al. 2008; Gera et al. 2012; Athe et al. 2014). An unequivocal impact of a rightward shift in the frequency distribution of serum folate levels in 6–10 years (1988–1994 to 1999–2000) in US have been reported with wheat flour fortified with folic acid (Pfeiffer et al. 2005).

WHO/FAO has advocated fortification of wheat flour fortification as a cost-effective strategy to address micronutrient deficiencies in LAMI countries and published guidelines on food fortification with micronutrient (WHO, FAO 2006). Over the years, many LAMI countries have focused on fortification of flour with iron and folic acid to address anaemia in general population. However, a systematic review of published data has documented weak evidence on the impact of this strategy in controlling anaemia prevalence but provided evidence in bringing down prevalence of ferritinemia (biomarker of iron stores) in women of reproductive age (Pachón et al. 2015). The authors argued that this may be due to certain variability in implementation of fortification strategies which needs careful assessment of regional and contextual and population dynamics.

9.6.3 Mandatory Vitamin A Supplementation for Children in India

The National Prophylaxis Programme against Nutritional Blindness started in 1970 was designed to prevent nutritional blindness due to vitamin A deficiency leading to keratomalacia in children. This strategy has significantly reduced severe deficencies of vitamin A (Bitot's spot/night blindness) below the public health significance level. Under this programme pharmacological oral dose of vitamin A was administered, which was designed and field tested as a single annual oral large dose of vitamin A (Swaminathan et al. 1970; WHO 2011a). A dose of 100,000 IU (30 mg retinol equivalents) in infants 6–11 months of age and 200,000 IU in children 12–59 months of age has been found to provide protection for 4–6 months.

9.6.4 Extra Nutritional Benefits of Vitamin A Supplementation in Children

Periodic surveys conducted in various parts of India have shown elimination of clinical signs of keratomalacia and a significant decrease in the prevalence of Bitot spots. However, based on the Cochrane review the scope of the program become broader and mega dose vitamin A programme has been advocated for reduction in childhood mortality (West and Sommer 1987; Imdad et al. 2022). Critical review of these benefits has been attributed to certain contextual issues such as scarce health care facilities and implementation of the strategy in areas where clinical deficiency is common (Kapil and Sachdev 2013). This example exemplifies the need for critical meta-analysis as a process in the evaluation and makes tall claims on the extra nutritional benefits for vitamin supplementation. These finding also underline the need for monitoring and withdrawing the interventions as and when the objectives have been met.

9.6.5 Vitamin and Mineral Fortification in India

Fortification programme with iodized salt has been considered as a public health success in India (Pandav et al. 2013). The prevalence of IDD was brought down by mandatory use of iodized salt in the country. In tune with the control of dual deficiencies of iron and iodine in India, ICMR-NIN, Hyderabad developed, field tested and implemented iron-fortified iodized salt double fortified salt (DFS) to tackle them simultaneously (Rao 1994; Ranganathan et al. 1996; Nair et al. 1998; Sivakumar and Nair 2002). Other technologies of DFS have also been developed and tested in India (De-Regil et al. 2008; Haas et al. 2014).

Global evidence (WHO, FAO 2006) and meager studies from India demonstrated that fortification of wheat flour with iron (Muthayya et al. 2012), milk with vitamin D (Khadgawat et al. 2013) and home fortification of foods with multiple micronutrient powders (WHO 2011b; Black et al. 2021) are all technologies with prospects for scale-up. In the recent years, the efficacy of use of rice fortified with iron- fortified rice kernels (FRK iron extruded rice) for 8 months on anaemia and iron deficiency among children attending mid-day meal programme in government schools was tested (Moretti et al. 2006; Radhika et al. 2011; WHO 2018). In view of the impact on micronutrient status FSSAI has set standards for iron, folic acid and vitamin B12 fortification of rice and wheat and vitamins A and D in milk and oil (FSSAI 2018).

Currently, in India, attempts are being made to assess the impact of supply of fortified rice through the PDS system in reducing the burden of anaemia. While DFS was mandated in Government led food and nutrition programmes like MDM in schools and ICDS feeding programmes in 2011, its performance is slow due to lack of scale up in its supply chain involving production, distribution, monitoring and long term implementation in the country. Further, the need for voluntary or mandatory fortification of milk and oil with vitamin A and D need to be regionally evaluated and implemented.

9.6.6 Home/Point-of Use Fortification with Micronutrient Powders (MNP)

Home or point-of-use fortification powders are an attractive strategy (WHO 2011b). In this strategy micronutrient powder (multiple micronutrients) is added to cooked food just before feeding children. The trials so far carried out in children have been found to be successful in reducing the prevalence of anaemia, iron deficiency and other micronutrient deficiencies. We have carried out a double-masked, cluster-RCT with the objective to assess whether this strategy compared with placebo fortification of preschool rice meals on cognitive development and whether impact vary by preschool quality (primary outcome) and blood parameters of anaemia and micronutrients (secondary outcomes). The MNP used was a combination of seven micronutrients (iron, zinc, vitamin A, folic acid, vitamin B12, vitamin C and riboflavin) below RDA level for children under 5 years of age. The intervention reduced anaemia ($<10\%$) and iron deficiency. The intervention increased preschoolers' expressive language and inhibitory control and reduced developmental disparities among preschoolers in low-quality preschools. Thus enhancing overall quality by implementing responsive care, learning, and nutrition along with MNP intervention in preschools may be necessary to improve preschoolers' cognitive development (Black et al. 2021). Similar results on biomarkers have been reported in the recent Cochrane review on this type of strategy, although the analysis found no significant impact on diarrhea among participants (De-Regil et al. 2017).

9.7   Challenges to Micronutrient Fortification in India

Micronutrient fortification in India has many challenges that require careful consideration. These include (1) clear understanding of the regional food preferences and cultural context and population dynamics, (2) technology of fortification, program design and implementation (in case of fortified rice segregation of fortified rice kernel stability of nutrients, cooking practices), (3) effectiveness of single vs. multiple micronutrients, (4) whether to consider a mandatory fortification approach (5) potential adverse effects of layering of multiple interventions (Kurpad et al. 2021), (6) changing lifestyle and use of micronutrient fortified multiple prepackaged foods (7) availability of a robust monitoring and evaluation with appropriate biomarkers/outcome measures in place (8) prioritizing and popularizing food environment conducive for dietary diversification with an intention of increasing food and nutrient synergy and bioavailability of iron (Nair et al. 2013; Roy Choudhury et al. 2021; Nair 2022; Konapur et al. 2022) and (9) formulating and field testing country specific MNP for children and MMS for pregnant mothers. There is also a need to consider and advocate new approaches such as 'food as a medicine approach'/food patterns to realize nutrition security and ameliorate health (Mozaffarian et al. 2022).

9.8   Major Knowledge Gaps and Research Opportunities

According to various surveys only a meager improvement in mean haemoglobin and a decrease in anaemia prevalence have been achieved among children under the age of 5 years and women (15–49 years) over a period of time (Stevens et al. 2013). Further improvements are needed in many regions of globe and in India to achieve the targets to halve anaemia prevalence by 2025 from 2011 levels. Addressing this needs holistic contextual assessment of all the risk factors of anaemia such as low bioavailability of dietary iron, deficiency of other hemopietic nutrients, sickle-cell anaemia and thalassaemia, malaria, parasitic infestation, HIV infection, and NCDs (Nair and Iyengar 2009).

We need accurate methods to design improved questionnaires and recall methods for assessing risk factors and to assess intake and deficiency stages of nutrients using digital technologies (AI based). Effective pharmaco-vigilance system and database creation to assess the actual composition of micronutrients in health supplements;

new biomarkers of intake using modern tools of nutri genomics, molecular imaging and systems biology network approaches are critical elements. Designing and conducting robust RCTs on the biological outcome of single, multiple micronutrients supplements and fortification among diverse populations that reflect the multiplicity of the country is a research priority.

9.9   Conclusions

Although, the short term supplementation and medium term food fortification interventions have made significant impact in controlling vitamin and minerals deficiencies in recent period, extra efforts directed towards achieving the long term and sustainable strategy of dietary diversification are needed. Integration of these approaches with all available maternal and child health policy/programs may ensure sustainable benefits. Nutrition education should also form an integral part of the above programs to improve awareness, acceptability and equity. More studies are needed to generate evidence from the LAMI countries to assess the direct impact of supplementation and fortification on other complex functional outcome measures such as morbidity and mortality.

References

Athe R, Rao MV, Nair KM (2014) Impact of iron-fortified foods on Hb concentration in children (<10 years): a systematic review and meta-analysis of randomized controlled trials. Public Health Nutr 17:579–586

Bhutta ZA, Ahmed T, Black RE, Cousens S, Dewey K, Giugliani E, Haider BA, Kirkwood B, Morris SS, Sachdev HPS (2008) What works? Interventions for maternal and child undernutrition and survival. Lancet 371(9610):417–440

Black MM, Fernandez-Rao, Nair KM, Balakrishna N et al (2021) A randomized multiple micronutrient powder point-of-use fortification trial implemented in Indian preschools increases expressive language and reduces anaemia and iron deficiency. J Nutr 151(7):2029–2042. https://doi.org/10.1093/jn/nxab066

Brown B, Wright C (2020) Safety and efficacy of supplements in pregnancy. Nutr Rev 78(10): 813–826. https://doi.org/10.1093/nutrit/nuz101

Codex (2005) CAC/GL 55 - 2005 Codex. www.fao.org/input/download/standards/10206/cxg_0 55e.pdf

Das JK, Salam RA, Kumar R, Bhutta ZA (2013) Micronutrient fortification of food and its impact on woman and child health: a systematic review. Syst Rev 23(2):67. https://doi.org/10.1186/2046-4053-2-67

De-Regil AM, Thankachan P, Muthayya S, Goud RB, Kurpad AV, Hurrell RF, Zimmermann MB (2008) Dual fortification of salt with iodine and iron: a randomized, double-blind, controlled trial of micronized ferric pyrophosphate and encapsulated ferrous fumarate in southern India. Am J Clin Nutr 88:1378–1387

De-Regil LM, Jefferds MED, Peña-Rosas JP (2017) Point-of-use fortification of foods with micronutrient powders containing iron in children of preschool and school-age. Cochrane Database Syst Rev 2023(11):CD009666. https://doi.org/10.1002/14651858.CD009666.pub2. Accessed 13 July 2023

El-Kadiki A, Sutton AJ (2005) Role of multivitamins and mineral supplements in preventing infections in elderly people: systematic review and meta-analysis of randomised controlled trials. BMJ 330:871

Fairfield K, Stamfer M (2007) Vitamin and mineral supplement for cancer prevention, issues and evidence. Am J Clin Nutr 85(1):289S–292S

FAO (1995) FAO technical consultation on food fortification: technology and quality control. FAO, Rome, pp 20–23. http://www.fao.org/docrep/W2840E/w2840e0b.htm#1

FAO (1997) Preventing micronutrient malnutrition: a guide to food-based approaches - a manual for policy makers and programme planners. International Life Sciences Institute, Washington

FSSAI (2018). https://www.fssai.gov.in/upload/uploadfiles/files/Gazette_Notification_Food_Forti fication_10_08_2018.pdf

FSSAI (2022). https://www.fssai.gov.in/upload/uploadfiles/files/Gazette_Notification/ Nutra_30_03_2022.pdf

Gera T, Sachdev HS, Erick B (2012) Effect of iron fortified foods on hematological and biological outcomes: systematic reviews of randomized controlled trials. Am J Clin Nutr 96:309–324

Ghosh S, Thomas T, Pullakhandam R, Nair KM, Sachdev HS, Kurpad AV (2023) A proposed method for defining the required fortification level of micronutrients in foods: an example using iron. Eur J Clin Nutr 77(4):436–446. https://doi.org/10.1038/s41430-022-01204-4

Gomes F, Agustina R, Black RE, Christian P, Dewey KG, Kraemer K, Shankar AH, Smith ER, Thorne-Lyman A, Tumilowicz A, Bourassa MW (2022a) Multiple micronutrient supplements versus iron-folic acid supplements and maternal anaemia outcomes: an iron dose analysis. Ann N Y Acad Sci 1512(1):114–125. https://doi.org/10.1111/nyas.14756

Gomes F, Black RE, Smith E, Shankar AH, Christian P (2022b) Micronutrient supplements in pregnancy: an urgent priority. Lancet Glob Health 10(9):e1239. https://doi.org/10.1016/S2214-109X(22)00308-4

Greenwald P, Anderson D, Nelson SA, Taylor PR (2007) Clinical trials of vitamin and mineral supplements for cancer prevention. Am J Clin Nutr 85(1):313–317

Haas JD, Rahn M, Venkatramanan S, Marquis GS, Wenger MJ, Murray-Kolb LE, Wesley AS, Reinhart GA (2014) Double-fortified salt is efficacious in improving indicators of iron deficiency in female Indian tea pickers. J Nutr 144:957–964

Haider BA, Bhutta ZA (2015) Multiple-micronutrient supplementation for women during pregnancy. Cochrane Database Syst Rev 2015:Cd004905

Heaney RP (2007) Bone health. Am J Clin Nutr 85(1):300S–303S

ICMR-NIN (2010) Indian Council of Medical Research. Nutrient requirements and recommended dietary allowances for Indians. In: A report of the expert group of the Indian Council of Medical Research. National Institute of Nutrition, Hyderabad

ICMR-NIN (2020) Indian Council of Medical Research/National Institute of Nutrition. Nutrient Requirements for Indians. Department of Health Research, Ministry of Health and Family Welfare, Government of India

Imdad A, Mayo-Wilson E, Haykal MR, Regan A, Sidhu J, Smith A, Bhutta ZA (2022) Vitamin A supplementation for preventing morbidity and mortality in children from six months to five years of age. Cochrane Database Syst Rev 3(3):CD008524. https://doi.org/10.1002/14651858. CD008524.pub4

Kapil U, Sachdev HPS (2013) Massive dose vitamin A programme in India - need for a targeted approach. Indian J Med Res 138(3):411–417

Khadgawat R, Marwaha RK, Garg MK, Ramot R, Oberoi AK, Sreenivas V, Gahlot M, Mehan N, Mathur P, Gupta N (2013) Impact of vitamin D fortified milk supplementation on vitamin D status of healthy school children aged 10-14 years. Osteoporos Int 24(8):2335–2343. https://doi. org/10.1007/s00198-013-2306-9

Konapur A, Gavaravarapu SM, Nair KM (2022) The 5 A's approach for contextual assessment of food environment. J Nutr Educ Behav 54(7):621–635. https://doi.org/10.1016/j.jneb.2022. 02.017

Kurpad AV, Sachdev HS (2022) Precision in prescription: multiple micronutrient supplements in pregnancy. Lancet Glob Health 10(6):e780–e781. https://doi.org/10.1016/S2214-109X(22) 00207-8

Kurpad AV, Ghosh S, Thomas T, Bandyopadhyay S et al (2021) Perspective: when the cure might become the malady: the layering of multiple interventions with mandatory micronutrient fortification of foods in India. Am J Clin Nutr 2021:1–6. https://doi.org/10.1093/ajcn/nqab245

Laxmaiah A, Nair KM, Arlappa N, Raghu P, Balakrishna N, Rao KM, Galreddy C, Kumar S, Ravindranath M, Rao VV, Brahmam GN (2012) Prevalence of ocular signs and subclinical vitamin A deficiency and its determinants among rural pre-school children in India. Public Health Nutr 15(4):568–577. https://doi.org/10.1017/S136898001100214X

Marks KJ, Luthringer CL, Ruth LJ, Rowe LA, Khan NA, De-Regil LM, López X, Pachón H (2018) Review of grain fortification legislation, standards, and monitoring documents. Glob Health Sci Pract 6(2):356–371. https://doi.org/10.9745/GHSP-D-17-00427

MoHFW (2019) Comprehensive National Nutrition Survey (CNNS). National report 2016–2018. MoHFW, Government of India, UNICEF and Population Council, New Delhi

Moretti D, Zimmermann MB, Muthayya S, Thankachan P, Lee TC, Kurpad AV, Hurrell RF (2006) Extruded rice fortified with micronized ground ferric pyrophosphate reduces iron deficiency in Indian school children: a double-blind randomized controlled trial. Am J Clin Nutr 84:822–829

Mozaffarian D, Blanck HM, Garfield KM, Wassung A, Petersen R (2022) A food is medicine approach to achieve nutrition security and improve health. Nat Med 28:2238–2240

Muthayya S, Thankachan P, Hirve S, Amalrajan V, Thomas T, Lubree H, Agarwal D, Srinivasan K, Hurrell RF, Yajnik CS, Kurpad AV (2012) Iron fortification of whole wheat flour reduces iron deficiency and iron deficiency anaemia and increases body iron stores in Indian school-aged children. J Nutr 142(11):1997–2003. https://doi.org/10.3945/jn.111.155135

NAAS (2022) Food fortification: issues and way forward, Policy Paper No. 111. National Academy of Agricultural Sciences, New Delhi, p 20

Nair KM (2019) Contextualizing the principles of iron fortification of foods in India, NFI Bulletin April

Nair KM (2022) Addressing iron bioavailability: core strategy to achieve aneaemia mukt Bharat. Bull Nutr Found India 43:4

Nair KM, Augustine LF (2016) Basis for current allowances of nutrients for food fortification in India. Nutr Bull 37(3):1–5

Nair KM, Iyengar V (2009) Iron content, bioavailability & factors affecting iron status of Indians. Indian J Med Res 130:634–645

Nair KM, Brahmam GNV, Ranganathan S, Vijayaraghavan K, Sivakumar B, Krishnaswamy K (1998) Impact evaluation of iron and iodine fortified salt. Indian J Med Res 108:203–211

Nair KM, Bhaskaram P, Balakrishna N, Ravinder P, Sesikeran B (2004) Response of haemoglobin, serum ferritin and serum transferrin receptor during iron supplementation in pregnancy: a prospective study. Nutrition 20:896

Nair KM, Brahmam GN, Radhika MS, Dripta RC, Ravinder P, Balakrishna N, Chen Z, Hawthorne KM, Abrams SA (2013) Inclusion of guava enhances non-heme iron bioavailability but not fractional zinc absorption from a rice-based meal in adolescents. J Nutr 143(6):852–858. https://doi.org/10.3945/jn.112.171702

Nair KM, Fernandez-Rao S, Nagalla B, Kankipati RV, Punjal R, Augustine LF, Hurley KM, Tilton N, Harding KB, Reinhart G, Black MM (2016) Characterisation of anaemia and associated factors among infants and pre-schoolers from rural India. Public Health Nutr 19(5): 861–871. https://doi.org/10.1017/S1368980015002050

Nair KM, Choudhury DR, Konapur A (2018) Appropriate doses of iron for treatment of anaemia amongst pregnant and lactating mothers; under 5 children. Children 6-10 years of age; adolescent girls and women of reproductive age groups. Indian J Community Health 30(Suppl):39–53

NFHS-3 (2006) International Institute for Population Sciences (IIPS) and ICF. 2006. National Family Health Survey (NFHS-3), 2005-06: India. IIPS, Mumbai

NFHS-4 (2016) International Institute for Population Sciences (IIPS) and ICF. National Family Health Survey (NFHS-4), 2015-16: India. IIPS, Mumbai

NFHS-5 (2021) International Institute for Population Sciences (IIPS) and ICF. National Family Health Survey (NFHS-5), 2019-21: India. IIPS, Mumbai

NHP (2017) Ministry of Health and Family Welfare, Government of India

NNMB (2012) Diet and nutritional status of rural population and prevalence of hypertension. Third repeat survey. NNMB Technical report series 26. National Institute of Nutrition (ICMR), Hyderabad

NNMB (2016) Report for the year 2016. Technical Report Series. National Nutrition Monitoring Bureau, National Institute of Nutrition (ICMR), Hyderabad

Pachón H, Spohrer R, Mei Z, Serdula MK (2015) Evidence of the effectiveness of flour fortification programs on iron status and anaemia: a systematic reviews. Nutr Rev 73:780–795. https://doi.org/10.1093/nutrit/nuv037

Pandav CS, Yadav K, Srivastava R, Pandav R, Karmarkar MG (2013) Iodine deficiency disorders (IDD) control in India. Indian J Med Res 138:418–433

Pfeiffer CM, Caudill SP, Gunter EW, Osterloh J, Sampson EJ (2005) Biochemical indicators of B vitamin status in the US population after folic acid fortification: results from the National Health and Nutrition Examination Survey 1999-2000. Am J Clin Nutr 82(2):442–450. https://doi.org/10.1093/ajcn.82.2.442

Prentice RL (2007) Clinical trials and observational studies to assess the chronic disease benefits and risks of multivitamin-mineral supplements. Am J Clin Nutr 85(1):308S–313S

Radhika MS, Nair KM, Hari Kumar R, Rao VM, Ravinder P, Gal Reddy C, Brahmam GNV (2011) Micronized ferric pyrophosphate supplied through extruded rice kernels improves body iron stores in children: doubleblind, randomized placebo-controlled midday meal feeding trial in Indian school children. Am J Clin Nutr 94:1202–1210

Ranganathan S, Reddy V, Ramamoorthy P (1996) Large scale production of salt fortified with iodine and iron. Food Nutr Bull 17:73–78

Ranjith A, Puri S, Vohra K, Khanam A, Bairwa M, Kaur R, Yadav K (2022) Ideal dose of iron in multiple micronutrient supplement: a narrative review of evidence. Cureus 14(9):e28688. https://doi.org/10.7759/cureus.28688. PMID: 36199654; PMCID: PMC9526876

Rao BSN (1994) Fortification of salt with iron and iodine to control anaemia and goitre. Development of a new formula with good stability and bioavailability of iron and iodine. Food Nutr Bull 15:32–39

Roy Choudhury D, Nair KM, Nagalla B, Vijaya Kankipati R, Ghosh S, Buwade J, Fernandez-Rao S (2021) Guava with an institutional supplementary meal improves iron status of preschoolers: a cluster-randomized controlled trial. Ann N Y Acad Sci 1492(1):82–95. https://doi.org/10.1111/nyas.14556

Roy M, Pitkin A (2007) Folate and neural tube defects. Am J Clin Nutr 85(1):285S–288S. https://doi.org/10.1093/ajcn/85.1.285S

Sarna A, Porwal A, Ramesh S, Agrawal PK, Acharya R, Johnston R, Khan N, Sachdev HPS, Nair KM, Ramakrishnan L, Abraham R, Deb S, Khera A, Saxena R (2020) Characterisation of the types of anaemia prevalent among children and adolescents aged 1–19 years in India: a population-based study. Lancet Child Adolesc Health 4(7):515–525

Sivakumar B, Nair KM (2002) Double fortified salt at crossroads. Indian J Pediatr 69:617–623

Stevens GA, Finucane MM, De-Regil LM, Paciorek CJ, Flaxman SR, Branca F, Peña-Rosas JP, Bhutta ZA, Ezzati M, Nutrition Impact Model Study Group (Anaemia) (2013) Global, regional, and national trends in haemoglobin concentration and prevalence of total and severe anaemia in children and pregnant and non-pregnant women for 1995-2011: a systematic analysis of population-representative data. Lancet Glob Health 1(1):16–25. https://doi.org/10.1016/S2214-109X(13)70001-9

Swaminathan MC, Susheela TP, Thimmayamma VS (1970) Field prophylactic trial with a single annual oral massive dose of vitamin A. Am J Clin Nutr 23:119–122

Traber MG (2007) Heart disease and single-vitamin supplementation. Am J Clin Nutr 85(1):293S–299S

Tuncalp Ö, Rogers LM, Lawrie TA, Barreix M, Peña-Rosas JP, Bucagu M, Neilson J, Oladapo OT (2020) WHO recommendations on antenatal nutrition: an update on multiple micronutrient supplements. BMJ Glob Health 5(7):e003375. https://doi.org/10.1136/bmjgh-2020-003375

UNICEF/WHO/UNU (1999) Composition of a multi-micronutrient supplement to be used in pilot programmes among pregnant women in developing countries. In: Report of an UNICEF/WHO/UNU workshop. UNICEF, New York. http://www.idpas.org/pdf/059CompositionofMult-MicronutrientSupplement.pdf

West KP and Sommer A (1987). Delivery of oral doses of vitamin A to prevent vitamin A deficiency and nutritional blindness. A state-of-the-art review. Nutrition Policy Discussion Paper No 2. Rome: United Nations Administrative Committee on Coordination, Subcommittee on Nutrition

WHO (2011a) Guideline: vitamin A supplementation in infants and children 6–59 months of age. World Health Organization, Geneva

WHO (2011b) Guideline: use of multiple micronutrient powders for home fortification of foods consumed by infants and children 6-23 months of age. World Health Organization, Geneva. http://whqlibdoc.who.int/publications/2011/9789241502047_eng.pdf

WHO (2018) Guideline: fortification of rice with vitamins and minerals as a public health strategy. WHO, Geneva

WHO, FAO (2006) Guidelines on food fortification with micronutrients. WHO Press, Geneva

Zimmermann MB, Andersson M (2021) Global perspectives in endocrinology: coverage of iodized salt programs and iodine status in 2020. Eur J Endocrinol 185:R13–R21

Demand-Supply of Agri-food Commodities in India

10

P. K. Joshi

Abstract

Indian agri-food system has transformed from a deficit to self-sufficient and surplus, and from import to export oriented. The production as well as consumption of different food commodities increased and diversified remarkably. The production and consumption diversified in favor of high-value commodities such as fruits, vegetables, dairy products, meat, and fish. The per capita consumption of cereals and coarse cereals declined overtime. There are projections that India will continue to be surplus in rice, wheat, milk, fruits, vegetables, fish but deficit in pulses and edible oils. Role of technologies and policies will play key role in increasing enough production to minimize import of pulses and edible oils.

Keywords

Food demand · Food consumption · Food supply · Food diversification

The paper is prepared for a Lead Lecture in the XVI Agricultural Science Congress on "Transforming Agri-food Systems for Achieving Sustainable Development Goals", scheduled to be held on 10-13 Oct 2023 in Kochi, Kerala.

P. K. Joshi (✉)
Formerly with International Food Policy Research Institute, New Delhi, India

10.1   Background

Overtime, Indian agri-food system has transformed impressively with several milestones of global success stories. Increasing food production was the primary policy objective of the government to achieve the food self-sufficiency and meet the growing demand for food. Technology-led revolution in agriculture dully supported by effective policies, inclusive institutions and required infrastructure has transformed Indian agri-food system from deficit to self-sufficient and from import to export-oriented. While production was remarkably rising, demand for food was also increasing and diversifying due to growing population, flourishing economy, expanding urbanization, unfolding globalization, and changing taste and preferences (Kumar and Joshi 2016).

The question is if India will continue its tempo of increasing food production to meet the growing demand or will it import food commodities by 2030? This paper presents the historical trends in food production and demand for food commodities and provides projections for 2032–33 under different scenario. The paper first presents the concept of agri-food system, which is followed by production and consumption of food commodities. It also provides demand and supply projections of food commodities by 2032–33. At the end, the paper briefly listed actions to be taken for a competitive and sustainable Indian agri-food system in future.

10.2   Agri-Food System

An agri-food system consists of three sub-systems: (1) production sub-system, (2) consumption sub-system, and (3) production-consumption linkage sub-system. The production sub-system includes all back-end activities of production such as seed, soil nutrients, water, machinery, labor, etc. The consumption sub-system includes government's social safety-net programs, diet diversity, food safety, food processing, etc. The middle one, which links production and consumption, includes financial and marketing institutions, logistics, and government policies (price policy, trade policy, incentives, and subsidies). Climate change and environment influence all the three sub-systems. The present agri-food system is largely unorganized, fragmented and highly inefficient. It ignores conservation and sustainability of natural resources such as soil and water, nutrition, and healthy diets. Several agri-food systems in different regions and agro-ecologies in the country are experiencing serious problem of degradation of soil and water resources, and acute undernourishment among women and children. There is a need for a sustainable and resilient agri-food system which should aim to promote production and consumption of nutritive, healthy, and affordable food without compromising with the natural resources and building resilience against any unforeseen eventualities. In this paper, we are covering the production and consumption sub-systems.

10.3 Production Sub-System

India has made remarkably progress in majority of agri-food commodities (Table 10.1). The food grains production increased substantially from 72.35 m tons in 1965–66 to a record level of 330.5 m tons in 2022–23. Rice and wheat have made substantial progress during 1960s, 1970s and 1980s due to all efforts were made by the government to ensure food security in the country. In rice and wheat, India not only became self-sufficient but also accumulated huge buffer stock for distribution under different social safety-net programs. India also started exporting rice and wheat after meeting the domestic requirements and government's commitments for the social safety-net programs. As of June 2020, rice and wheat stock with the Food Corporation of India was as high as 84 m tons. Such a huge buffer stock helped the government to provide free ration (5 kg per head) to about 800 million poor population under the *Prime Minister's Garib Kalyan Anna Yojana* during Covid 19 pandemic. In addition, India exported rice and wheat worth of US\$ 9.96 billion and US\$ 2.12 billion, respectively in 2022.

Other food commodities, such as maize, coarse cereals and pulses made impressive progress during the last two decades. Maize production witnessed sharp increase due to single cross hybrids as well as its cultivation in non-traditional areas. Production of pulses increased all time high due to constituting Pulse Mission by conversing all the government programs, including supplying improved seeds, higher minimum support prices and ensuring procurement, to reduce import of pulses. Coarse cereals also received due importance due to International Year of Millets 2023.

There was a remarkable increase in the production of cotton and sugarcane during the last two decades. Cotton production was mainly due to large-scale adoption of Bt cotton varieties, and sugarcane production was due to improved and high-yielding varieties, which not only increased yields but also sugar content.

More impressive trends were noted in production of horticultural commodities, milk, fish and poultry. For example, milk production escalated from 20 m tons in 1965–66 to 221 m tons in 2020–21. During the same period, fish production went-up from 1.37 m tons to 16.2 m tons and horticulture production to 342.33 m tons in 2021–22 from a low of 96 m tons in 1990–91. Production of fruits sharply increased

Table 10.1 Production of different crops during 1966–67, 2000–01 and 2022–23 (million tons)

Crop	1966–67	2000–01	2022–23
Rice	30.4	84.9	135.5
Wheat	10.4	69.7	112.7
Maize	4.8	12.0	35.9
Coarse cereals	21.2	32.7	54.7
Pulses	8.4	11.1	27.5
Total foodgrains	**74.3**	**196.8**	**330.5**
Oilseeds	6.4	18.4	40.9
Cotton	5.3	9.5	34.3
Sugarcane	92.8	297.2	494.4

Source: Compiled by the author

from 28.63 m tons in 1990–91 to 107.10 m tons in 2020–21, and of vegetables from 58.5 m tons to 204.6 m tons during the same period.

The main reasons for such an impressive increase in production of majority of commodities were (1) adoption of improved and high-yielding varieties; (2) impressive use in inorganic fertilizers; (3) massive investment in canal irrigation and incentives in use of groundwater for irrigation; (4) effective policies, including announcement of minimum support prices and assured procurement, especially for rice and wheat; and (5) development of effective and large network of markets and banking institutions. In addition, establishment of National Dairy Development Board (NDDB), National Horticulture Development Board, and Fisheries Development Board significantly contributed in increasing production of horticultural commodities, milk and fish. Role of cooperatives played a key role in augmenting milk production, while contract farming was responsible for an impressive increase in production of poultry meat and eggs.

10.4 Consumption Pattern of Food Commodities

Historical trends in consumption of food commodities have been drawn from Joshi et al. (2016). Earlier studies on consumption of food commodities reveal (a) share of food in total expenditure has fallen from 62.7% in 1983 to 46% in 2011; and (b) consumption of food items has diversified, the diversity index increased from 0.43 in 1983 to 0.55 in 2011 (Joshi and Khadka 2022). The changes in consumption pattern of food commodities are presented in Table 10.2.[1]

The table shows that per capita demand for cereal-based commodities has declined, while those of high-value commodities (such as fruits, vegetables, milk, meat, eggs and fish) have risen over time. The changing dietary pattern is leading to imbalance in nutrient intake, with the consumption of fat rising while those of calories, protein and iron are falling (Joshi et al. 2016).

Demand for food commodities is generally driven by their prices and income of the consumers. Demand is also influenced by the prices of substitute commodities, and tastes and preferences of the consumers. Globalization of food is also influencing the consumption pattern; consumers are consuming more of processed commodities and increasing intake of western diets. Table 10.3 presents the expenditure (proxy of income) and own price elasticity, which shows the response of consumers towards different food commodities.

Three inferences can be drawn from the table:

(a) Own price elasticities are negative for all the commodities, indicating that increase in the prices of any commodity will have negative impact on its

---

[1]The limitation of this section is that the latest data on consumption pattern of different food commodities is available only up to 2011. There has been no survey on household consumption of food commodities.

Table 10.2 Changes in consumption pattern of different food commodities (kg/capita/annum)

Food commodity	1983	2011	Change (%)
Rice	81.09	80.1	−1.22
Wheat	55.22	47.74	−13.55
Coarse cereals	31.74	5.55	−82.51
Total cereals	168.05	133.39	−20.62
Pulses	11.82	9.97	−15.65
Edible oil	4.52	8.07	78.54
Vegetables	47.97	56.25	17.26
Fruits	3.31	11.93	260.42
Milk & milk products	44.99	64.91	44.28
Sugar	11.40	10.01	−12.19
Meat, fish & eggs	5.36	7.49	39.74

Source: Joshi et al. (2016)

Table 10.3 Expenditure and own price elasticities of consumers

Food commodity	Expenditure elasticity	Own price elasticity
Rice	0.026	−0.291
Wheat	0.083	−0.379
Coarse cereals	−0.148	−0.246
Pulses	0.206	−0.456
Edible oils	0.259	−0.492
Sugar	0.064	−0.343
Vegetables	0.256	−0.499
Fruits	0.368	−0.620
Milk	0.377	−0.605
Meat, fish and eggs	0.651	−0.837

Source: Kumar and Joshi (2016). Refer the paper by Kumar and Joshi (2016) for income group wise elasticities

demand. The magnitude of own price elasticities of fruits & vegetables, milk, meat, fish and eggs is much higher than cereals and pulses, indicating that the former group is more sensitive to the prices.

(b) The expenditure/income elasticities ate positive for all commodities, except coarse cereals. It means with the rise in income of the consumers, the demand for food commodities will increase, with the exceptions of coarse cereals.

(c) The magnitude of own price elasticities of a commodity is more than the expenditure/income elasticities. This indicates that equal increase in the prices and income will have negative impact on the demand of food commodities.

10.5 Demand and Supply Projections

A Working Group Report of NITI Aayog on "Demand and Supply Projections Towards 2033", projected demand for and supply of different food commodities for 2032–33 (Table 10.4).

Table 10.4 Aggregate demand and supply projections by 2032–33 (MMT)

Food commodity	Demand	Supply	Net surplus/deficit
Rice	120.84	151.6	30.76
Wheat	113.46	138.8	25.34
Coarse cereals	67.48	61.7	−5.78
Total cereals	301.78	352.3	50.52
Pulses	35.23	33.9	−1.33
Food grains	337.01	386.2	49.19
Oilseeds	99.59	59.9	−39.69
Milk & products	292.15	329.7	37.55
Fruits	203.55	202.6	−0.95
Vegetables	360.77	362.8	2.03

Source: NITI Aayog (2018)

Table reveals that India will be surplus in most of the food commodities with exception of coarse cereals, pulses, oilseeds, and little bit of fruits by 2032–33. With surplus in majority of the food commodities, India needs to make a policy shift from deficit management to surplus management. For surplus commodities, more focus is to be towards export without adversely affecting the domestic prices. It will also require effective logistics, including transport, warehouses, cold storage, and adequate processing facilities. For food commodities, deficit at present and in future, there is a need to increase their production through improved technologies, effective policies and need-based institutions.

10.6 Conclusions

The future of Indian food-system appears to be quite impressive in achieving the Sustainable Development Goals (SDGs), especially Goal 1 (No Poverty). A recent report from the NITI Aayog shows that Multidimensional Poverty Index has declined from 24.85% in 2015–16 to 14.96% in 2019–21 (NITI Aayog 2023). This indicates that as many as 135 million people came-out of Multi-dimensional Poverty during the same period. However, there is slow progress in achieving Goal 2 (Zero Hunger); malnourishment among children (stunting, wasting and underweight) under 5 years has marginally fallen from 38.4% to 35.5%, 21.0% to 19.3% and 35.8% to 32.1% respectively between 2015–16 and 2019–21 (Government of India 2021). Malnourishment among women aged 15–49 years has reduced from 22.9% to 18.7%. To accelerate the progress in achieving Goal 1 and 2, efforts are needed to further strengthen and make food-system more efficient, competitive, and sustainable. This will require (a) more investment in agriculture, including agricultural research and infrastructure development, (b) improve input-use and water-use efficiency, (c) conserve natural resources, (d) make agriculture climate resilient, (d) strengthen domestic and global value chains, and (e) make agriculture globally competitive. In addition, role of innovative institutions, such as cooperatives, farmer producer organizations, contract farming, and self-help groups, will play key role in empowering small holders to harness the global opportunities. On consumption side, especially achieving Goal 2, the on-going social safety-net programs may continue with some modification. Efforts are also needed to improve dietary diversity, reduce post-harvest losses, promote millets and bio-fortified food commodities, improve food safety, and popularize nutrition education. With these efforts, India will accomplish Sustainable Development Goals, and make Indian agri-food system more efficient, competitive, nutritive, and sustainable without adversely affecting the environment.

References

Government of India (2021) Malnutrition free India. Ministry of Women and Child Development, Government of India, New Delhi

Joshi PK, Shinoj P, Kumar P (2016) Dynamics of food consumption and nutrient insecurity in India. Proc Indian Natl Sci Acad 82(5):1587–1599

Joshi P, Khadka S (2022) Remandating Indian agriculture: pathways for transformation. In: Chand R, Joshi P, Khadka S (eds) Indian agriculture towards 2030: pathways for enhancing farmers' income, nutritional security and sustainable food and farm systems. Springer, Singapore

Kumar P, Joshi PK (2016) Food demand and supply projections to 2030: India. In: Brouwer F, Joshi PK (eds) International trade and food security: the future of Indian agriculture. CABI, UK

NITI Aayog (2018) Demand and supply projections towards 2030: the working group report. NITI Aayog, Government of India

NITI Aayog (2023) India: multi-dimensional poverty index: a Progress review 2023. NITI Aayog, Government of India

R. Hemalatha

Abstract

This study examines the dietary patterns in India, taking into account various factors such as age, activity levels, income, and rural-urban settings. The findings reveal significant disparities in food consumption across different population groups. While cereals are consumed in excess by most groups, there is a substantial lack of intake of protective foods like pulses, legumes, milk, nuts, vegetables, and fruits.

The analysis demonstrates that rural populations tend to consume higher quantities of cereals compared to their urban counterparts. Even the urban dwellers do not come close to the recommended levels of intakes; though, they exhibit higher consumption of protective foods compared to the rural people. However, the urban population have higher intake of added fats and oils, added sugars and foods from outside sources.

These imbalanced dietary patterns raise concerns about their impact on the genesis of non-communicable diseases (NCDs). Inadequate consumption of protective foods, coupled with excessive intake of cereals and fats, lead to nutrient imbalance in both urban and rural populations, which may potentially contribute to the increased risk of diabetes, coronary heart disease, hypertension, stroke and other NCDs. Analysis showed association of low intake of vegetables and fruits with a higher risk of diabetes, while low intake of milk and milk products was linked to hypertension.

To address these dietary imbalances and reduce the risk of nutrition-related diseases, it is crucial to promote a diverse and balanced diet that includes protective foods. Public health interventions should target specific groups, such as adult females from low socioeconomic backgrounds, to improve their access to

R. Hemalatha (✉)
ICMR-National Institute of Nutrition, Hyderabad, Telangana, India

© National Academy of Agricultural Sciences, under exclusive license to Springer
Nature Singapore Pte Ltd. 2023
K. C. Bansal et al. (eds.), *Transformation of Agri-Food Systems*,
https://doi.org/10.1007/978-981-99-8014-7_11

nutritious foods. Additionally, raising awareness about the importance of incorporating a variety of protective foods into daily meals and reduction of foods with high fat, sugar and salt (HFSS foods) can help mitigate the rising epidemic of NCDs in the Indian population. However, both the rural (2012) and the urban data (2016) are old. More recent food consumption data might show a different picture with increasing purchasing capacity and growing availability of convenience foods and HFSS foods.

Keywords

Nutritional assessment · Dietary patterns · Nutrition · Dietary habits

11.1   Background

Rapid urbanization leading to a shift in dietary habits of large sections of the population towards foods that are high in energy, fats, free sugars or salt/sodium, and a decrease in intake of sufficient quantities of fruits, vegetables and dietary fibers like in the form of whole grains has resulted in increasing prevalence of double burden of malnutrition within communities, families and within individuals. This has also resulted in a steep increase in non-communicable diseases (NCDs) and related conditions in several sections of the population. Diet is a strong determinant of nutritional status of an individual which in turn is correlated with health. However, for a country like India, the dietary spectrum is extremely wide making it quite difficult to implement national-level nutritional assessment studies and then to design dietary recommendations that can cater to such high divergence of dietary habits. Here, we have laid out suggestions for achieving a balanced diet through 'My Plate for the Day' developed by the ICMR-National Institute of Nutrition; and a snapshot analysis of current dietary practices, alongside macronutrients and energy for different sections of the population including children are presented using the 2012 and 2016 dietary data of two large-scale surveys conducted in India (National Institute of Nutrition, Indian Council of Medical Research. National Nutrition Monitoring Bureau (NNMB) 2012; National Institute of Nutrition, Indian Council of Medical Research. National Nutrition Monitoring Bureau (NNMB) 2017). The energy intake and food consumption pattern of the population are also compared with the ICMR-NIN recommendations (Hemalatha 2023; Anon 2023).

11.2   'My Plate for the Day'

'My Plate for the Day' by the ICMR-NIN recommends not more than 45% of the total energy (kcal or En) from cereals (Fig. 11.1 and Table 11.1). Whereas from pulses, eggs and flesh foods the total recommended energy percentage is around 14–15%. While milk and milk products intake should be 250–300 ml/day that contributes to 10–11% of total energy per day. As for vegetables, including green

Fig. 11.1 ICMR-NIN "My Plate for the Day"

leafy vegetables (GLV) and tubers (excluding potato), the recommendation is around 400 g, which is about 8% energy per day. About 100 g fruits (3% energy) and 30 g nuts and oil seeds (up to 8% energy) are recommended for a 2000 Kilo-calories diet in a day (Table 11.1). Energy from total fats should be limited to 30 percent of total En; with at least 15 percent coming from fats integrated with foods such as nuts, oilseeds, certain millets and sea foods. For quantities of suggested food groups for balanced diets for all age groups, please refer to the-Revised Short Summary Report of the 2020 ICMR-NIN Expert Group on Nutrient Requirement for Indians (Anon 2023).

Though this model plate is not a representation of any therapeutic diet, regular intake of foods in the mentioned proportions, coupled with regular physical activity has the potential to reduce the risk of non-communicable diseases such as diabetes, hypertension, heart attack, stroke, cancer, arthritis and will support immune function. Although 'My Plate for the Day' provides a satisfactorily rounded recommendation for a balanced nutrition, there is a massive need for several such

Table 11.1 Calories (energy, Kcal), protein and fats from different Food Groups ('MY PLATE FOR THE DAY') for a 2000 Kcal diet

Food groups	Foods to be consumed (g)/day	Percent of total E/day	Total E (kcal)/ day	Total Protein (g)/day	Total fat (g)/ day	Carbohydrate (g)/day
Cereals (incl. Nutri-cereals)	260	44	~876	~25	~5	~178
Pulses[a]	85	14	~274	~19	~3	~42
Milk/curd (ml)	300	11	~216	~10	~13	~16
Vegetables[b] green leafy vegetable (GLV)	400	8	~184	~10	~2	~21
Fruits[c]	100	3	~56	~1	~1	~11
Nuts & seeds	30	8	~155	~5	~12	~6
Fats & Oils	27	12	~243	–	~27	–
Total	1200	–	~2000	14% E	28% E	–

Source: ICMR-NIN My Plate
NOTE: The above quantities for different food components are recommended for people consuming approximately a 2000 Kcal diet. For people with different energy requirements, the intake of cereals and pulses must be increased or decreased while maintaining the 3:1 ratio of cereal to pulse. The recommended quantities for rest of the food groups will more or less remain unalter
[a]Eggs/fish/meat can substitute a portion of pulses
[b]Prescribed amount of vegetables (excluding potato) may be consumed either in cooked form/ salad
[c]Prefer fresh fruits (avoid juices)

recommendations each tuned for different subsections of the population having different nutritional needs that cannot be met with a one-for-all approach.

11.3 Food Group Consumption among Rural and Urban Population

Consumption of different 'Food groups' and estimated 'Nutrients' across different age groups in the urban and rural population pan-India are given in 'What India Eats' (WIE) (Hemalatha et al. 2023). A brief summary is presented below.

Mean intake of 'Cereals' was higher for all age groups in the rural survey when compared to data from the urban population. A remarkable difference of over 100 grams per day cereals intake was noted for rural girls (16–18 years), rural 18–50 years men- moderate work and rural 50 + to 60 years men as compared to data from their urban counterparts. When data was stratified to look at percent difference in intake of 'Cereals' between Urban and Rural population, several additional rural groups other than the ones mentioned above showed at least a 25% higher intake of cereals when compared to their corresponding urban groups.

However, mean intake of 'pulses' was lower by 10 g across all the rural population, but for one age group (1–3 years) compared to the urban. This was also reflected in the 'Cereal to Pulse' ratio for each age-group. Comparing data from all rural and urban age groups, it was observed that every age-group in the rural data had a lower 'Cereal to Pulse' ratio than the corresponding urban age group. The recommended cereal: pulse ratio is 3:1 for vegetarians and 4 or 5:1 for those who consume recommended quantity of flesh foods.

Similarly, milk and milk products, flesh foods, eggs, vegetables and fruits consumption were higher for urban population of all age groups compared to the rural population. However, both urban and rural populations were consuming lesser than the levels recommended by ICMR-NIN.

Overall consumption of nuts and oilseeds were lower among children and adults across rural and urban population. However, other food group that is food eaten from outside (including HFSS) was uniformly higher across all age-groups in the urban compared to intake data from rural India. It was evident from this data that intake of food from outside was almost twice as much in urban India as compared to rural India across all age groups. Added sugar and added oils consumption will be high when HFSS foods are taken into consideration. As far as possible, sugar should be avoided at least for children up to 2-year age, but sugar intake was found to be around 10 grams for children below 2.

11.4 Energy and Protein Intake among Rural and Urban Population

Median 'Energy' values for Urban children from 4 to 6 years and 7 to 9 years and both Boys and Girls from 10 to 12 years (Urban) and Boys (Urban) aged 13 to 15 years were around 100 kcal more than corresponding rural groups. All other age groups only showed minimal differences in median 'Energy' values across urban and rural data.

The median quality protein intake data for all the different age-groups in the urban data was higher when compared to their corresponding rural age-groups (except the 18–50 years Women—Moderate Work—rural group). The highest difference in protein intake (ranging from 3.3 to 3.9 gm more) were seen for the urban groups 10–12 years (both Boys and Girls) and Boys 13–15 years and Lactating Women (Sedentary Work) compared to their corresponding rural groups. Percent difference values for intake additionally revealed that the rural group Children 4–6 years consumed 10% lesser protein than their urban counterparts and this was the highest percent difference in 'Protein' intake across all age groups.

11.5 Percent Calories (Energy-En) from Different Food Groups among Urban and Rural Adults (Fig. 11.2a, b)

Mean calorie intake of adults in urban areas was 2129 kcal/day with 308 g of Carbohydrate (CHO), 62 g of Fat and and 58 g of quality Protein intake respectively per day; while in rural areas, the mean calorie intake was 2152 Kcal/day with 372 g

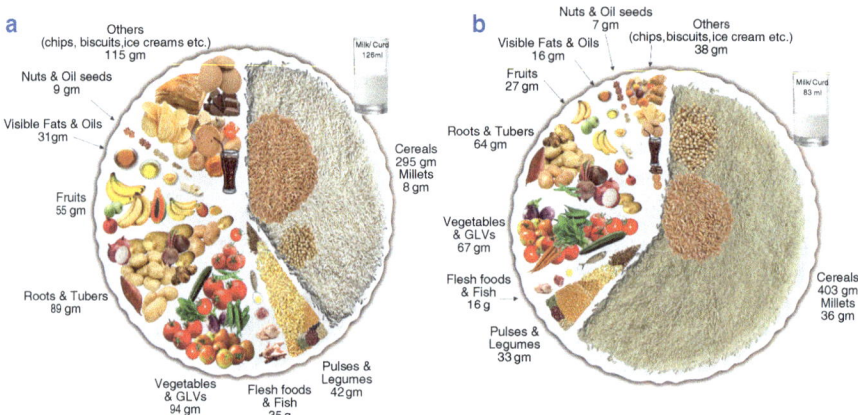

Fig. 11.2 (**a**) Urban plate. (**b**) Rural plate

of carbohydrates and 37 g of Fat and 54 g of quality Protein. Cereals contributed to 1040 Kcal/day of the total energy/ calorie (hence forth referred as En) intake, while visible fats and pulses contributed to 266 and 135 Kcal/day respectively in urban areas. In contrast, the total energy intake from cereals was much higher (1508) Kcal/ day, and considerably lower from Fats and Pulses in rural areas (fats, 146 Kcal/day and pulses, 106 Kcal/day respectively). Whereas, Milk and Milk Products contributed with wide variations in urban (258 Kcal/day) and rural areas (170 Kcal/day). The % En derived from different food groups showed that refined cereals contributed to 49% of En per day in urban areas and 69% of En in rural areas, which was far more than the recommended level of not more than 45% energy from cereals. Milk and milk products, and pulses contributed 12% and 6.4% of En per day respectively, in urban areas, which was slightly higher than rural areas (8% En and 5% En respectively). Together, pulses, meat, poultry and fish contributed to 8.8% of the total energy per day in urban areas and 6% of En in rural areas as against the recommended intake level of 14% of total energy from these foods. As for energy from vegetables (including roots & tubers) and fruits, it was only 4.4% En in urban areas and 3.4% En in rural areas as against the minimum requirement of 8–11% En per day from vegetables and fruits. Similarly, whole nuts and oilseeds formed only 2.2% En in urban areas and 1.7% En in rural areas as against recommendation of 8% En per day. Other foods (which include chips, biscuits, chocolates, sweets, juices, etc.,) contributed to 10% En per day in urban areas, while it was low, but substantial in rural areas (4%).

The trend showed that 'refined cereals' and 'fats & oils' were contributing to much higher energy values than the 'My plate' recommendations among the urban and rural population. Whereas, intake from the other groups containing protective foods such as legumes, milk, nuts, vegetables and fruits was significantly below the recommendation, which may potentially increase the risk of non-communicable diseases.

Overweight and Obesity among Different Age/Activity-Groups The proportion of overweight and obesity among different age groups was performed using the 'Weight-for-Height' z-score for individuals up to 5 years of age; 'BMI-for-z score' for individuals 6–18 years of age and BMI score for individuals above 18 years of age. We analyzed the data using Asian BMI categories for overweight and obesity. As expected, children under 5 years of age showed a slightly higher proportion of children with overweight in the urban settings compared to corresponding rural children. However, the proportion of obesity was similar between urban and rural children up to 5 years of age. There was a steady increase in proportion of children who were overweight and obese with increasing age between the ages 6–18 years with a steep increase in girls after 12 years and in boys after 15 years. Overall, the rural data showed 2.5% overweight and obesity among boys and girls of 12–18 years old, while the urban children of same age had four-fold higher level of overweight and obesity.

Overweight and obesity was higher in the urban population compared to the rural; and a sharp increase in the proportion of overweight and obesity was observed among the rural and urban population aged 18–49 years. Mostly a trend of reduction in overweight and obesity proportions was seen among adults with increasing activity status among both rural and urban groups using the Asian BMI standards. The highest overweight and obesity percentages were observed among urban sedentary adult females (18 to 49 years old) with 30.6% individuals overweight and 14.1% obese among them when compared to 30.4% overweight and 6.6% obese in urban sedentary males (18-49 years).

Hypertension and Diabetes The overall prevalence of hypertension (HTN) was 32.8% and 21.9% in urban and rural regions respectively, while the prevalence of diabetes was 25.4% in urban and 7.4% in rural areas. An increase in the risk of hypertension was observed with low intake of milk and milk products. And low intake of vegetables and fruits was observed to augment diabetes. Higher intake of foods such as sweets, ice-creams, carbonated beverages, fried snacks and packaged foods was associated with risk of diabetes. Healthy diet and adequate physical activity are the only strategies for halting or preventing the development of type 2 diabetes, hypertension, coronary heart disease, stroke etc. Small, judicious changes in dietary intakes will result in huge benefits in the health and nutrition of our population.

11.6 Highlights from the Surveys

A common observation across most population groups was that there was an over-excess intake of cereals than recommended levels and less intake of protective foods such as legumes, milk, nuts, vegetables and fruits. The consumption pattern was different across rural and urban India as highlighted below:

- Age-group wise data showed a higher mean intake of 'Cereals' in all age groups among the rural when compared to the urban population.
- Some rural groups consumed over 25% more 'Cereals' than their urban counterparts.
- The trend was reversed for the 'Pulses' food group. In addition, urban population consumed close to double or more in terms of quantity of 'Eggs and flesh foods' compared to rural population. Mean intake of 'Milk & Milk products' was also higher for all urban age-groups.
- Most urban age groups also consumed far higher 'Added sugar and added fats/ oils than their rural counterparts.
- Mean intake of 'Nuts and Oilseeds' was generally higher in most age-groups in the urban population.
- The mean intake of 'Vegetables' was higher for majority of the age-groups of urban population.
- Intake of food from outside was also twice as much in urban India as compared to rural India across all age groups.

11.7   Conclusion

Optimum nutrition is the cornerstone of good health; and improved nutrition can address many of the societal, environmental, and economic challenges faced by a country. However, to understand the population's health and nutrition to effectively and efficiently improve and sustain optimal health, the collection of accurate and relevant information periodically and systematically is required. Particularly, dietary monitoring is an important requirement to plan for nutrition actions. Further, the development of effective nutrition policies including food fortifications and nutrition norms of the supplementation programs such as the Integrated Child Development Scheme (ICDS), the Mid-day Meal program (MDM), depend on the availability of robust data from routine systematic surveys. Currently the only data available are the rural (2012) and the urban data (2016), which are old and are not representative of the current nutrition scenario of the population. In the given context, a national level dietary survey, "Diet and Biomarker Survey in India (DABS-I)", has been rolled out to fulfil the above needs of the country in line with the National Nutrition Policy recommendations.

This comprehensive survey will have stratified national level diet intake data using state of the art techniques to measure individual daily food/nutrient intakes, biomarkers of nutrition and health indicators. The survey also proposes measurement of haemoglobin levels by the gold standard venous blood method, which may provide more accurate estimate of burden of anaemia in the country. The study's main purpose is to support all National Nutrition programs in the country by providing the much needed dietary data across all age groups along with nutrition and health biomarkers among urban and rural populations of all states/UTs of India. The survey data is expected to help in data-driven policy making process, particularly for ongoing subsidies, supplementary nutrition programs and for developing regulation of food policy in the country. The survey findings will also help develop appropriate strategies/interventions to mitigate the rising overweight/ obesity complications in India.

In addition to the dietary data (food consumption data), for achieving health, elimination of all forms of malnutrition, to reach the SDGs and to reduce NCDs, there is an urgent need to create awareness among households for inculcating healthy dietary practices and improve intake of locally grown and available protective foods.

Also, while not everyone has equal access to diverse, micronutrient rich foods such as fresh fruits, vegetables, pulses, legumes and nuts, foods that are high in salt, sugars, saturated fats and trans fats have become inexpensive and are more widely available. Increased availability of inexpensive staple cereal crops has reduced hunger, but at the expense of diet diversity, displacing local ingredients and protective foods.

Although 'My Plate for the Day' provides a satisfactorily rounded recommendation for a balanced diet for adults, keeping in view the different subsections of the population having different nutritional needs we have developed the "Dietary Guidelines for Indians" (DGI), which will be published in a couple months.

The dietary guidelines are a way to translate scientific knowledge on nutrients into specific advice for individuals. The guidelines focus on a food-based approach to achieve optimal nutrition and promote nutritionally adequate diets and healthy lifestyles throughout all stages of life.

These guidelines are intended to provide information to the general population as well as specific age, gender, activity, and physiological groups, in order to promote health benefits. They are designed to be practical, dynamic, and flexible, taking into account the prevailing situation and considering social, economic, agricultural, and environmental conditions.

The use of these guidelines may require adaptation to different contexts, but they are meant to be utilized by medical and health professionals, nutritionists, dietitians, and other stakeholders. The guidelines are considered scientifically evidence-based information that aligns with the goals stated in the National Nutrition Policy, as well as national policies on Agriculture and Health.

In summary, the DGI aims to educate individuals and professionals about nutrition, guide food choices, and contribute to the achievement of national nutrition and health goals.

Acknowledgements Dr. B. Senthil Kumar, Sr. Technical Officer is acknowledged for statistical analysis of the data. Mrs. G. Neeraja, Technical Officer for helping and supporting the Statistician in collating the dietary data. Also, the services of Mr.S.Devendran, Senior Technician-3 (Artist) is hereby acknowledged and appreciated for designing the 'My Plate', 'Rural and Urban Plates'.

All the Scientists and Technical Staff of Division of Community Studies, NIN who were involved in the supervision and quality control of the data collection and analysis and Project staff of National Nutrition Monitoring Bureau (NNMB) for data collection are greatly acknowledged.

References

Anon (2023) Revised short summary report of the 2020, ICMR-NIN Expert Group on Nutrient Requirement for Indians, Recommended Dietary Allowances (RDA) and Estimated Average Requirements (EAR)

Hemalatha R (2023) Promotion of "my plate for the day" and physical activity among the population to prevent all forms of malnutrition and NCDs in the country. ICMR-NIN, Hyderabad. (policy brief)

Hemalatha R, Gupta SS, Raghavendra C, Neeraja G, Senthil Kumar B, Laxmiah A, NNMB Team (2023) What India Eats-2: food consumption pattern and nutrient intake among adults and children. ICMR-NIN, Hyderabad

National Institute of Nutrition, Indian Council of Medical Research. National Nutrition Monitoring Bureau (NNMB) (2012) Diet and nutritional status of rural population, prevalence of hypertension and diabetes among adults and infant and young child feeding practices-3rd Repeat Survey. Report No.26. Hyderabad. Available from: https://www.nin.res.in/downloads/NNMB_Third_Repeat_Rural_Survey%20%20%20Technicl_Report_26%20(1).pdf

National Institute of Nutrition, Indian Council of Medical Research. National Nutrition Monitoring Bureau (NNMB) (2017) Diet and nutritional status of urban population in India and prevalence of obesity, hypertension, diabetes and hyperlipidaemia in urban men and women. Report No.27. Hyderabad. Available from: https://www.nin.res.in/downloads/NNMB%20Urban%20Nutrition%20Report%20-Brief%20%20%20report.pdf

Balancing Human Demand and Ecological Supply for Sustainable Agriculture

P. Das

Abstract

It is supposed that the first organized agriculture developed in West Asia, somewhere in the present sites of *Jermo*, a small village, in the fertile crescent of Mesopotamia, covering an area of 12,000 to 16,000 m^2, dated to 7090 BC, (by carbon-14) consisting of some 25 houses, and about 150 people. They mainly dwelled in mud-thatched houses and practised a primitive way of living. Today's agriculture serves food for 2.3 billion Houses inhabited by 8.02 billion people, cultivating 6000–7000 species of crops. The global agriculture market saw a 9.4% CAGR (Compound Annual Growth Rate) worth $12,245.63 billion in 2022 to $13,398.79 billion in 2023.

Integrated solutions must be devised for better land and water governance, requiring both technical and managerial innovation. The priorities must be set in order to drive transformation and scale it up. Investment and assistance for agriculture need to be diverted to benefit environmental and social advantages that is derived from water and land management. Of India's 328.72 million mha total geographical area (TGA), over 97.85 million mha experienced land degradation in 2018–2019. India is committed to neutralizing land degradation by 2030.

The paper discusses various aspects of challenges of soil and water, the status of water availability, demand and future requirements, and balancing ecological footprint and biocapacity, which is a widely recognized measure of sustainabilityin one hand, and the status of ecological deficit/reserve and biocapacity creditors on the other. The application of Sustainability Science in terms of transdisciplinary approach requiring consideration of the need for transformational function were also discussed. The major farming systems of

P. Das (✉)
ICAR, New Delhi, India

© National Academy of Agricultural Sciences, under exclusive license to Springer
Nature Singapore Pte Ltd. 2023
K. C. Bansal et al. (eds.), *Transformation of Agri-Food Systems*,
https://doi.org/10.1007/978-981-99-8014-7_12

South Asia, the properties of sustainable agroecosystems, its structural and functional properties and indicating factors were critically analysed for optimum management of natural resources including various approaches that globally evolved; in terms of (a) clear ideologies, (b) environmental, economic, and societal objectives and (c) the approaches evolved in their own right over time. In addition, the supporting activities and similar environmentally friendly practices were also discussed. Finally, the issues that need to be addressed for better appreciation and adoption of sustainable agriculture were outlined.

Keywords

Human demand · Ecological supply · Sustainability · Ecological footprint · Sustainability · Ecological deficit · Ecological reserve · Transdisciplinary · Transformational

12.1 Introduction

It is believed that the first systematized agriculture developed in West Asia, somewhere in the present sites of *Jermo*, a small village, in the fertile crescent of Mesopotamia, covering an area of 12,000 to 16,000 m^2, dated to 7090 BC, (by carbon-14) consisting of some 25 houses, in all, about 150 people. They mainly dwelled in mud-thatched houses and practised a primitive way of living (Wikipedia 2023). Today's agriculture serves food for 2.3 billion Houses inhabited by 8.02 billion people, cultivating 6000–7000 species of crops. The global agriculture market saw a 9.4% CAGR (Compound Annual Growth Rate) worth $12,245.63 billion in 2022 to $13,398.79 billion in 2023 (The Business Research Company 2023).

India Shares (2019) 2.44% of the World's Area, 11.28% of the World's Arable Land, 17.73% of the World's Population, 4.00% of Water, 12.70% of Cattle, 56.70% of Buffaloes, and 14.50% of Goats. The the country employs 152 million people (the financial year 2021 to produce 21.60% of rice, 10.94% of total cereals, 25.44% of total pulses production, and 15.70% of milk).

12.2 State of the World's Land and Water Resources System for Food

The Synthesis Report 2021 of the state of the world's land and water resources system for food and agriculture suggests its present State, The challenges, and The Responses and actions (FAO 2021).

1. Integrated solutions must be devised for better land and water governance, and technical and managerial innovation is necessary to mitigate land degradation and water scarcity.

2. For scaling up the integrated solutions, they must be prepared at all levels and packaged as packages or programs of technical, institutional, governance, and financial support.
3. Adopting new technologies and management approaches can aid in reviving poor soils, and combat drought and water scarcity.

12.2.1 Land Degradation in India

A deteriorating trend in land quality due to direct or indirect causes is referred to as "land degradation." Anthropogenic activities have led to climate change, affecting biological productivity, ecological truthfulness, or human value. Forest degradation is defined as land degradation that takes place in forest land; however, it also occurs in non-forest areas. A subset of land degradation progressions known as "soil degradation" immediately impacts the soil (Olsson et al. 2019).

The article prepared by the Indian Space Research Organization (ISRO), and Desertification and Land Degradation Atlas of India exposed that:

1. Some 97.85 million hectares (mha) (29.7%) of India's total geographical area (TGA) of 328.72 mha underwent land degradation during 2018–2019 (Anon 2021).
2. The country is devoted to achieving land degradation neutrality by 2030. It has identified achieving land degradation neutrality as a means to recuperate biodiversity.
3. Additionally, the Government has launched numerous schemes or programs in this direction, including the National Afforestation Program, Green India Mission, and Watershed Development Component of *Pradhan Mantri Krishi Sinchayee Yojana* (PMKSY), which pays for the restoration of 26 million hectares of degraded land (Anon 2023).

12.2.2 Safeguarding Water Resources in India

Evapotranspiration accounts for about 53.3% of the total precipitation loss, leaving the nation a deficit of 1869 BCM of water. As a result of topographical boundaries and an unequal distribution of water resources over time and location, around 40% of the bourgeoning is inaccessible. As a result, the country's potential for using water is projected to be 1123 BCM, made up of 433 BCM of groundwater and 690 BCM of surface water (NITI 2023a) given by Ministry of Water Resources (MoWR), and National Commission for Integrated Water Resources Development (NCIWRD). The irrigation prerequisite estimated by NCIWRD is on the lower side as compared to that estimated by the Standing Sub-Committee because NCIWRD presumed that the irrigation proficiency will increase to 60% from the current level of 35 to 40% (Government of India, Ministry of Water Resources 2006).

With a set of 28 Key Performance Indicators (KPI) covering irrigation status, drinking water, and other water-related sectors, NITI Aayog has created a Composite Water Management Index. The index would be a supportive tool for monitoring performance in the water industry and encouraging timely remedial action for remedial management of water resources (NITI 2023b).

12.3 Balancing Demand Side (Ecological Footprint) and Supply Side Biocapacity

A widely recognized measure of sustainability, the Ecological Footprint provides an integrated, multiscale approach to tracking the use and overuse of natural resources, and the consequent impacts on ecosystems (Mancini et al. 2018) and biodiversity (Galli et al. 2014).

The Ecological Footprint is an account-based system of indicators whose underlying context is the recognition that Earth has a finite amount of biological production that supports all life on it (Wackernagel et al. 2018a, b).

The productivity of a country's biological resources, such as its cropland, grazing land, forest land, fishing grounds, and built-up land, is measured by its biocapacity on the supply side. If left untapped, these places can also be used to absorb the garbage we produce, particularly the carbon emissions from the combustion of fossil fuels (Footprintnetwork 2023).

The ecological footprint (EF) calculates the "global hectares" (gha) of biologically productive land needed in various scales or productive areas (croplands, grazing lands, forested areas for timber, marine areas for fisheries, residential land, and forested land to purify the atmosphere) (Hayden 2023).

Ecological Deficit/Reserve: An ecological deficit happens when a population's ecological footprint exceeds the biocapacity of the space that the population has access to. **Biocapacity Creditors** are those countries where the Biocapacity is Greater than the Footprint and are classified into four groups as >150%, 100%–150%, 50%–100%, and 50%–0%. Similarly, **Biocapacity Debtors** are those Countries where the Footprint is Greater than Biocapacity and are also classified into four groups as >150%, 100%–150%, 50%–100%, and 50%–0%.

For our current population to be sustained, 1.75 Earths would be needed. By 2050, if the recent trends continue, there will be required three Earths. How Do We Advance from Here? (World Population History n.d.).

Out of the top 10 Countries with Ecological Footprint (2022), India is in the third position after China and USA, analyzed based on the data available from Global Footprint Network, York University, and Footprint Data Foundation, 2022 (Global Footprint Network 2022; Anon n.d.), respectively.

As far as the Countries with **Biocapacity Reserve** (% of Biocapacity Exceeds Ecological Footprints) are concerned, there are 51 countries in the World in which none of the countries of South Asia found the place. On the other hand, there are 135 countries with Biocapacity Deficit with all the South Asian countries including

India. The analysis has been made based on the data available from Global Footprint Network, York University, and Footprint Data Foundation, 2022 (Anon n.d.).

12.4 Sustainability Science (SS)

Sustainability from sustain, "to nourish, maintain, prolong" (The Science of Sustainability 2023). The SS is thought of as a young, practical, and solution-focused discipline whose goal is to manage environmental, social, and economic problems in the context of institutional, historical, and cultural views.

- **The discipline faces challenges in better understanding, the issues affecting sustainability and** moving forward with solutions by using an integrated, all-encompassing, and participative approach.
- **Focus on dynamic communications between nature and society:** the systemic dynamics, susceptibility, and resilience of complex social-ecological systems.
- **Transdisciplinarity approach:** There is a need to incorporate the various tiers of reality knowledge into a reductionist or comprehensive perspective; and.
- **Transformational function**: creation of collaborative and coordinated solutions to sustainability issues (Sala et al. 2012).

12.4.1 Categories of Farming Systems

Based on a number of key factors, including (1) the available natural resource base; (2) the prevalent mode of farm work and household subsistence, including relationship to markets; and (3) the level of production, as well as the applicability of these standards to each of the six principal emerging world areas, 72 Farming Systems: were identified of eight broad categories of farming systems:

1. Irrigated farming techniques that produce a variety of food and cash crops.
2. Systems for growing rice in wetlands that rely on seasonal rainfall and irrigation.
3. In humid regions, rainfed farming systems with particular main crops or mixed crop livestock systems.
4. Rainfed agriculture on highland and hill terrain,
5. Rainfed farming systems in dearth areas,
6. Dualistic (mixed large commercial and smallholders) farming systems,
7. Coastal artisanal fishing systems, and.
8. Urban-based farming systems.

Each category's systems, with the exception of the dualistic systems, are dominated by smallholder agriculture. These eight categories have distinguishing characteristics, such as: (1) the availability of water resources, such as irrigated, rainfed, moist, or dry; (2) the climate, such as tropical, temperate, or cold; (3) the altitude of the landscape, such as highland or lowland; (4) farm size, such as large scale; (5) the intensity of production, such as intensive, extensive, or sparse; (6) the primary source of income, such as root crops, maize, tree crops, artisanal fishing, or pastoral activities; (7) dual crop lifestyles, such as cereal-root and rice-wheat (note that the term mixed denotes crop-livestock integration); and (8) location, such as forest-based, coastal, or urban-based (Dixon et al. 2001a).

12.4.2 Major Farming Systems of South Asia

It is revealed from Table 12.1 that there are 10 major farming systems in South Asia, the largest being, in terms of coverage of population, is Rice-Wheat (33%), followed by Rainfed Mixed (30%) and Rice (17%). However, in terms of largest area coverage, is Rainfed Mixed (29%), followed by Rice-Wheat (19%), and Highland Mixed (12%); Rice in sixth position covers 7% of the area along with Sparse (Mountain).

12.5 Properties of Sustainable Agroecosystems

Productivity, stability, sustainability, equity, and autonomy are the five system attributes that the Southeast Asian Universities Agroecosystem Network (SUAN) has used to evaluate the functioning of agroecosystems (Marten 1988). However, Gliessman (2005) identified 12 such properties while comparing Natural Ecosystems, Modern Agroecosystems, and Sustainable agroecosystems (Table 12.2).

Catherine Hill 2012 while dealing with 'Decent rural livelihoods and rights in a green economy environment' classified the properties of Agro-ecosystems into two major groups as Functional Properties and Structural Properties. Functional Properties have been further divided into Efficiency, Resilience, and Structural

Table 12.1 Major farming systems of South Asia

Farming systems	Land area (% of region)	Agric popn (% of the region)
Rice	7	17
Coastal artisanal fishing	1	2
Rice-wheat	19	33
Highland mixed	12	7
Rainfed mixed	29	30
Dry Rainfed	4	4
Pastoral	11	3
Sparse (arid)	11	1
Sparse (mountain)	7	<1
Tree crop	Dispersed	1
Urban based	<1	1

Source: (Dixon et al. 2001b; Anon n.d.)

Table 12.2 Properties of natural ecosystems compared with modern and sustainable agroecosystems. (Adapted from Gliessman (2005), Anon (n.d.))

Sl. No.	Property	Natural ecosystem	Modern Agroecosystem	Sustainable Agroecosystem
1	Productivity	Medium	High	Medium (possibly high)
2	Species diversity	High	Low	Medium
3	Functional diversity	High	Low	Medium–high
4	Output stability	Medium	Low–medium	High
5	Biomass accumulation	High	Low	Medium–high
6	Nutrient recycling	Closed	Open	Semi-closed
7	Trophic relationships	Complex	Simple	Intermediate
8	Natural population regulation	High	Low	Medium–high
9	Resilience	High	Low	Medium
10	Dependence on external inputs	Low	High	Medium
11	Human displacement of ecological processes	Low	High	Low–medium
12	Sustainability	High	Low	High

Table. 12.3 Agro-ecosystems' functional and structural properties and indicators

Functional properties		
Efficiency of resources Under **normal** conditions based on: • Productivity • Commercial yield per unit of input • Life quality of producers and consumers	**Resilience** To environmental and macro-economic risk under **disturbed** conditions in terms of: •Productivity • Commercial yield per unit of input • Life quality of producers and consumers	
Structural properties		
Connectedness Based on: •Transboundary pollution and environmental aspects • Dependency on monetary and imputs • Participation and social integration	**Coherence** Based on: • Ecological balance of water, soil, habitat, nutrient, energy • Economic integration • Household manpower	**Diversity** Based on: • Biological diversity • Diversification of income • Knowledge

Source: (Hill 2012; Anon n.d.)

Properties into Connectedness, Coherence, and Diversity, the details of which are given Table 12.3.

12.5.1 Effect of Agricultural Sustainability, Technologies and Practices

Table 12.4 summarizes the implementation of agricultural sustainability technology and practices and their effects on 286 projects across 57 countries. Based on the results of 12.56 million farmers with a coverage of 36.95 million hectares, there has been an average increase in yield by 79.2 %, with a geometric mean of 64.0%, even with wider variation in the Farming Systems as well as the number of farmers and area covered in different categories of Farming Systems (only in rice there were three reports where yields decreased).

According to the study, these sustainable agroecosystems also have advantages that support the development of social capital, natural capital, and human capabilities (Ostrom 1990; Pretty 2003).

Examples of positive side recorded in numerous developing countries include:

- Enhancements to natural capital that result in improved carbon sequestration and increased agrobiodiversity contain increased soil water retention, raised water tables (with more drinkable water during the dry season), and less soil erosion.
- increases in social capital, together with local social organizations that are more plentiful and powerful, new guidelines and standards for the management of common resources, and improved ties to external policy institutions.

Table 12.4 Technologies and Practices for agricultural sustainability and their effects on 286 projects across 57 countries

Sl. No.	FAO categories of Farm System[a]	No. of Farmers into sustainable Agriculture	Area Under Sustainable Agriculture (ha)	% Increase Crop Yields[b] (Average)
1	Smallholder irrigated	177,287	357,940	129.8 (±21.5)
2	Wetland rice	8,711,236	7,007,564	22.3 (±2.8)
3	Smallholder rainfed humid	1,704,958	1,081,071	102.2 (±9.0)
4	Smallholder rainfed highland	401,699	725,535	107.3 (±14.7)
5	Smallholder rainfed dry/cold	604,804	737,896	99.2 (±12.5)
6	Dualistic mixed	537,311	26,846,750	76.5 (±12.6)
7	Coastal artisanal	220,000	160,000	62.0 (±20.0)
8	Urban-based and kitchen garden	207,479	36,147	146.0 (±32.9)
9	All projects	12,564,774	36,952,903	79.2 (±4.5)

[a]Farm classes from (Dixon et al. 2001b)
[b]Yield data from 360 crop-project undertaken; reported as % increase, Standard errors in supports
Source: Pretty, Jules (2007). Agricultural sustainability: concepts, principles, and evidence. Philosophical Transactions of Royal Society B, doi:10.1098/rstb.2007.2163 (Pretty 2007a)

– improvements to human capital comprising of increased local experimentation and problem-solving skills, a reduction in malaria cases in rice-fishing regions, raised status for women and beforehand marginalized groups, improved child health and nutrition, especially during dry seasons, a reversal of migration, and more local employment (Pretty 2007b).

12.6 Approaches to Sustainable Natural Resource Management

There are 14 Approaches to sustainable natural resource management along with 54 principles and 121 practices (Oberč and Arroyo Schnell 2020a).

1. These include Agroecology, a new way of considering agriculture and its relationship with society (Silici 2014). Today's visions of agroecology integrate transdisciplinary knowledge, farmers' practices, and social movements, while recognising their mutual interdependence (based on the definition from the FAO's High-Level Panel of Experts on Food Security and Nutrition (FAO 2017)).

2. A system of agriculture known as "nature-inclusive agriculture" (NIA) is economically viable and provides a groundwork for sustainable commercial operations, including care for ecological services and biodiversity on or neighbouring the farm. Nature-inclusive agriculture aspires to become mainstream, whereas agroecology attentions much more conceptually on the local food chain and context. This is a significant dissimilarity between the two approaches (van Doorn et al. 2016).

3. Permaculture is the deliberate creation and upkeep of harmonious landscape and human integration that meets their requirements for food, energy, shelter, and other material and non-material needs in a sustainable manner (Mollison 1988).

4. In biodynamic agriculture, farmers use living solutions for fertility and pest management as a substitute of synthetic pesticides and fertilizers, and they set aside at least 10% of their entire farmland for biodiversity (Demeter 2012).

5. Organic farming does not employ inputs that have negative impacts but rather rely on ecological processes, biodiversity, and cycles that are tailored to local conditions. In order to benefit the environment as a whole, organic agriculture blends tradition, creativity, and science. It also encourages the coexistence of all those involved (International Federation of Organic Agriculture Movements (IFOAM) 2005).

6. Conservation agriculture, which has its focuses on the soil and derives from the Latin word "conserve" which means "to keep together," seeks to "keep the soil together" as a living ecosystem that facilitates food production and aids in combating climate change (Kassam et al. 2018).

7. Robert Rodale first used the term in the early 1980s (Gold and Potter Gates 2007). In that it aims to improve and sustain the health of the soil by refilling its organic matter and enhancing its fertility and productivity, regenerative

agriculture builds on conservation agriculture in that way. The health of the soil is its main concern.

8. Given the growing urgency with which climate change must be handled and its objective of becoming climate-neutral by 2050, the EU is eager to scale up the practice of carbon farming (Gillman 2019; Nijman 2019).

9. An integrated strategy called "climate-smart agriculture" (CSA) seeks to concurrently achieve three goals: improved productivity, enhanced resilience, and reduced emissions (The World Bank n.d.).

10. High Value for Nature Farming mandates that because it is at the lower end of the farming intensity scale, a variety of wildlife species that are housed at intensively farmed land are supported by both the agricultural margins and the productive land (Oppermann et al. 2012).

11. Low External Input Agriculture does not mean restricting external inputs like fertilizer and pesticides, but rather emphasizing better agronomic methods, integrated pest management, manual labour, and general farm management to preserve yields (FAO 2019).

12. Resource efficiency as opposed to production efficiency: Utilizing a whole-systems approach, circular agriculture integrates crops and livestock while making the best use of available resources, including side streams (de Boer and van Ittersum 2018).

13. A method to agricultural production known as "ecological intensification" incorporates the management of ecosystem services provided by biodiversity into production systems (Bommarco et al. 2013) and.

14. Sustainable intensification is a strategy that "increases yields without having a negative impact on the environment and without using more land for cultivation" (Royal Society 2009).

Sustainable natural resource management was formulated with various approaches that evolved globally. These were characterised as (a) clear principles, (b) environmental, economic, and social objectives and (c) had evolved as approaches in their own right over some time (i.e., agroecology or sustainable intensification). These methods frequently work in a range of conditions and production kinds, or they may be designed for the entire farm or system. In the majority of situations, they already have practitioners and occasionally a market or label attached to them (for example, organic farming). The fact that they are all decisions that farmers can make and that will largely determine how they manage their farm in the long run despite their potential diversity in scope (more "overarching" such as agroecology or sustainable intensification, or more "focused" such as permaculture or perhaps high nature value farming) (Oberč and Arroyo Schnell 2020b).

12.6.1 Supporting Activities and Similar Environmentally Friendly Practices

Supporting Activities (7):	Similar environmentally friendly practices (11):
1. Genetic improvement 2. Precision farming 3. Mixed farming systems 4. Integrated farming tool 5. Pasture-based and free-range farming 6. Landscape and ecosystems approach 7. Supporting socio-economic activities	(1) Crop rotation; (2) cover and companion crops; (3) mixed crop and intercropping; (4) reduction of synthetic pesticide and mineral fertilizer use; (5) no or minimal tillage; (6) lower livestock densities, managed grazing, free range; (7) crop diversification, (8) mixing farming and forestry, (8) mixed crop and animal farming, (9) nutrient balancing, (10) recovery and reuse, (11) inclusion of landscape elements such as hedgerows and flower strips

Source: (Oberč and Arroyo Schnell 2020b)

12.7 Issues Need to Be Addressed

1. There is a need for a common vision for sustainable agriculture in future.
2. Different approaches have several important commonalities and diversities, which is a strength in itself.
3. For implementation, the choice of approach depends on local contexts and specific priorities.
4. The challenge for policymaking is to create the (market or regulatory) environment according to local contexts.
5. Need for common metrics to ascertain and monitor the environmental performance of various approaches, which currently need to be improved.
6. The demand for more labour contributions.
7. Productivity and profitability related issues.
8. Yield gaps are a frequently mentioned problem for several techniques, affecting their viability from an economic standpoint.
9. Scalability and uptake issues.
10. Public support or private investments in the transition to sustainable land use may help to reduce potential expenses and/or a decline in profitability.
11. Difficulties in taking into account all important environmental factors at the same level, including soil, water, biodiversity, and temperature (Gupta et al. 2021).

12.8 To Conclude

Jules Pretty (2008) in his paper entitled, 'Agricultural sustainability: concepts, principles and evidence' indicated that several countries have given subregional support to agricultural sustainability, such as the states of Santa Caterina, Parana´ and Rio Grande do Sul in southern Brazil supporting zero-tillage, catchment management and rural agribusiness development and some states in India supporting participatory watershed and irrigation management. More nations have changed specific aspects of their agricultural policies. Examples include China's support for integrated ecological demonstration villages, Kenya's catchment approach to soil conservation, Indonesia's ban on pesticides and program for farmer field schools, Bolivia's regional incorporation of agricultural and rural policies, Sweden's support for organic agriculture, Burkina Faso's land policy, Sri Lanka's and the Philippines' requirement that water users' groups be formed. Several agri-environmental programs have been launched in Europe and North America during the past 10 years, albeit their success has been uneven (Dobbs and Pretty 2004; Kleijn et al. 2001; Marggraf 2003; Carey et al. 2005; Feehan et al. 2005; Herzog et al. 2005; Meyer-Aurich 2005).

References

Anon (2021) Status of land degradation and desertification. Environment. https://journalsofindia. com/status-of-land-degradation-and-desertification/?print=pdf

Anon (2023) G20 India: 'Arrest Land Degradation, Restore Ecosystem', discussed at 1st Working Group Meeting focused on environment, 13 February 2023

Anon (n.d.) *Ibid*

Bommarco R, Kleijn D, Potts SG (2013) Ecological intensification: harnessing ecosystem services for food security. Trends Ecol Evol 28(4):230–238. In ecological intensification in EU agriculture policy implications of research findings from project LIBERATION. Ecological intensification in EU agriculture (fao.org)

Carey P, Manchester SJ, Firbank LG (2005) Performance of two agri-environment schemes in England: a comparison of ecological and multi-disciplinary evaluations. Agric Ecosyst Environ 108:178–188. https://doi.org/10.1016/j.agee.2005.02.002

de Boer IJM, van Ittersum MK (2018) Circularity in agricultural production. Scientific basis for the Mansholt lecture 2018. Wageningen University and Research. https://edepot.wur.nl/470625

Demeter (2012) Biodynamic preparations. https://www.demeter.net/what-is-demeter/biodynamicpreparations

Dixon JA, Gibbon DP, Gulliver A (2001a) SUMMARY farming systems and poverty improving farmers' livelihoods in a changing world. FAO and World Bank, Rome and Washington DC

Dixon J, Gulliver A, Gibbon D (2001b) Farming systems and poverty. FAO, Rome

Dobbs T, Pretty JN (2004) Agri-environmental stewardship schemes and 'multifunctionality' rev. Agric Econ 26:220–237. https://doi.org/10.1111/j.1467-9353.2004.00172.x

FAO (2017) Agroecology and Ecosystem based Adaptation (EbA). Introduction to ecosystem-based adaptation (EbA) in the agricultural sectors: context, approaches and lessons. http://www.fao.org/3/CA1409EN/ca1409en.pdf

FAO (2019) The state of the World's biodiversity for food and agriculture. In: Bélanger J, Pilling D (eds) FAO commission on genetic resources for food and agriculture assessments. Rome, FAO, p 572; http://www.fao.org/3/CA3129EN/CA3129EN.pdf

FAO (2021) The state of the world's land and water resources for food and agriculture—systems at breaking point. Synthesis report 2021. FAO, Rome. https://doi.org/10.4060/cb7654en

Feehan J, Gillmor DA, Culleton N (2005) Effects of an agri-environment scheme on farmland biodiversity in Ireland. Agric Ecosyst Environ 107:275–286. https://doi.org/10.1016/j.agee.2004.10.024

Footprintnetwork (2023) Global footprint network-advancing the science of sustainability. Ecological Footprint. https://www.footprintnetwork.org/our-work/ecological-footprint/. Accessed 17 Apr 2023

Galli A, Wackernagel M, Iha K, Lazarus E (2014) Ecological footprint: implications for biodiversity. Biol Conserv 173:121–132

Gillman Steve (2019) EU wants to scale carbon farming to help achieve 2050 climate ambitions. IEG Policy. https://iegpolicy.agribusinessintelligence.informa.com/PL221802/EU-wants-to-scale-carbon-farming-tohelp-achieve-2050-climate-ambitions

Gliessman SR (2005) Agroecology and agroecosystems. In: Pretty J (ed) The Earthscan reader in sustainable agriculture. Earthscan, London

Global Footprint Network (2022) York University, and Footprint Data Foundation

Gold MV, Potter Gates J (2007) Tracing the evolution of organic/sustainable agriculture. A selected and annotated bibliography. Alternative farming systems information center. US Department of Agriculture. https://www.nal.usda.gov/afsic/tracing-evolution-organicsustainable-agriculture-tesa1980

Government of India, Ministry of Water Resources (2006) Report of the working group on water resources for the XI five year plan (2007–12) New Delhi, December 2006

Gupta Niti, Pradhan Shanal, Jain Abhishek, Patel Nahya (2021) Sustainable agriculture in India 2021- What We Know and How to Scale Up Council on Energy, Environment and Water (CEEW) Report April 2021. CEEW-FOLU-Sustainable-Agriculture-in-India-2021-20Apr21. pdf. Accessed 15 Jan 2023

Hayden Anders (2023) Ecological footprint. https://www.britannica.com/science/natural-resource

Herzog F, Dreier S, Hofer G, Marfurt C, Schupbach B, Spiess M, Walter T (2005) Effect of ecological compensation on floristic and breeding bird diversity in Swiss agricultural landscapes. Agric Ecosyst Environ 108:189–204. https://doi.org/10.1016/j.agee.2005.02.003

Hill, Catherine (2012) Working Paper 2 GEA–access Decent rural livelihoods and rights in a green economy environment coordination: Paolo Groppo, natural resources management and environment department. In: Greening the economy with agriculture. Extract from the FAO council document CL 143/18: status of preparation of FAO contributions to the 2012 United Nations Conference on Sustainable Development: Governance for Greening the Economy with Agriculture

International Federation of Organic Agriculture Movements (IFOAM) (2005) Definition of organic agriculture. https://www.ifoam.bio/fr/organic-landmarks/definition-organicagriculture

Kassam A, Friedrich T, Derpsch R (2018) Global spread of conservation agriculture. Int J Environ Studies 76:29. https://doi.org/10.1080/00207233.2018.1494927

Kleijn D, Berendse F, Smit R, Gilessen N (2001) Agri-environment schemes do not effectively protect biodiversity in Dutch agricultural landscapes. Nature 413:723–725. https://doi.org/10.1038/35099540

Mancini MS, Galli A, Coscieme L, Niccolucci V, Lin D, Pulselli FM, Bastianoni S, Marchettini N (2018) Exploring ecosystem services assessment through ecological footprint accounting. Ecosyst Serv 30:228–235

Marggraf R (2003) Comparative assessment of agri-environment programmes in federal states of Germany. Agric Ecosyst Environ 98:507–516. https://doi.org/10.1016/S0167-8809(03)00109-9

Marten GG (1988) Productivity, stability, sustainability, equitability and autonomy as properties for agroecosystem assessment. Agric Syst 26(1988):291–316

Meyer-Aurich A (2005) Economic and environmental analysis of sustainable farming practices—a Bavarian case study. Agric Syst 86:190–206. https://doi.org/10.1016/j.agsy.2004.09.007

Mollison B (1988) Permaculture: a Designer's manual. Tagari Publications, Tyalgum

Nijman P (2019) Call for participation, carbon farming. Interreg–North Sea region carbon farming. https://northsearegion.eu/carbon-farming

NITI (2023a) NITI Aayog, National Institution for Transforming India. Water. Government of India. https://social.niti.gov.in/water-index

NITI (2023b) NITI Aayog, National institution for transforming India, Government of India, April, 2023. op.cit

Oberč BP, Arroyo Schnell A (2020a) Approaches to sustainable agriculture. Exploring the pathways towards the future of farming. IUCN EURO, Brussels

Oberč BP, Arroyo Schnell A (2020b)˙ op.cit

Olsson L, Barbosa H, Bhadwal S, Cowie A, Delusca K, Flores-Renteria D, Hermans K, Jobbagy E, Kurz W, Li D, Sonwa DJ, Stringer L (2019) Land degradation. In: Shukla PR, Skea J, Buendia EC, Masson-Delmotte V, Pörtner H-O, Roberts DC, Zhai P, Slade R, Connors S, van Diemen R, Ferrat M, Haughey E, Luz S, Neogi S, Pathak M, Petzold J, Pereira JP, Vyas P, Huntley E, Kissick K, Belkacemi M, Malley J (eds) Climate change and land: an IPCC special report on climate change, desertification, land degradation, sustainable land management, food security, and greenhouse gas fluxes in terrestrial ecosystems. https://doi.org/10.1017/9781009157988.006

Oppermann R, Beaufoy G, Jones G (2012) High nature value farming in Europe: 35 European countries –experiences and perspectives. C. Verlaggegionalkultur, Germany. p 544. ISBN:978-89735-657-3

Ostrom E (1990) Governing the commons: the evolution of institutions for collective action. Cambridge University Press, New York, NY

Pretty J (2003) Social capital and the collective management of resources. Science 302:1912–1915. https://doi.org/10.1126/science.1090847

Pretty J (2007a) Agricultural sustainability: concepts, principles, and evidence. Philos Trans Royal Soc Lond B Biol Sci 363:447. https://doi.org/10.1098/rstb.2007.2163

Pretty Jules (2007b) op.cit

Pretty J (2008) Agricultural sustainability: concepts, principles and evidence. Philos Trans R Soc Lond B Biol Sci 363(1491):447–465. https://doi.org/10.1098/rstb.2007.2163; https://www.ncbi.nlm.nih.gov/pmc/articles/PMC2610163/. Published online 2007 Jul 25

Royal Society (2009) Reaping the benefits: science and the sustainable intensification of global agriculture. The Royal Society, London

Sala S, Farioli F, Zamagn A (2012) Progress in sustainability science: lessons learnt from current methodologies for sustainability assessment: part 1. Int J Life Cycle Assess 18:1653. https://doi.org/10.1007/s11367-012-0508-6; Progress-in-sustainability-science-Lessons-learnt-from-current-methodologies-for-sustainability-assessment-Part-1.pdf (researchgate.net). Accessed 26 Dec 2022

Silici L (2014) Agroecology: what it is and what it has to offer. IIED Issue Paper. IIED, London. ISBN: 978–1–78431-065-3. https://pubs.iied.org/pdfs/14629IIED.pdf

The Business Research Company (2023) The business research company: report of agriculture market size, trends and global forecast to 2032. https://www.thebusinessresearchcompany.com. Accessed 20 Apr 2023

The Science of Sustainability (2023) The Science of Sustainability.pdf. Accessed 11 Jan 2023

The World Bank (n.d.). https://www.worldbank.org/en/topic/climate-smart-agriculture

van Doorn A et al (2016) Food-for-thought: natuurinclusieve landbouw. Wageningen University & Research, Wageningen. https://doi.org/10.18174/401503

Wackernagel M, Galli A, Hanscom L, Lin D, Mailhes L, Drummond T (2018a) Chapter 16: ecological footprint accounts: principles. In: Bell S, Morse S (eds) Routledge handbook of sustainability indictors. Routledge International Handbooks; Routledge, Abingdon, pp 244–264

Wackernagel M, Galli A, Hanscom L, Lin D, Mailhes L, Drummond T (2018b) Chapter 33: ecological footprint accounts: criticisms and applications. In: Bell S, Morse S (eds) Routledge handbook of sustainability indictors. Routledge International Handbooks; Routledge, Abingdon, pp 521–539

Wikipedia (2023). https://en.wikipedia.org/wiki/Jarmo. Accessed 20 Apr 2023

World Population History (n.d.). https://worldpopulationhistory.org/carrying-capacity/

Extension Strategies for Climate Resilient Agriculture in Eastern Himalayan

Anupam Mishra

Abstract

Mountains are among the most vulnerable habitats on the planet Earth harbouring rich biodiversity. Climate change caused by anthropogenic forces has emerged as a major problem worldwide in terms of socioeconomic and environmental sustainability. Agriculture is generally regarded as a very vulnerable business to climate change, with extreme weather events posing an immediate threat to food security and livelihood sustainability. Globally, agricultural practices must adapt quickly to assure future food security in view of dual challenges of climate change and increasing population.

To make climate adaptation planning more robust, assessing resilience in different regions is vital for the development and scaling up of appropriate, site-specific interventions and policies that strengthen the resilience of agricultural system. Adaptation is critical for mitigating the harmful effects of climate change, particularly in the agricultural sector. This paper seeks to draw insights based on a multi-scalar and multi-indicator assessment analysis and debate through profiling of resilience in the Eastern Himalayan Regions of India, based on the development of a climate-resilient agriculture.

Farmers' views and response to drastically changing climatic conditions are perceived as critical policy strategies that can possibly mitigate the adversities of climate change and simultaneously feed the increasing population. In this context, this paper systematically analyses farmer perception and adaptation strategies in India's Eastern Himalayan region, from the jurisdiction of the Central Agricultural University, Imphal, Manipur, located across the seven sister states of North East India. Farmers in this region acknowledge climate change events such as temperature rise, unpredictable and decreased rainfall. They are receptive to a

A. Mishra (✉)
Central Agricultural University, Imphal, Manipur, India

© National Academy of Agricultural Sciences, under exclusive license to Springer Nature Singapore Pte Ltd. 2023
K. C. Bansal et al. (eds.), *Transformation of Agri-Food Systems*,
https://doi.org/10.1007/978-981-99-8014-7_13

wide range of adaptation strategies, which are incremental and systemic. Transformational adaptations such as substantial changes in land use, cropping systems
and adopting natural farming need further nudging by all stakeholders to ensure
their acceptance by farmers as a long term intervention against climate change.

The paper reiterates the role of extension strategies in influencing adaptation of
climate smart agriculture practices relevant to Eastern Himalayan region of India
and emphasizes on the role of policy makers, agriculture institutions and farmers
in efficient execution of these strategies in the target areas.

Keywords

Agricultural practices · Climate change · Extension strategies · North Eastern
India

13.1 Introduction

Agriculture contributes 16.5% of India's GDP and employs approximately 42.30%
of the labour force, including 71% of women workers, especially those living in rural
areas (World Bank 2019). With about 66% rural population, India remains primarily
a rural economy, and agriculture continues to be the primary livelihood of many
people. Smallholder farmers dominate India's agriculture sector with an estimated
87% (126 million) of India's total 146.4 million land holders. Roughly 69% of
India's small holders are marginal farmers, who own less than one hectare of land.
But, small and marginal farmers together accounted for only about 47% of the total
operating area (2015–2016), indicating severe land inequities. Rising fragmentation
of land is another important cause of concern for Indian agriculture. The average
land holding size has steadily decreased from 2.28 ha in 1970–1971 to 1.08 ha in
2015–2016 (DoAC and FW 2019). These are unsustainable levels, causing farmers
to abandon their land in search of greater prospects elsewhere, leading to accelerated
unplanned urbanization. As a result of a shortage of physical and human capital,
enormous swaths of productive land are either left fallow or are utilized at a very low
production (NITI Aayog 2016).

Climate change predictions made through 33 global climate models suggest that
the rise in lowest temperature in India is projected to be greater than that in
maximum temperature. The rise will be higher in winter (October–April) than *kharif*
(June–September). Minimum temperatures are expected to rise by 0.946–4.067 °C
during 2020–2080 compared to the baseline period of 1976–2005 in *kharif* and by
1.096–4.652 °C in *rabi*. The rise in maximum temperatures during the
corresponding period is expected to be 0.741–3.533 °C and 0.882–4.01 °C. Temperature rises are expected to be greater in northern India than in southern India.
Minimum and maximum temperature variability is expected to be substantially
greater during *rabi* than *kharif*. Rainfall will increase by 2.3–3.3% and 4.9–10.1%
by 2020 and 2050, respectively, during *kharif*, and by 12% and 12–17% during the
corresponding period in *rabi* season as compared to the baseline period of

1976–2005 (Naresh Kumar et al. 2019). Extreme weather events such as unseasonal rains, droughts, and floods will become more common across the country as a result of these changes. Over the previous century, the sea-surface temperature has increased by about 1.0 °C and is likely to rise more in the next years (Krishnan et al. 2020). This will result in a 300-mm sea-level increase in the Indian Ocean by the end of the century, compared to 1986–2005 estimates. Since the 1950s, the geographical and temporal frequency of droughts has increased. Some of these droughts have been more severe in recent decades, and their frequency and amplitude are expected to grow further. An increase in severe tropical cyclones along India's coastline is also expected. Agriculture also accounts for around 16% of total greenhouse gas (GHG) emissions in India. Agriculture sector in India emits 417.2 Mt. CO_2 of total GHG emissions, 54% of which comes from enteric fermentation, 19% from agricultural soils, 18% from rice cultivation, 7% from manure management and 2% from field burning and agricultural residues (MoEFCC 2018).

Farming households in India, Africa and other low-income countries are adopting a range of production methods to alleviate the detrimental effects of climate change on their livelihoods. They are planting early maturing cultivars and drought-tolerant crops, practicing crop diversification and changing land management practices to adapt to climate change (Williams et al. 2019; Bawakyillenuo et al. 2016). Farmers must overcome numerous barriers including lack of financial resources, lack of local participation in policy decision making and insufficient information on climate change characteristics and other climate information to implement such adaptation practices (Yiran and Stringer 2017). Adapting to climate change therefore, demands a transformation in people's behaviour, expertise and abilities to aid in resilience (Nnadi et al. 2013). In most of the instances, both informal and formal institutions aid in such learning.

Extension services are one example of a formal institution that play a critical role in assisting small-scale agriculture and achieving national and household food security across the blobe (Rickards et al. 2018). Agricultural extension services are known to enhance farmers' agricultural knowledge and skills, disseminate new technology and change farmers' attitude (Lamm et al. 2023), additionally also promote socioeconomic growth through human and social capital development, facilitate market access and work with farmers to achieve effective management of natural resources (Landini et al. 2017).

In the context of climate change, agricultural extension has a pivotal role to play in coordinating the distribution of innovative extension approaches and improving knowledge of the best possible local adaptations that may be used to manage climate risks (Afsar and Idrees 2019), in addition to supporting farmers in avoiding maladaptation (Antwi-Agyei et al. 2018; Juhola et al. 2016). While earlier research has demonstrated that extension services play an important role in improving farmers' welfare and productivity as well as equipping them to deal with the consequences of climate change, there are considerable information gaps addressing extension officers' capacity-building requirements. Rather than scientifically exploring how agricultural extension agents may be more effective, the emphasis has traditionally been on aiding farmers in adapting the technologies. Assessing the capacity building needs of agricultural extension agents to fulfill this essential mandate is very critical to well equip them so that they can assist smallholders in effectively managing and adapting to climate change, improving ability in building efforts to generate knowledge, skills, and expertise in order to enhance their analytical capacity that may assist in improving sustenance and productivity of agricultural systems (IFPRI 2005).

The focus of this article is to explore and figure out how extension techniques in India's Eastern Himalayan Region might help smallholder farmers navigate and manage the consequences of climate change on agriculture. It explores answers to vital questions such as: (1) what information sources are used by agriculture extension workers in India's Eastern Himalayan region? (2) What are the capacity-building needs for agricultural extension workers in order to effectively communicate climate knowledge to construct resilient agricultural systems? and (3) Identify the main hurdles of extension outcomes to good climate change adaptation?

13.2   The Eastern Himalayan Region: Agro-Climatic Zone (EHR-ACZ)

The EHR-ACZ extends between 21° 56′ 26.296″ and 29° 27′ 41.59″ North latitude and 87° 59′ 14.024″ to 97° 24′ 43.345″ East longitude and comprises the States of Arunachal Pradesh, Assam, Manipur, Meghalaya, Mizoram, Nagaland, Sikkim, Tripura and hill regions of West Bengal (Fig. 13.1). The region accounts for 8.3% of India's geographical area and has a human population of 456 million (Jain et al. 2013). It is known for its challenging and complex mountainous terrain, diverse climatic areas such as arctic, subarctic, temperate, subtropical and warm tropical, and a diverse flora and fauna. The climate in the EHR-ACZ region is usually humid subtropical with hot and humid summer, severe monsoon, and mild winter with annual mean temperature and humidity ranging from 5 to 30 °C and 70% to 85%, respectively (Jhajharia et al. 2009; Sharma et al. 2009). The EHR-ACZ is located at the intersection of the Indo-Chinese, Indo-Malayan and Indian bio-geographical realms (Nath et al. 2019) and contains a variety of forest types including moist, dense, evergreen, semi-evergreen and temperate forests (CEPF: Critical Ecosystem Partnership Fund 2005), as well as several endemic and globally threatened species (Malczewski 1999). The forest cover in the EHR-ACZ region stands at 1,70,541 square kilometres, accounting for 65.05% of its geographical area (IFSR: India State of Forest Report 2019). The major characteristics of EHR are presented in Table 13.1.

13.3   Extension Strategies: The Basics

Agricultural extension disseminates knowledge from local and world-wide studies to farmers by expediting knowledge transfer and assisting farmers in becoming better managers. It helps farmers make better decision that increases output, potentially leading to agricultural progress and greater profits (Anderson and Feder 2007).

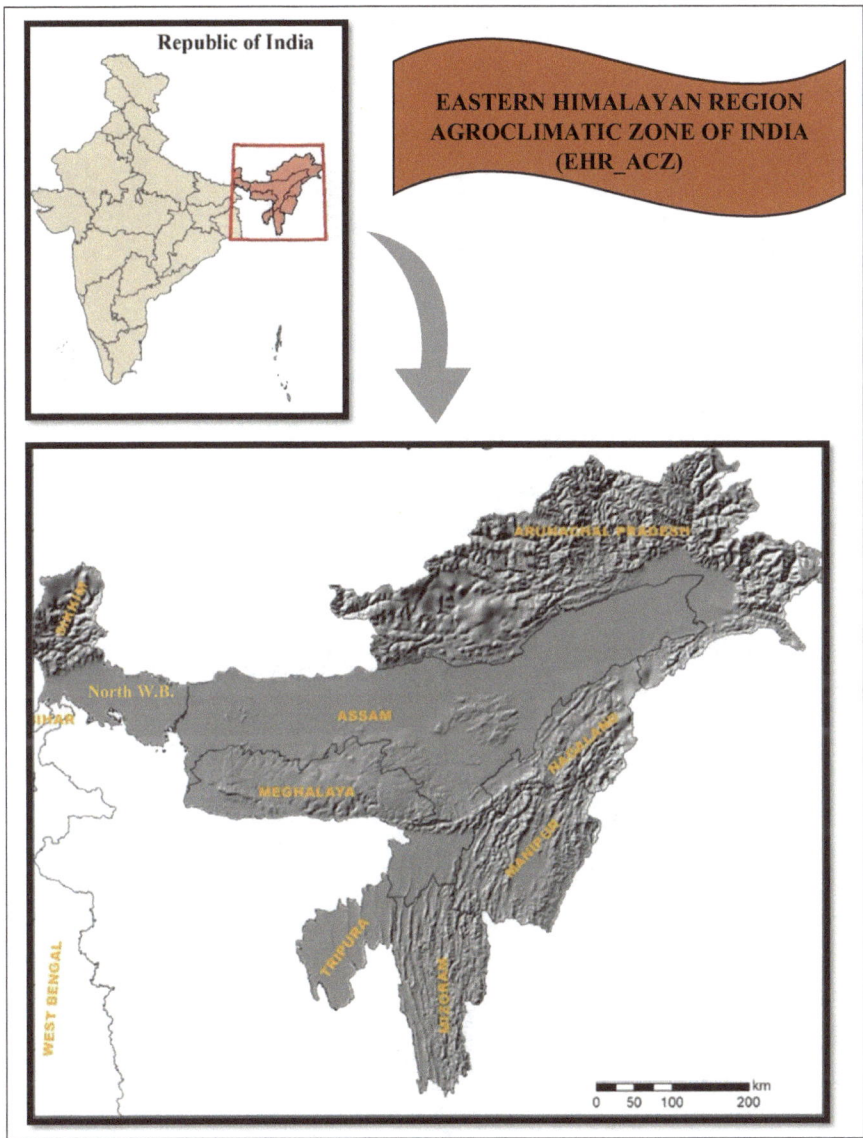

Fig. 13.1 Physical and Political Map of Eastern Himalayan Region of India. (Adapted from: (Nath et al. 2021)

Agricultural extension services are extensively accessible around the world, with more than 90% of public extension agents employed in developing nations. China with an estimated 6 lakh extension agents leads the world, while India has 90,000, Indonesia 54,000 and Ethiopia 46,000 extension workers. In comparison, the United States has less than 8000 extension agents, who are backed by 3000–4000 experts,

---

(cleaning)

Table 13.1 Major characteristics of the Eastern Himalayan Region-Agroclimatic Zone (EHR-ACZ)

Number of districts under EHR-ACZ	Distribution of states	Climate	Major soils	Annual rainfall (mm)	Annual temperature, °C (Min, Max)	Area[a] (km²)	Rural population[b] (million)
89	Arunachal Pradesh, Assam, Manipur, Meghalaya, Mizoram, Nagaland, Sikkim, Tripura, West Bengal	Per humid to humid	Red, sandy laterite, acidic, alluvial, red loamy, terai	2313	18.43, 28.45	2,74,942 (8.36)	43.83 (5.26)

Source—Singh et al. (2021)

Note: *Area and population estimates were derived from district-level data received from the Census of India (2011); rainfall and temperature data (1991–2015) were taken from the India Meteorological Department (IMD), Ministry of Earth Sciences, Government of India*
[a]Figures in parenthesis include share of ACZ in the total geographical area of the country
[b]Figure in the parenthesis include percentage share of ACZ in the total rural population of the country

who work at Land-Grant institutions or Agricultural experiment stations and are responsible for designing extension materials and programmes for agents (Swanson and Davis 2014).

The field of Agriculture Extension has evolved over the years from a traditional emphasis on technology transfer and farm management information provided by the public sector to a broader public and private advisory service mode, addressing topics such as marketing, environmental sustainability, pest diagnostics and risk management. For a multitude of reasons, extension systems have become increasingly pluralistic throughout time, relying on numerous delivery modes and deriving economic support from alternate financing sources that may inclide both public and private entities (Davis and Franzel 2018). Recent advances in information communication technology (ICT) have lessened barriers to information dissemination, gently shifting extension activities away from message delivery towards adapting advances in science to make them more accessible to farmers (World Bank 2017).

This has two vital elements:

- Practical knowledge dissemination, such as better seeds, soil quality, equipment, water management, crop protection, agricultural practices and livestock management options, and
- Application of this knowledge in farms.

Extension is an important aspect of agricultural research and development, in addition to helping rural development. Institutions of academic excellence emphasize on the scientific aspects of developing usable technology. Farmers' acceptance and implementation of these technologies is the focus of extension. The two vital elements mentioned above must work in tandem to achieve the expected goals.

Three key sources of agricultural extension are:

- **The public sector.**
  Government Ministries and Agriculture Departments, and Agricultural research centers, State and Central Agricultural Universities.
- **The private non-profit sector.**
  Non-governmental organisations (NGOs), Foundations, Community groups and Associations on a local and worldwide scale; Bilateral and multilateral aid programmes; and other non-commercial organisations.
- **The private for-profit sector.**
  Commercial organisations (such as input makers and distributors), Commercial farmers or Farmer' group-operated entities in which farmers are both users and suppliers of agricultural information, Agro-marketing and processing companies; Trade organisations and Private consultancy and Media firms.

13.3.1 Agricultural Extension in the Developing World

Countries with developed agricultural sectors, such as the USA, Canada, Australia, and Denmark have their robust extension services. This is lacking in the developing world, where agricultural extension has many times failed the expectations of the farmers in need.

The World Bank-designed 'Training and Visit' (T&V) system played a crucial role in India's Green Revolution, but it also fell short in certain key areas. It was inadequate for diverse farming systems in rainfed locations. It also struggled to handle changing issues such as increasing sustainability, diversification and linking farmers to markets. As a result, poverty persists in the sector, with far-reaching consequences. These include the younger generations relocating to cities, putting a burden on inadequate urban infrastructure and triggering food insecurity.

13.4   Novel Extension Strategies to Mitigate Climate Change

Numerous institutional, technological, socioeconomic and infrastructure limitations restrict an array of viable climate adaptation alternatives. Identifying these options and their adaptation is crucial for finding possible opportunities to tackle the challenges posed by climate change (Eisenack et al. 2014). Limited access to agricultural extension services has been identified as a major obstacle by farmers in adapting climate resilience strategies in earlier studies. To combat new and developing difficulties in agricultural markets, fresh thinking about technology and sustainability is required. Extension now needs to focus on

- Participatory approaches that shape demand-driven services.
- Multiple extension service providers.
- Development of agricultural innovation systems strategies.
- Decentralization of extension advisory services through technology.

The institutions supported by government have been the conventional means of research and extension services in the developing countries. However, it is becoming obvious that these institutions have their own limitations in terms of reach, cost and priorities. To ensure development and to feed rising populations, small-scale farmers must have an active role in the services they require. Furthermore, hurdles in engaging the corporate and non-governmental organisations in the extension programmes must be addressed and novel public-private partnerships must be developed.

13.5 Innovative Extension Models for Climate Resilience in Agriculture

A few innovative models that can be adapted in Eastern Himalayan Region are as follows:

1. **Village Knowledge Centers:** The Village Knowledge Centers (VKC) serve as a conduit to disseminate information, knowledge and skills to the end users in rural areas by utilizing the best-fit Information and Communication Technologies (ICTs). Eight VKCs have been constructed as a part of the climate adapt initiatives by M.S. Swaminathan Research Foundation (MSSRF) with assistance from NIBIO and the Norwegian Embassy. The VKC's goal is to give the community access to a sustainable single window knowledge platform that provides location-specific climate smart agro advisories and technologies to farmers for informed decisions as well as demand responsive, comprehensive knowledge and information to the community at large, based on their knowledge needs.

2. **Extension model adopted under NICRA:** The objectives of the program are as follows
   (a) The creation and use of improved production and risk management technologies.
   (b) To exhibit site-specific technological packages for coping with climate hazards on farmers' land.
   (c) To increase the expertise of researchers in climate-resilient agriculture and other interested parties.

The components of the scheme are as follows:
Strategic analysis of mitigation and adaptation.
Technology dissemination in 100 susceptible districts to handle the current climate variability.
Competitive sponsored research to close major gaps.
Capacity building.

13.6 National Climate Change Adaptation Programs

(a) One of the eight missions within the National Action Plan on Climate Change (NAPCC), the National Mission of Sustainable Agriculture was adopted in 2010 to encourage theme wise management of resources.
(b) To solve the problems with water resources and offers a long-term solution that envisions **Per Drop More Crop**, the *Pradhan Mantri Krishi Sinchayee Yojana* (PMKSY) was introduced in the year 2015. It encourages micro/drip irrigation thereby promoting water saving in farmers' field.
(c) The Government of India established the Green India Mission in 2014 under the auspices of National Action Plan on Climate Change (NAPCC) with the primary

goal of protecting, enhancing and restoring India's dwindling forest covers, thereby decreasing the harmful consequences of climate change.

(d) *The Paramparagat Krishi Vikas Yojana* was carried out in collaboration with the Indian Council of Agricultural Research and government stakeholders to fully use the adaption of climate-smart practices and technologies.

(e) The Soil Health Card program was established by the Government of India (GOI) to analyze soil samples from clusters and advise farmers on the fertility of their land. Neem-Coated Urea (NCU) was additionally introduced to reduce the overuse of urea fertilizers, preserving soil health and feeding plants with nitrogen.

13.7 Plausible Mechanism of Linkage of Extension Strategies with SDGs

Figure 13.2 elucidates the mechanism by which extension advisory services promote adaptation of climate resilience among the targeted farmers ensuring enhanced income which subsequently results in promoting food and nutrition security. The main mechanism flow depicted in Fig. 13.2 indicates that the direct causal process starts from extension agents' knowledge and skill impartation that empower farmers with knowledge and skill to adopt climate resilience practices in agriculture leading

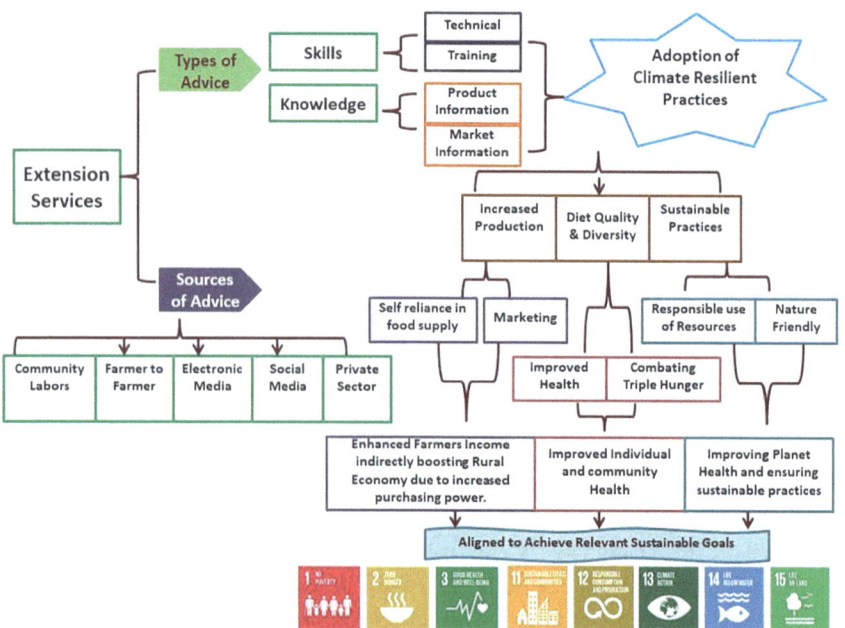

Fig. 13.2 The process flow of plausible mechanism aligning extension services and SDGs

to increase in farmers' crop and animal production. Thus, farmers have the option to consume their own products and sell the remaining that provides additional income. The income generated improves the purchasing power of the farmer and enables him/her to invest in food and non-food related expenses. Therefore, an increase in production, income generated plus supplements received via Agricultural extension and advisory services (AEAS) training positively correlates with inclusive economic development, reduce extreme poverty, ensure food security and ultimately help achieving the SDG1 (No Poverty), SDG 2 (Zero Hunger), SDG 3 (Good Health and Well Being), SDG 11 (Sustainable Cities and Communities), SDG 12 (Responsible Consumption and Production), SDG 13 (Climate Action), SDG 14 (Life Below Water) and SDG 15 (Life on Land).

13.8 Extension Strategies: Success Stories in Eastern Himalayan Region

13.8.1 The m4Agri Project at Central Agricultural University, Imphal, Manipur

Multiple sectors of world economy have been influenced by the boom in the Information and Communication Technology (ICT) in terms of improved efficiency and enhanced productivity. Agriculture sector is also witnessing tremendous use of ICT application in all fields of its operation. Daum et al. (2022) observed that in recent years, ICTs had become one of the main driving tools used by farmers to manage the essential factors of production (land, labour, capital and soil) in agriculture. Decent cell phone accessibility to farmers in India allows the targeted delivery of timely messages related to routine farming operations by extension workers creating a robust advisory system. Resource poor farmers in developing countries continue to be at a disadvantage in terms of mass media hardware such as computers and smart phones, but extension and advisory systems are rapidly gaining clients for a diverse range of mass media programming among large and small commercial farmers. Simple messages like those that can be communicated in personal interactions are disseminated more efficiently as the number of farmers addressed are in mass and it involves low cost. Improved weather and disease predictions, market information, pest and disease diagnoses and suggestions, and timely reminders of essential agricultural practices are just a few examples of what extension staff bring to farmers in remote locations. Some of these approaches for disseminating extension information have been studied for their efficiency (Singh Naveen et al. 2018; Larochelle et al. 2019; Daum et al. 2022).

The ICTs have the potential to strengthen the partnership between extension, research and farmers. These may be used as a channel for research institutions to disseminate fresh research findings to extension workers, who can then interact with farmers (Ayisi Nyarko and Kozári 2021). Farmers may also utilise the same ICT platform to provide feedback on new technologies from their field experiences to extension workers, who will then transfer the information to research institutes for necessary action. Furthermore, ICT tools like Whats App or Facebook group have the in-built platform on which farmers, extension workers and researchers can get on-board to share ideas, knowledge and valuable information with potential for resolution of problems in real time. Annor-Frempong et al. (2006) advocated that ICTs are among the modern tools that facilitate rapid information delivery and knowledge sharing among farmers, extension agents and other stakeholders such as research institutions (Purnomo and Lee 2010). Globally, various ICT projects and apps have been developed to improve communication among extension workers and other stakeholders in the agriculture and allied sectors.

Central Agricultural University (CAU), Imphal, Manipur (India) in partnership with Digital India Corporation (DIC) is implementing an ICT based project m4Agri since 2015. It is operational in seven states of EHR-ACZ. The project entitled "Mobile Based Agro Advisory System for North-East India (m4agriNEI)" was implemented through its colleges in Meghalaya at Barapani (College of Post Graduate Studies and Agricultural Sciences, Umiam, Barapani) and Tura (College of Community Science, Dobasipara, West Garo Hills, Tura, Meghalaya) with the objective to empower the farmers by providing 'right information at right time in local languages- Khasi and Garo, respectively'.

Based on the success of the project in Meghalaya, the entire M4agriNEI program with complete IT platform, farmers' database and project manpower was taken over by the Government of Meghalaya (GoM). The GoM has set up a unique farmer centric innovative project titled 1917-iTEAMS with the Meghalaya Institute of Entrepreneurship (MIE) as its Project Management Unit (PMU), in collaboration with DIC (Digital India Corporation) for the benefit of farmers of Meghalaya and which is the first of its kind disruptive, cloud based, IT driven farmer centric dedicated supply chain logistics, marketing, evacuation and advisory service platform.

Currently the project is operational in six constituent colleges of CAU, Imphal spread over six North Eastern States. It has registered 37,416 farmers who have made 1,24,332 calls to the project for seeking advice from experts in agriculture and allied fields that include, Animal Husbandry, Fisheries, Horticulture, Agriculture Engineering and Forestry (Fig. 13.3).

The six colleges implementing the project located at six different states cater to various aspects of the farmer's need. The state-wise registrations done by the farmers, total calls received by the project team, advisories delivered on demand, messages pushed, need-based workshops conducted and the locations covered are depicted in Fig. 13.4. To make the project farmer friendly care has been taken to name the project as per local dialect and the extension advisories are provided in local language for efficient dissemination and adoption of information by the local farmers.

Fig. 13.3 m4agri Project - a success story implemented by Central Agricultural University, Imphal as an extension strategy for benefit of farmers in North Eastern India

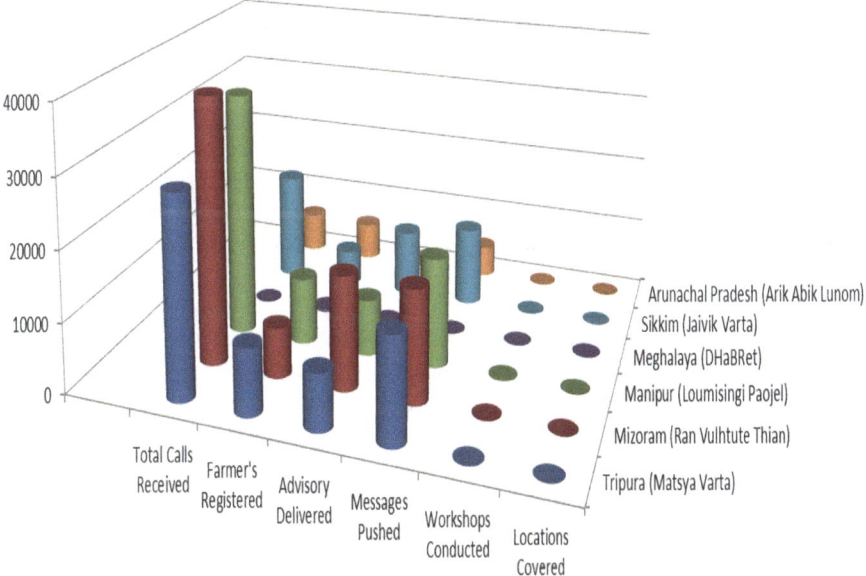

Fig. 13.4 ICT based m4agri project implemented by CAU, Imphal and Digital India. Corporation in North Eastern India

13.9 Suggestions for Policy Makers

Advent of ICTs have catalyzed the evolution of diverse extension and consultancy services involving institutional and private players working in tandem. This plurality encourages flexibility, stimulates desired services and reduces some of the economic burden on the government sector. However, it may jeopardize collaboration to ensure that the needs of all farmers are satisfied irrespective of their land holding. It is challenging to ensure that extension providers are scrutinized while also ensuring that the public sector extension service is neither underfunded or destroyed. Eliminating public extension may result in inefficient responses to farmer demands due to resource constraints, stifling environmental improvement. In this coordination, different duties of extension service providers, including government led extension services, private stakeholders, NGOs and FPOs must be considered. Decentralization can assist with accountability and meeting the needs of diverse target groups, but it also necessitates the allocation of resources for monitoring and, notably, assessment. Evaluation can be used to analyse what works and what doesn't, as well as which mix of tactics is the most cost-effective for the country. Extension activities frequently have checks and balances that ensure credibility of the services, but they usually focus on monitoring rather than measuring practicality of the adopted solution. Many times extension activities are notoriously characterized by insufficient learning and creative thinking leading to misconstrued judgments regarding monitoring, evaluation and impact.

Decentralizing programme planning and management with strong client involvement that links governmental extension to sources of fresh knowledge is crucial, while ensuring the system barely political enough to assure resources but not so political that those in power start taking political mileage of it. When farmers regard them as government eyes and ears rather than information sources, overly politicised institutions lose trust and, eventually, monetary support. Because of the financial load of T&V, India, which has the world's second largest extension system, has seen extension migrate from a centralized T&V system during 1970s to 1990s to a decentralized approach more recently led by the Agricultural Technology Management Agency (ATMA). As governmental extension systems continue to face economic challenges, private-public partnerships (PPPs) are one avenue for mitigating their impact, whether extension information is included in products or advice. One model for enhancing inclusivity that ensures "last mile" access to desired farmer groups who are small holders is to promote linkages between extension and farmer organizations.

It is essential for institutional and private extension stakeholders to fully adapt and justifiably use the application of the latest ICT. These novel approaches have the ability to expand and speed up the dissemination of information to farmers, supplement existing extension techniques, and be cost-effective to use. They can also aid in issue diagnosis and assist extension systems discover what is effective. Robust dissemination of ICT-based extension information may reduce gender bias by providing information to women-headed limited-resource households if female access to ICT is at par to that for men. Subject to the availability of access, women are frequently as interested in using social networks online or offline as men. Farmer organizations and virtual networks are becoming increasingly crucial in the spread of technology. These networks must be used by extension services (World Bank 2019). Products like Whats App and other cell-phone-based networking and messaging applications, for example, are gaining popularity even in underdeveloped countries and their usage in extension services is expanding quickly for aspects like agricultural pest diagnosis and advice. As farmer's access to computers gains momentum in developing nations, in-depth decision-making skills may be conveyed via webinars and other distant learning techniques.

Effective in-service training is a necessity for every extension programme, but it is especially vital now since ICT technologies are changing at the same rate as agricultural advances. Maintaining strong links to the most recent advances created via public and commercial research is critical to successful extension. Several agricultural extension initiatives have been created and deployed to assist farmers. Every model has strengths and disadvantages, and the effectiveness of these innovations varies by country and depending on sociocultural characteristics and institutional support mechanisms. No particular innovation is universally applicable. Extension and advising services will always be an important component of agricultural development programmes. Extension service providers such as public, private and civil society organisations need to develop vibrant and dynamic coordination so that the extensions strategies to mitigate climate change are efficiently adapted and implemented by farmers globally, in general and the Eastern Himalayan Region of India in particular.

References

Afsar N, Idrees M (2019) Farmers' perception of agricultural extension services in disseminating climate change knowledge. Sarhad J Agric 35(3):942–947

Anderson JR, Feder G (2007) Agricultural extension. In: Evenson R, Pingali P (eds) Handbook of agricultural economics, vol 3. North Holland Press, Amsterdam, pp 2343–2378

Annor-Frempong F, Kwarteng J, Agunga R, Zinnah MM (2006) Challenges of infusing information and communication technologies in extension for agricultural and rural development in Ghana. J Ext Syst 22(2):69

Antwi-Agyei P, Dougill AJ, Stringer LC, Codjoe SNA (2018) Adaptation opportunities and maladaptive outcomes in climate vulnerability hotspots of Northern Ghana. Clim Risk Manag 19:83–93

Ayisi Nyarko D, Kozári J (2021) Information and communication technologies (ICTs) usage among agricultural extension officers and its impact on extension delivery in Ghana. J Saudi Soc Agric Sci 20(3):164–172. https://doi.org/10.1016/j.jssas.2021.01.002; Accessed 29 Jun 2023

Bawakyillenuo S, Yaro JA, Teye J (2016) Exploring the autonomous adaptation strategies to climate change and climate variability in selected villages in the rural northern savannah zone of Ghana. Local Environ 21(3):361–382

Daum T, Ravichandran T, Kariuki J, Chagunda M, Birner R (2022) Connected cows and cyber chickens? Stocktaking and case studies of digital livestock tools in Kenya and India. Agric Syst 196:103353

Davis K, Franzel S (2018) Extension and advisory services in 10 developing countries: a cross-sectional analysis. USAID, feed the future DLEC project, September. https://www.digitalgreen.org/wp-content/uploads/2017/09/EASin-Developing-Countries-FINAL.pdf. Accessed 26 Jun 2023

DoAC & FW (2019) Agriculture census: 2015–16. New Delhi, Agriculture Census Division, Department of Agriculture, Co-operation & Farmers Welfare, Ministry of Agriculture and Farmers Welfare, Government of India

Eisenack K, Moser SC, Hoffmann E, Klein RJ, Oberlack C, Pechan A, Rotter M, Termeer CJ (2014) Explaining and overcoming barriers to climate change adaptation. Natl Clim Chang 4(10):867–872

IFPRI FORUM (2005) Building local skills and knowledge for food security. International Food Policy Research Institute. www.ifpri.org. Accessed 25 Jun 2023

IFSR: India State of Forest Report (2019) Forest Survey of India, Ministry of Environment, Forest and Climate Change, Government of India

Jain SK, Kumar V, Saharia M (2013) Analysis of rainfall and temperature trends in Northeast India. Int J Climatol 33(4):968–978

Jhajharia D, Shrivastava SK, Sarkar DSAS, Sarkar S (2009) Temporal characteristics of pan evaporation trends under the humid conditions of Northeast India. Agric For Meteorol 149(5): 763–770

Juhola S, Glaas E, Linnér BO, Neset TS (2016) Redefining maladaptation. Environ Sci Policy 55:135–140

Krishnan R, Sanjay J, Gnanaseelan C, Mujumdar M, Kulkarni A, Chakraborty S (2020) Assessment of climate change over the indian region: a report of the Ministry of Earth Sciences (MoES), Government of India. Springer, Singapore

Lamm KW, Lamm AJ, Davis K, Sanders C, Powell A, Park J (2023) Extension networks and dissemination of horticultural advancements: development and validation of a professionalization instrument. Horticulturae 9(2):245

Landini F, Vargas G, Bianqui V, MI YR, Martínez M (2017) Contributions to group work and to the management of collective processes in extension and rural development. J Rural Stud 56:143–155

Larochelle C, Alwang J, Travis E, Barrera VH, Dominguez JM (2019) Did you really get the message? Using text reminders to stimulate adoption of agricultural technologies. J Dev Stud 55(4):548–564

Malczewski J (1999) GIS and multicriteria decision analysis. John Wiley & Sons

MoEFCC (2018) India: second biennial update report to the United Nations framework convention on climate change. Ministry of Environment, Forest and Climate Change, Government of India

Naresh Kumar S, Islam A, Swarooparani DN, Panjwani S, Sharma K, Lodhi NK, Chander S, Sinha P, Khanna M, Singh DK, Bandyopadhyay SK (2019) Seasonal climate change scenarios for india: impacts and adaptation strategies for wheat and rice. p 56. ICAR-IARI Pub TB-ICN: 233/2019

Nath AJ, Tiwari BK, Sileshi GW, Sahoo UK, Brahma B, Deb S, Gupta A (2019) Allometric models for estimation of forest biomass in North East India. Forests 10(2):103

Nath AJ, Kumar R, Devi NB, Rocky P, Giri K, Sahoo UK, Pandey R (2021) Agroforestry land suitability analysis in the Eastern Indian Himalayan region. Environ Chall 4:100199

NITI Aayog (2016) Report of the Expert Committee on land leasing. NITI Aayog, Government of India

Nnadi FN, Chikaire J, Ezudike KE (2013) Assessment of indigenous knowledge practices for sustainable agriculture and food security in Idemili South Local Government Area of Anambra State, Nigeria. J Res Develop Manag 1:1

Purnomo S, Lee YH (2010) An assessment of readiness and barriers towards ICT program implementation: perceptions of agricultural extension officers in Indonesia. Int J Educ Develop ICT 6(3):19–36

Rickards L, Alexandra J, Jolley C, Frewer T (2018) Final report: review of agricultural extension. Australian centre for International Agricultural Research (ACIAR). Accessed 25 Jun 2023

Sharma E, Chettri N, Tse-ring K, Shrestha AB, Jing F, Mool P, Eriksson M (2009) Climate change impacts and vulnerability in the Eastern Himalayas

Singh Naveen P, Bhawna A, Khan MA (2018) Micro-level perception to climate change and adaptation issues: a prelude to mainstreaming climate adaptation into developmental landscape in India. Nat Hazards 92:1287–1304. https://doi.org/10.1007/s11069-018-3250-y; Accessed 29 Jun 2023

Singh NP, Anand B, Singh S, Srivastava SK, Rao CS, Rao KV, Bal SK (2021) Synergies and trade-offs for climate-resilient agriculture in India: an agro-climatic zone assessment. Clim Chang 164:1–26

Swanson BE, Davis K (2014) Status of agricultural extension and rural advisory services worldwide: summary report. Global Forum for Rural Advisory Services, Lindau

Williams PA, Crespo O, Abu M (2019) Adapting to changing climate through improving adaptive capacity at the local level–the case of smallholder horticultural producers in Ghana. Clim Risk Manag 23:124–135

World Bank (2017) ICT in agriculture: connecting smallholders to knowledge, networks, and institutions. Updated Edition, Washington, DC. http://documents.worldbank.org/curated/en/522141499680975973/pdf/117319-PUB-Date-6-27-2017-PUBLIC.pdf. Accessed 29 Jun 2023

World Bank (2019) World Development Indicators. The World Bank

Yiran GA, Stringer LC (2017) Adaptation to climatic hazards in the savannah ecosystem: improving adaptation policy and action. Environ Manag 60(4):665–678

Transforming the Indian Livestock Sector

14

Abhijit Mitra, Amit Kumar Tripathy, Raj Kumar Singh,
and Kamal Malla Bujarbaruah

Abstract

India is committed to addressing global challenges in line with the 2030 Agenda
for Sustainable Development and its Sustainable Development Goals (SDGs)
over the next 7 years. The SDGs aim to eliminate poverty and hunger, protect the
environment, and ensure prosperity for all. Achieving these goals requires creating awareness, understanding each economic sector's potential contribution, and
building consensus among stakeholders. While livestock, including Animal
Source Foods (ASFs), has been vital for human progress and food security, it
faces challenges such as poverty, malnutrition, soil degradation, and climate
change. The livestock sector can play a significant role in addressing these
challenges. Traditionally, discussions focused on sustainable production, but
the UN's 2030 Agenda has broadened the conversation to emphasize the sector's
contribution to achieving the SDGs. By offering valuable insights, this article
serves as a concise overview of the progress made and challenges faced in
transforming the Indian livestock sector for sustainable development.
Policymakers, researchers, and stakeholders can utilize these insights to drive
the adoption of sustainable practices and ensure a thriving Indian livestock sector.

Keywords

Transforming · Indian livestock sector · SDGs

A. Mitra (✉)
Department of Animal Husbandry and Dairying (DAHD), Ministry of Fisheries, Animal Husbandry
and Dairying, Government of India, New Delhi, India

A. K. Tripathy · R. K. Singh
One Health Support Unit, DAHD, GoI, New Delhi, India

K. M. Bujarbaruah
ICAR, Assam Agricultural University, Jorhat, Assam, India

14.1   Introduction

India contributes three-quarters of the total output in the South Asia region, and it has demonstrated robust growth, reaching approximately 7% in 2022/23 (Indian Economy Continues to Show Resilience Amid Global Uncertainties n.d.), despite recent challenges. Similar growth is anticipated in 2023/24. Notably, India's agriculture sector has consistently shown strong annual growth of over 3% (World Development Indicators 2023). With this progress, India is on track to become the fastest-growing economy among the world's largest emerging markets and developing economies in the near future. However, considering this potential, it is essential to assess the impacts of food systems to identify both potential future risks and promising policy options. These options can enhance early warning systems, immediate response capabilities, and overall resilience building in India. The track to achieve sustainable Development Goals (SDGs) by 2030 and progress in tackling the problem having achieved considerable gains has to be relooked in the face of post-COVID-19. Pandemic-induced economic disruptions, macroeconomic issues, and climate change provide an opportunity to emphasize SDGs.

By 2050, India's population is projected to reach 1.65 billion (World Popul Rev 2023), indicating a 16% increase from the current figure of 1.43 billion. This growth will be accompanied by a continued and accelerated process of urbanization, with approximately 55% of the population residing in urban areas. Furthermore, income levels are expected to rise significantly, reaching around 401,839 INR per capita in 2050, a substantial increase from the 53,331 INR per capita recorded in 2010–2011 (Indian Council of Agricultural Research, New Delhi 2015).

14.2   Context

The longstanding livestock debate, centered on increasing efficiency to feed a growing global population, has evolved with the United Nations' 2030 Sustainable Development Goals agenda. It now emphasizes not just sustainable livestock production but also how the sector can better support the attainment of the SDGs (Box).

Assessing SDGs vis-à-vis the Indian livestock sector; SDG 1, 2, 3, 4, 5, 6, 7, 8, 10, 12, 13, and 17 all fall in place and offer the necessary roadmap.

- SDG-1 – Zero Poverty offers the catalytic role to achieve livelihood objectives by enhancing the five capitals - human, social, natural, physical, and financial including the resilience and consumption smoothing strategy against external shocks.
- SDG-2 – Hunger Eradication, contributes at different levels and from different entry points by increasing the direct consumption of healthy and nutritious animal-source foods (ASFs).
- SDG-3 – Healthy Lives, warns us about the action desired today for antimicrobial resistance (AMR) to ensure health and well-being for people of all ages.
- SDG-4 – Quality Education (inclusive and equitable quality education) – promoting lifelong learning opportunities for all can help relate to ASFs to improve children's cognitive and physical development as well as school attendance and performance.
- SDG-5 – Gender Equality, enables women to meaningfully operate in, and benefit from, the livestock sector, policies and programmes and thereby remove all obstacles and constraints in their way.
- SDG-6 – Given the large and growing water footprint associated with livestock production, improving water-use efficiency throughout the production system is important and thereby ensures access to safe water sources and sanitation for all.
- SDG-7 – seeks access to affordable, reliable, sustainable, and modern energy sets in the priority to be offered to turning animal manure into biogas to eliminate a leading source of methane, a powerful driver of global warming.
- SDG-8 – Given the remarkable growth rate predictions, the Indian livestock sector has tremendous potential to create jobs and reduce inequality, by promoting inclusive and sustainable economic growth, employment, and decent work for all.
- •SDG-10 – is closely correlated to the ¬rest of the SDGs (elimination of poverty) and while there has been progress on poverty reduction over the past decades, the world continues to suffer from substantial inequalities. To reach both SDG-1 and SDG-10, efforts to foster growth need to be complemented by equity-enhancing policies and interventions.
- SDG-12 – the SGD-12 targets highlight the importance of information, especially to consumers - critical for the livestock sector as demand for ASF is growing fast. Hence an effort in food supply chains, with the participation of all stakeholders, to reduce the amount of meat, milk and eggs wasted by consumers and the food industry or lost in the production process which ultimately will translate to major sustainability gains.
- SDG-13 – aims to strengthen resilience and adaptive capacity to climate-related hazards and natural disasters and an adaptation to the adverse impacts of climate change. Climate change impacts livestock directly (for example through heat stress and increased morbidity and mortality) and indirectly (e.g., through quality and availability of feed and forages, and animal diseases). Smallholder livestock keepers and pastoralists are among the most vulnerable to climate change. At the same time, the livestock sector contributes significantly to climate change.
- SDG-17 – calls for multi-stakeholder partnerships between various actors to help provide financial, knowledge and institutional support to spur progress across different sectors. By working together in partnership, all stakeholders can help achieve transformative change. Leading partnerships with recognized work on sustainable livestock development mechanism(s)/include One Health, Global Agenda for sustainable livestock (GASL), Livestock Environmental Assessment and Performance (LEAP) Partnership, Global Pastoralists Knowledge Hub, Multi-stakeholder Feed Safety Partnership, Tripartite partnership of FAO, World Health Organization (WHO), and World Organisation for Animal Health (WOAH – erstwhile referred to as OIE), Global Alliance for Livestock Veterinary Medicine (GALVmed), Livestock Global Alliance, Dairy Asia – to name a few.

The future of the livestock sector presents significant opportunities alongside its formidable challenges. This sector has the potential to enhance millions of lives by ensuring a steady supply of meat, milk, eggs, and dairy products, promoting the direct consumption of animal-source foods, boosting income generation and employment, and fortifying the resources that rural households rely on for their livelihoods. Additionally, it can contribute to enhancing children's cognitive and physical development, school attendance, and performance, empower rural women, improve natural resource efficiency, expand access to clean and renewable energy, and support sustainable economic growth. Furthermore, it can generate fiscal revenue and foreign exchange, facilitate value addition and industrialization, stimulate smallholder entrepreneurship, address inequality gaps, encourage sustainable consumption and production practices, enhance household resilience to climate-related shocks, and unite diverse stakeholders in achieving these objectives.

Demographic and economic shifts in India will result in a growing need for calorie consumption. However, the proportion of calories derived from vegetable sources is projected to decrease to 84%, while the contribution from animal sources will double to 16% (Indian Council of Agricultural Research, New Delhi 2015). The demand for meat, fish, and eggs is projected to surge from 11.64 million tonnes to 35.52 million tonnes (by 205%) (NITI Ayoga 2018). Moreover, the demand for milk and its products is expected to grow 3.7 times faster than that of food grains (NITI Ayoga 2018).

India holds a prominent position among the top twenty agricultural exporters globally, with a 2.1% share in global agricultural exports (PIB 2021). The nation ranks as the fourth largest exporter of buffalo meat and holds the top position as exporter of sheep and goat meat (APEDA 2021a) and additionally, it makes significant exports of poultry and dairy products (APEDA 2021b; APEDA 2021c). These numbers underscore the significant contribution of India's livestock sector to global food security, in addition to meeting its domestic demand (Fig. 14.1).

14.3   Scope and Potential

In 2022, India experienced a wide range of natural disasters almost daily, according to a report by the Centre for Science & Environment (CSE). These disasters resulted in significant damages, including the loss of 2755 lives, damage to 1.8 million hectares of crop area, destruction of over 400,000 houses, and the death of nearly 70,000 livestock (Centre for Science and Environment 2022). The irregular monsoon patterns added to the challenges by causing food price fluctuations, inflation concerns, and complications in managing monetary policies.

The World Livestock (WoLi) report, titled "Transforming the Livestock Sector through Sustainable Development Goals," (FAO 2018) focuses on how the livestock sector interacts with each of the Sustainable Development Goals (SDGs). It examines potential synergies, trade-offs, and complex interconnections within the sector. The report emphasizes the need for an integrated approach to sustainable

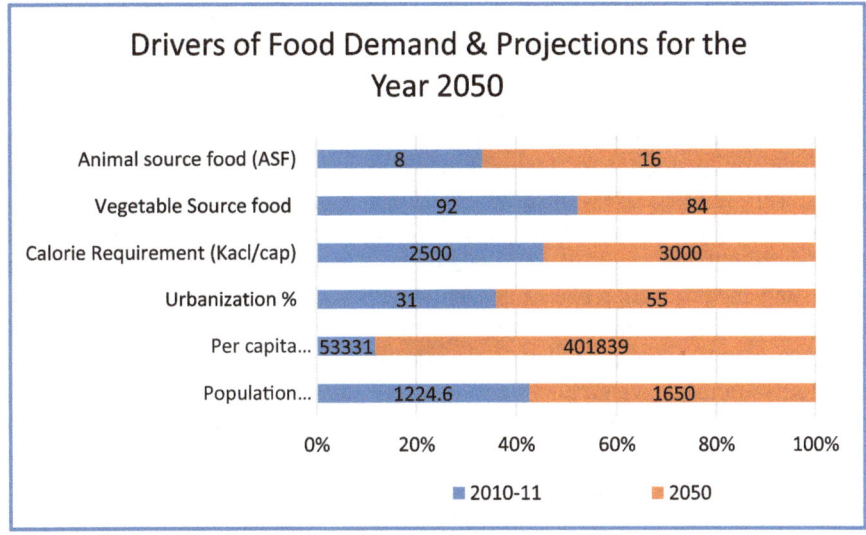

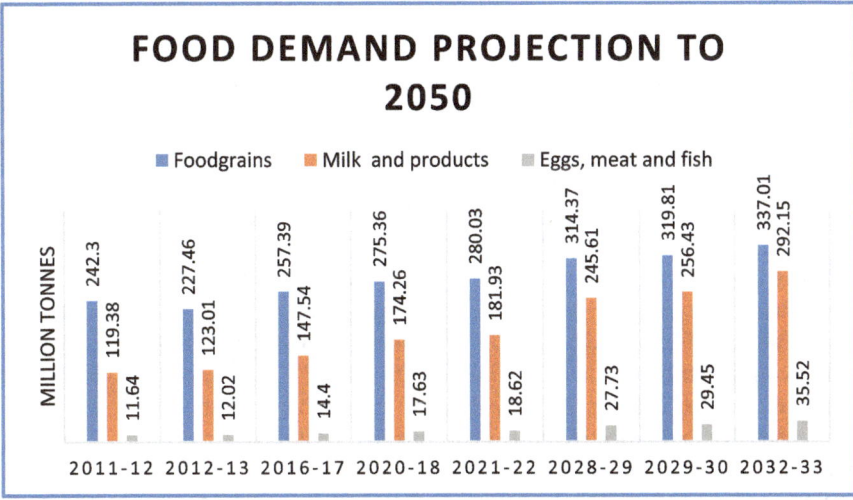

Fig. 14.1 Food demand projections to 2050

livestock development and highlights the challenge of translating the SDGs into specific and targeted national policies (FAO 2018).

The drivers in the Indian context for transformation through the SDGs include the following:

Enhancing economic growth through the multiplier effects of the livestock sector: Livestock systems not only significantly contribute to the national economy but also represent one of the fastest-growing segments within Indian agriculture. They constitute 31% of the Gross Value of Output in the Agriculture and Allied sector (GVOA) (Department of Animal Husbandry and Dairying Ministry of Fisheries, Animal Husbandry and Dairying Government of India 2023). Among these, milk production holds the largest share, accounting for 20% of GVOA (Department of Animal Husbandry and Dairying Ministry of Fisheries, Animal Husbandry and Dairying Government of India 2023). The potential for boosting economic growth through the livestock sector extends beyond mere production and can be achieved through vertical and horizontal multiplier effects. However, it is important to note that the Indian livestock sector is highly fragmented (Nanda Kumar et al. 2022a), with varying levels of labor productivity observed between processing and production phases and among different types of production systems, including commercial and subsistence ones.

Turning rapid livestock sector growth into accelerated poverty reduction: Small-scale farmers face limitations in leveraging their available resources to generate income. To effectively translate the swift growth of the livestock sector into poverty alleviation, the following strategies can be pursued: (1) expanding the livestock sector within the economy, (2) increasing livestock sector growth, (3) involving poor communities actively, (4) improving producers' access to production factors, (5) facilitating workers' access to employment opportunities, and (6) providing consumers with competitive prices, safer food, and high-quality diets.

Unleashing the potential of the livestock sector to combat hunger and malnutrition: To attain these goals, it is essential to prioritize specific actions, such as boosting the productivity and income of small-scale food producers, fostering the development of sustainable and resilient food systems, preserving the diversity of genetic resources, guaranteeing the effective operation of food markets, and reducing the reliance on antimicrobials through improved access to quality veterinary services and sound animal husbandry practices.

Ensuring healthy lives by proactively preventing animal diseases: To safeguard the health and well-being of both animals and humans, it is imperative to prevent animal diseases. Achieving this requires a collaborative effort among experts in animal production and health, public health officials, and the commercial sector, including the feed industry. A "One Health" approach is essential for developing a comprehensive and proactive strategy to mitigate health risks associated with livestock that can affect human health.

Promoting a balanced consumption of animal-source foods for the enhancement of children's cognitive development, school attendance, and academic performance: These foods offer high-quality, energy-dense protein and readily absorbable micronutrients that are more easily obtained compared to plant-based sources. Insufficient intake of these essential micronutrients during pregnancy and childhood can lead to health issues affecting both growth and educational progress. Embracing dietary diversity can contribute to improved cognitive function, micronutrient levels, physical development, academic achievements, and better vaccine responses, while also bolstering immunity against opportunistic infections. The inclusion of animal-source foods in mid-day meals can serve as a valuable tool to encourage school enrollment and attendance.

Promoting active involvement and decision-making authority for women in the livestock sector: The fact that men predominantly hold leadership positions in agricultural cooperatives and producer associations, which play a crucial role in shaping government development plans and policies, has repercussions. This gender imbalance affects the sensitivity of these plans and policies towards gender-related issues and their overall impact on women and girls. Therefore, fostering women's participation and decision-making influence within the livestock sector can contribute to the reduction of gender disparities in rural areas and ensure that women gain equal access to productive resources and services.

Enhancing water-use efficiency in livestock production is critical to address water scarcity: The runoff and leakage of nutrients or residues from concentrated sources of livestock waste pose a significant threat to both freshwater supplies and coastal ecosystems. If not managed effectively, these nutrient runoffs and high concentrations of nitrogen and phosphorus can harm surrounding ecosystems, estuaries, and coastal fisheries. Furthermore, discharges from manure and slurry pits and outflows from abattoirs and food processing facilities contribute to water contamination unless properly treated. Given the substantial water footprint of livestock production, improving water-use efficiency and providing comprehensive guidance throughout the production process are essential steps to ensure access to safe water sources and sanitation.

Transforming animal manure into clean and renewable energy: The ongoing "Energy Revolution," seeks to replace environmentally damaging coal and oil with sustainable alternatives. This transition is anticipated to be one of the most significant achievements of the twenty-first century. By converting livestock manure into biogas, there is the potential to provide a substantial domestic source of renewable fuel to over a billion people. This could grant them access to cost-effective, dependable, and environmentally sustainable energy solutions.

Improving the delivery of ecosystem services through sustainable management of grasslands and increased feed-use efficiency: While the livestock sector has been associated with negative impacts such as biodiversity loss, land degradation, and deforestation; it also plays a vital role in providing essential services that contribute to the protection, restoration, and sustainable utilization of terrestrial ecosystems. This includes efforts to combat desertification, reverse land degradation, and curb biodiversity erosion. The ultimate environmental impact of the livestock sector, whether positive or negative, hinges on the specific production and management practices employed. Livestock production can be a significant asset in supporting initiatives like sustainable rangeland management, wildlife preservation, and the enhancement of soil fertility and nutrient cycling.

While the Sustainable Development Goals (SDGs) and their targets are aspirational, it is imperative to integrate the role of livestock in these goals into national planning processes, policies, and strategies. This integration should take into account the unique circumstances of each nation. To enhance the alignment of livestock policy and practices with sustainable development strategies, the Policy Framework serves as a valuable tool to strengthen the impact of livestock policy analysis in achieving the 2030 Agenda. It's objectives include building the capacity of

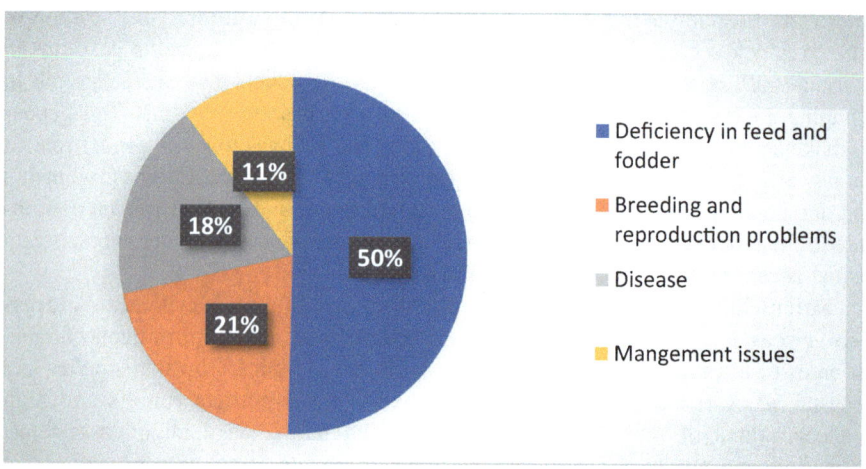

Sustainable and Efficient Production

Fig. 14.2 Factors affecting livestock productivity and their contribution to total loss

stakeholders to assess how the livestock sector contributes to the SDGs, identifying areas of synergy, as well as trade-offs, pinpointing opportunities for change, generating analytical evidence to evaluate existing programs, and promoting the use of methods and tools to monitor the sector's contribution to the SDGs.

Considering the difficulties, changing economy and population trends, and the rising need for ASFs; it is important to transform the Indian Livestock sector to align with the SDGs. To achieve sustainable and efficient production in the Indian Livestock sector; it is crucial to address constraints such as feeding, breeding, health, and management issues. Currently, Indian animals have lower average yields of milk and meat compared to the global average by 20–60%. Among the factors affecting productivity, the deficiency of feed and fodder contributes to 50.2% of the total loss, followed by breeding and reproduction problems (21.1%), diseases (17.9%), and management issues (10.5%) (Indian Grassland and Fodder Research Institute (Indian Council of Agricultural Research) Gwalior Road, Jhansi -284 003 2013) (Fig. 14.2). These challenges must be systematically addressed to improve productivity and ensure sustainable production (Indian Grassland and Fodder Research Institute (Indian Council of Agricultural Research) Gwalior Road, Jhansi -284 003 2013).

14.4 Quality Feed and Fodder Availability

India, despite occupying only 2.29% of the world's land area, supports 17% of the global human population and 10.70% of the livestock population. This places immense strain on land, water, and resources. Consequently, there is a significant shortage of feed and fodder, with deficits in green fodder (35.6%), dry crop residues (10.95%), and concentrate feed ingredients (44%) (India's agricultural and processed food products exports have grown at a steady pace in the last decade notwithstanding challenges 2021). This has left approximately 500 million animals in the country without adequate feed and fodder security (NITI Ayog 2018). The availability of quality forage crop seeds is limited, meeting only 15–20% of the national requirement through an unorganized seed market (India's agricultural and processed food products exports have grown at a steady pace in the last decade notwithstanding challenges 2021). To address the feed and fodder shortage, the National Livestock Mission (NLM) has focused on producing 36,400 MTs of quality fodder seeds in 2022–2023 (Department of Animal Husbandry and Dairying Ministry of Fisheries, Animal Husbandry and Dairying Government of India 2023). Furthermore, India has significant potential to utilize around 40 million hectares of wastelands, degraded lands, and forest margins for forage resources out of the total available 122 million hectares (India's agricultural and processed food products exports have grown at a steady pace in the last decade notwithstanding challenges 2021). In addition, the establishment of animal feed manufacturing plants under the Animal Husbandry Infrastructure Development Fund (AHIDF) has been undertaken. However, proper scientific management, regulation, and institutionalization at the local level are crucial for sustainable utilization. A comprehensive feed and fodder policy should be implemented, ensuring year-round availability of fodder at reasonable prices in all states, particularly in fodder-deficient regions. Technological interventions such as converting horticultural waste into fodder, hydroponics/aeroponics for cost-effective fodder production, and the promotion of hay and silage making can significantly contribute to the solution.

14.5 Breeding and Reproduction

India, a global leader in livestock population and milk production, faces challenges with low ruminant productivity, particularly in cattle. The Indian cattle had an average annual productivity of 1777 kg per animal in 2019–2020, compared to the global average of 2699 kg (PIB 2021). The lack of good quality animals is one contributing factor. However, there has been a remarkable improvement in productivity for all categories of animals, including indigenous, non-descript cattle, buffaloes, and crossbred cattle, with a 16.74% increase between 2014–2015 and 2021–2022 (Department of Animal Husbandry and Dairying Ministry of Fisheries, Animal Husbandry and Dairying Government of India 2023). The *Rashtriya Gokul Mission* (RGM) has played a vital role in enhancing milk production and bovine productivity in India. In 2021–2022 alone, an impressive 98 million of Artificial Insemination (AI) procedures were performed across all the states. The government has established five Sex-sorted Semen Stations capable of producing upto one million doses annually. Under this mission, government semen stations have successfully generated three million doses of sex-sorted semen, with an additional 3.3 million doses contributed by Milk Federations, NGOs, and private semen stations (Department of Animal Husbandry and Dairying Ministry of Fisheries, Animal Husbandry and Dairying Government of India 2023). To meet future demands and improve livestock productivity, India needs to expand the coverage of AI services and widely adopt the use of sex-sorted semen and in vitro fertilization (IVF).

In contrast to traditional progeny testing program, Genomic Selection (GS) offers the advantage of identifying animals with high genetic potential at a younger age, significantly accelerating genetic progress compared to traditional methods that require several years for assessing genetic worth. Considering the lack of suitable chip having sufficient information of indigenous breeds of cattle, under RGM, the DAHD established the National Bovine Genomic Centre. Projects from Indian Council of Agricultural Research—National Bureau of Animal Genetic Resources (ICAR-NBAGR) and National Dairy Development Board (NDDB) were approved. Subsequently, DAHD coordinated to converge the efforts made by different agencies in development of genomic chip such as ICAR- NBAGR, NDDB, BAIF Development Research Foundation and National Institute of Animal Biotechnology (NIAB) for development of reliable genomic chip for production traits and initiating genomic selection in the country. Two medium density chips related to milk productivity are being developed by pooling of data available with aforementioned agencies one for priority indigenous breeds of cattle and one for buffaloes for initiating full scale genomic selection in the country. The commercial common chip will be available and chip will be validated through creation of referral population (genotyped and phenotyped).

Livestock breeding is a state subject. Nevertheless, to ensure the effective implementation and coordination, the existing overarching National Livestock Breeding Policy is being reviewed so that it can serve as a guideline to complement the state animal breeding policies. These measures will contribute to increasing the number of productive milch animals in India's national herd, ensuring sustainable livestock productivity, and meeting future demands.

14.6   Animal Health and Welfare

Globally, diseases affecting food-producing animals have had a significant impact, leading to a substantial 20% reduction in production levels. In India specifically, the past 3 years (2020–2022) have witnessed numerous outbreaks of preventable diseases, including 749 outbreaks of Haemorrhagic Septicaemia, 384 outbreaks of Foot-and-Mouth Disease (FMD), 26 outbreaks of Brucellosis, 55 outbreaks of *peste des petits ruminants* (PPR), and 26 outbreaks of Classical Swine Fever (CSF) (Department of Animal Husbandry and Dairying Ministry of Fisheries, Animal Husbandry and Dairying Government of India 2021, 2022, 2023).

These outbreaks have resulted in significant economic losses, with annual estimates reaching Rs 52,550 million for Haemorrhagic Septicaemia (Singh et al. 2014b), Rs 200,000 million for FMD (Venkataramanan et al. 2006), Rs 204,000 million for Brucellosis (Singh et al. 2015), Rs 88,950 million for PPR (Singh et al. 2014a), and Rs 4290 million for CSF (Singh et al. 2016). As a result, farmers face a direct loss of nearly Rs 550,000 million each year due to these fully preventable diseases, which can be controlled through vaccination (Phand and Gummagolmath

2021). To combat this issue, the Government of India has implemented the Livestock Health & Disease Control Programme (LHDCP). While good progress has been made, there is still room for improvement. Currently, 84% of the target population has been vaccinated against FMD, and 48% has been vaccinated against Brucellosis (Department of Animal Husbandry and Dairying Ministry of Fisheries, Animal Husbandry and Dairying Government of India 2023). Vaccination efforts are also underway for PPR and CSF. In addition, the establishment of 4332 Mobile Veterinary Units (MVUs) (Department of Animal Husbandry and Dairying Ministry of Fisheries, Animal Husbandry and Dairying Government of India 2023) is a crucial step in enhancing vaccine coverage and providing essential services such as basic treatment, diagnostics, and extending veterinary services directly to farmers at their doorstep.

In recent decades, five out of six public health emergencies of international concern (PHEIC), as declared by the World Health Organization (WHO), originated from animals. The COVID-19 pandemic underscored the importance of pandemic preparedness and response (PPR) with a One Health approach, emphasizing the significance of animal health security. Therefore, the primary goal of this investment is to minimize the likelihood of a pathogen emerging from animals, whether domesticated or wildlife, transmitting into the human population and posing risks to their health, nutritional well-being, and the livelihoods of vulnerable communities. The Department is inititaing a program for Animal Health System Support for Improved One Health (AHSSOH) with World Bank support wherein animal health infrastructure and capacities would be strengthened for effective implementation of One Health concept at the national level. The project would target integrated disease surveillance for zoonotic diseases linking to human and wildlife systems, thereby creating a truly integrated system. With an objective to effectively address the issues and to have an enabaling policy, DAHD has launched the Animal Pandemic Preparedness Initiative (APPI), first of its kind in the world. These initiatives focus on enhancing veterinary services and infrastructure, improving disease surveillance mechanisms, enabling early detection and response, and building the capacity of professionals. Simultaneously, they aim to increase the awareness among farmers through community outreach efforts. By prioritizing these measures, India is preparing itself to respond to future animal pandemics, including zoonotic diseases, comprehensively and efficiently. India is taking significant steps towards mitigating the impact of diseases affecting food-producing animals, preventing economic losses, and safeguarding public health. Continued investment and sustained efforts in animal health and disease control will further improve the country's preparedness and response capabilities while ensuring the well-being of livestock and supporting the livelihoods of farmers.

14.7 Value Chain Development

Agri-food value chains have generally demonstrated resilience when faced with various shocks, although this resilience varies depending on the specific context and the type of crisis, whether it is related to climate, pandemics, or other factors. Policy interventions have sometimes played a direct role in bolstering this resilience, such as exempting food service industries from lockdowns during the COVID-19 pandemic. In India, the livestock and livestock product marketing sector is predominantly unorganized and fragmented (Nanda Kumar et al. 2022b; Sen and Muthukumar 2022; Singh et al. 2012). Differences in the structure of value chains can influence their resilience. Smaller informal firms may be more susceptible to vulnerabilities and constraints compared to larger formal enterprises. Therefore, interventions should be tailored not only to the nature of the crisis but also to the specific context, type of value chain, and ideally, the size of the affected enterprise.

The adoption of improved and novel technologies is crucial in enhancing value chain resilience. It requires proactive investments to ensure widespread access to practical knowledge and shock-resistant technologies, including climate-smart practices and relevant information and communication tools (ICT). Regulatory and business environments need to support the development and dissemination of crisis-responsive innovations.

Gender considerations are vital since women often bear a disproportionate burden during crises. Ensuring that women maintain access to productive opportunities at various stages of value chains can mitigate the impact on food security and livelihoods. This may involve enabling women to leverage digital agriculture and financial innovations, as well as providing training in food safety and other relevant practices. Additionally, women tend to deplete their savings more rapidly than men during crises, highlighting the need for further research on women's coping mechanisms and ways to enhance them. Livestock products, including milk, meat, eggs, and poultry, are primarily produced in decentralized systems, often located far from consumer markets. However, their perishable nature necessitates efficient marketing and processing to maximize their value. A report from 2015 highlighted significant post-harvest losses in livestock products in India. The losses were found to be 0.92% for milk, 2.71% for meat, 7.19% for eggs, and 6.74% for poultry meat (Jha et al. 2015). These losses can be attributed to factors such as limited access to national markets, resulting in localized surpluses and wastage.

The implementation of the Animal Husbandry Infrastructure Development Fund is playing a significant role in developing the overall livestock ecosystem and addressing these challenges. This fund supports eligible projects aimed at establishing dairy and meat processing and value addition infrastructure, breed improvement technology and breed multiplication farms, animal waste management for wealth creation, and veterinary vaccine and drug production facilities.

In the dairy sector, Milk Producer Companies (MPCs) and the three-tier cooperative system play crucial roles and exhibit a strong platform for innovation, entrepreneurship, and market-oriented strategies. The three-tier cooperative system has also contributed significantly to the growth and development of the dairy sector, covering about 17.263 million farmers through 0.196 million village-level dairy cooperative societies. These cooperative milk unions procure an average of 46.196 million Kg of milk per day, playing a vital role in the milk collection and supply chain (Department of Animal Husbandry and Dairying Ministry of Fisheries, Animal Husbandry and Dairying Government of India 2023).

However, addressing food loss and waste requires improvements in infrastructure, transportation, and cold chain facilities. Promoting organized marketing channels, strengthening cooperatives, and supporting small-scale producers can reduce post-harvest losses and enhance efficiency. Policy support is essential for developing processing and marketing infrastructure to ensure safe ASF production. Modern facilities are also needed to handle lean season demand and manage surpluses in the livestock sector. By addressing these aspects, the livestock and livestock product marketing sector in India can become more sustainable, contributing to the overall development of the agricultural and food systems in the country.

14.8   Pastoral/Extensive Livestock Production System: Transforming Food Systems

Pastoralism represents more than just a means of livelihood; it embodies a cultural system intricately intertwined with its natural surroundings. Pastoralists possess a wealth of local and indigenous knowledge. The utilization of pastoral production systems can have a positive impact on biodiversity and contribute to species conservation in various ways. Mobile pastoral practices, such as transhumance routes involving seasonal migrations between lowlands and mountains, can create bio-corridors and aid in seed dispersal, thereby enhancing biodiversity across landscapes. Mobile livestock herding also fosters fertile areas within grasslands and plays a crucial role in reducing fire risks in vulnerable ecosystems (PASTER n.d.).

A sustainable pastoral production system not only sustains livelihoods and food production but also provides a range of ecosystem services. These services include the conservation and restoration of biodiversity and the supply of animal-source food and byproducts to markets. Pastoralism holds significant potential in addressing the Sustainable Development Goals (SDGs), especially in the context of increasing global climate variability and uncertainty. Engaging with pastoralism aligns with virtually all the SDGs (FAO 2021). Sustainable pastoralism seeks to harness natural synergies and harmonize food production (e.g., meat and dairy) with ecological processes, thereby reducing input costs and detrimental externalities, including harm to people, climate, and the environment. Food production from pastoralism, such as meat and dairy, serves as an essential source of protein and animal byproducts for subsistence or sale. Moreover, it tends to have a relatively low carbon and environmental footprint since it does not rely on fossil fuels, agrochemicals, or imported feed. Historically, pastoralists have maintained mutually beneficial relationships with sedentary farming communities. Pastoralists rely on farmers for grain and as customers for their products, while their livestock contribute to fertilizing farmers' fields and weed control.

Globally, food systems are responsible for a significant share of deforestation, freshwater consumption, and terrestrial biodiversity loss. Concurrently, the health of soil and below-ground biodiversity, which provides the majority of our food calories, has been overlooked in the industrial agricultural revolution of the past century. Ensuring the sustainability and resilience of our food systems would make a substantial contribution to advancing global agendas related to land, biodiversity, and climate.

14.9 Environmental Sustainability

Livestock farming in India is a major contributor to global methane emissions, despite having a smaller share of the global livestock population. Livestock accounted for 63% of the GHG emission from the agriculture sector and 10% of the total GHG emissions from various sectors in the country (Chakraborty et al. 2022). Buffalo and indigenous cattle are the primary sources of these emissions (Sharma 2020). India has committed to reducing the intensity of its GHG emissions 45% by 2030 compared to 2005 levels, as part of its "Intended Nationally Determined Contributions (INDC)" submitted to the UNFCCC (PIB 2022) (Fig. 14.3).

To address the environmental impact of livestock farming, India has implemented the Ration Balancing Program (RBP). This program has shown promising results in reducing costs and GHG emissions, and managing fodder requirements. It has resulted in a 13.7% reduction in enteric methane emissions per kilogram of milk produced in lactating cows and buffaloes (The World Bank 2020). The RBP has been implemented in 2.86 million milch animals across 18 states, covering 33,374 villages (World Bank 2010). Scaling up the program to cover all milch animals could

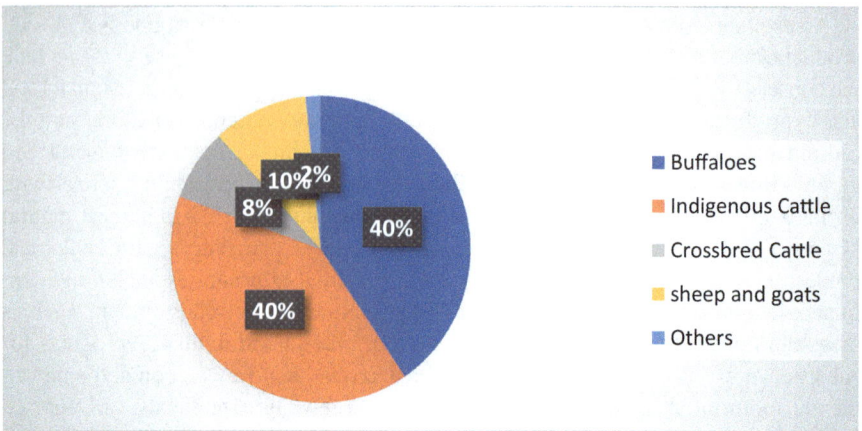

Fig. 14.3 Species wise share of GHG emission in India

lead to substantial reductions in costs, GHG emissions, and fodder needs. Despite the success of increasing milk production, around 37% of the growth can be attributed to an increase in livestock numbers rather than improved productivity (Chand 2017). This overreliance on expanding the livestock population for milk output growth is not sustainable in the long run.

In addition to methane emissions, water usage in dairy farming is a significant concern for sustainability. Dairy farming requires substantial amounts of water for various purposes, including crop cultivation, feed production, and animal needs. Traditional feeding practices resulted in higher water footprints, while a balanced ration consisting of green fodder, dry fodder, and concentrate feed ingredients led to a 14% reduction in water footprint, from 1236 to 1062 litres per kilogram of milk (NDDB 2020). To ensure sustainable water use in the dairy sector, it is crucial to extensively map the water footprint and develop strategies for efficient water management throughout the dairy value chain.

The livestock sector, similar to other agricultural sectors, encounters the task of enhancing efficiency, which involves minimizing disease and death rates through the enhancement of animal nutrition and healthcare systems. Climate change mitigation strategies offer substantial environmental benefits, and specific technical measures can potentially reduce the environmental impact of livestock by 14% to 41% (Gerber et al. 2013). These measures typically involve improving resource efficiency, thereby reducing greenhouse gas (GHG) emissions per unit of output. They also play a significant role in bolstering food security. The path forward involves the development of novel livestock feeding and genetics approaches that empower farmers and herders to increase food production with fewer animals and lower GHG emissions. These interventions will assist small-scale food producers in meeting the growing demand for livestock products while simultaneously decreasing GHG emissions per unit of animal-source foods produced.

14.10 Livelihoods and Nutritional Security

The livestock sector plays a crucial role in addressing poverty and promoting economic growth, supporting approximately 20.5 million individuals who depend on it for their livelihoods. This sector's contribution to the achievement of the "No Poverty" and "Economic Growth" goals cannot be understated. Moreover, it employs around 8.8% of the population, making it a key driver of inclusive economic development (Department of Animal Husbandry and Dairying Ministry of Fisheries, Animal Husbandry and Dairying Government of India 2023). Compared to similar growth in the crop sector, the expansion of the livestock sector has a greater potential to alleviate poverty (Birthal and Taneja 2006; Mellor 2004; Saxena et al. 2020). Additionally, it offers unique opportunities for female empowerment, as two-thirds of India's female rural workforce are engaged in livestock rearing (Kumar and Singh 2008). When women gain control over assets and participate in household decision-making, not only does their well-being improve, but it also positively impacts household food security, child nutrition, and education, aligning with the objectives of zero hunger and gender equality (Quisumbing and Maluccio 2003; Smith et al. 2003; World Development Indicators 2023). The Government of India has successfully empowered women in the livestock sector through cooperative movements, with 5.06 million female members now constituting 30% of livestock cooperative memberships (Department of Animal Husbandry and Dairying Ministry of Fisheries, Animal Husbandry and Dairying Government of India 2023). This notable achievement demonstrates the government's commitment to reducing inequalities and promoting gender equality within the sector. Initiatives like the NLM, RGM, AHIDF, and Animal Husbandry Start-Up Challenge further contribute to creating a favourable environment for the development of the livestock sector and generating employment opportunities in rural and urban settings. Overall, the sustainable growth and development of the livestock sector have the potential to positively impact livelihoods, reduce poverty, promote gender equality, ensure food security, and improve health outcomes, making a significant contribution to multiple SDGs.

14.11 Conclusion

The transformation of the Indian livestock sector in line with the SDGs is crucial to meet the growing demand for ASFs, ensure sustainable production, enhance animal health and welfare, develop a resilient value chain, and promote environmental sustainability. By addressing key areas such as sustainable and efficient production, feed and fodder availability, breeding and reproduction, animal health and welfare, value chain development, and environmental sustainability; India can attain a more sustainable and inclusive livestock sector. This transformation requires comprehensive policies, technological interventions, research and development, infrastructure improvement, and support for cooperative systems. By prioritizing these efforts; India can ensure food security, improve livelihoods, reduce production losses, and contribute to the overall development of the agricultural and food systems in the country.

The Department of Animal Husbandry & Dairying (DAHD), Government of India is fully concerned about transforming the Indian Livestock Sector to achieve the SDG goals and the goals for Amrit Kaal. Accordingly, the DAHD has initiated several programs starting from National Livestock Mission, earmarking AHIDF, implementing RGM Project, and largest-ever vaccination program against important livestock diseases with 100% central assistance under National Animal Disease Control Program. The adoption and field implementation of innovative strategies and technologies including in vitro fertilization, sex-sorted semen, and identification of elite germplasm using genomic chip have acted as game changers. Launching APPI and implementation of AHSSOH will create an enabling policy as well as strengthen the core facilities. With these and other programs already launched, a positive trend in milk, meat and egg production has been visible and with the escalated thrust on increasing the production and productivity of the livestock sector further; the sector—in all likelihood—shall be able to achieve the SDGs and other targets of Amrit Kaal. The Department is always open to suggestions to touch upon newer areas for the growth and development of the livestock sector as it is the sector that ensures the economic sustainability of the farmers and industrial setups.

References

APEDA (2021a) Buffalo meat. https://apeda.gov.in/apedawebsite/SubHead_Products/Buffalo_Meat.htm#:~:text=India%20is%20the%20fourth%20largest,%2C%20Malaysia%2C%20Egypt%20and%20Indonesia. Accessed 30 May 2023

APEDA (2021b) Dairy products. https://apeda.gov.in/apedawebsite/SubHead_Products/Dairy_Products.htm. Accessed 30 May 2023

APEDA (2021c) Poultry products. https://apeda.gov.in/apedawebsite/SubHead_Products/Poultry_Products.htm. Accessed 30 May 2023

Birthal PS, Taneja VK (2006) Livestock sector in India: opportunities and challenges, presented at the ICAR-ILRI workshop on 'smallholder livestock production in India' held during January 24-25, 2006 at NCAP, New Delhi

Centre for Science and Environment (2022) ANNUAL REPORT knowledge-based activism 2021–22. In CENTRE FOR SCIENCE AND ENVIRONMENT. https://www.cseindia.org/static/page/CSE-annual-report-2022.pdf. Accessed 18 Jul 2023

Chakraborty M, Solanki S, Dutt R, Krishnaswamy (2022) Major emitting sources from the livestock sub-sector: trends in India and mitigation opportunities. ghgplatform-india.org. https://www.ghgplatform-india.org/wp-content/uploads/2022/09/GHGPI-PhaseIV-Briefing-Paper-on-Major-Emitting-Sources-from-the-Livestock-Sub-sector-Vasudha-Foundation-Sep22.pdf. Accessed 26 Jun 2023

Chand R (2017) NITI Aayog, strategy paper for doubling farmers' income goal of the government, 2017

Department of Animal Husbandry and Dairying Ministry of Fisheries, Animal Husbandry and Dairying Government of India (2021) Annual Report 2022–2023. https://dahd.nic.in/sites/default/files/AnnualRep-2021.06.21.pdf. Accessed 30 May 2023

Department of Animal Husbandry and Dairying Ministry of Fisheries, Animal Husbandry and Dairying Government of India (2022) Annual report 2022–2023. https://dahd.nic.in/sites/default/files/AnnualEnglish.pdf. Accessed 30 May 2023

Department of Animal Husbandry and Dairying Ministry of Fisheries, Animal Husbandry and Dairying Government of India (2023) Annual report 2022–2023. https://dahd.nic.in/sites/default/files/FINALREPORT2023ENGLISH.pdf. Accessed 30 May 2023

FAO (2018) World livestock: transforming the livestock sector through the sustainable development goals. Rome, p 222. https://doi.org/10.4060/ca1201en.Licence; CC BY-NC-SA 3.0 IGO

FAO (2021) Pastoralism–making variability work. FAO Animal Production and Health Paper No. 185. Rome. https://doi.org/10.4060/cb5855en

Gerber PJ, Steinfeld H, Henderson B, Mottet A, Opio C, Dijkman J, Falcucci A, Tempio G (2013) Tackling climate change through livestock—a global assessment of emissions and mitigation opportunities. Food and Agriculture Organization of the United Nations (FAO), Rome

India's agricultural and processed food products exports have grown at a steady pace in the last decade notwithstanding challenges (2021). https://pib.gov.in/PressReleasePage.aspx?PRID=1786508. https://pib.gov.in

Indian Council of Agricultural Research, New Delhi (2015) Vision 2050. https://icar.org.in/files/Vision-2050-ICAR.pdf. Accessed 30 May 2023

Indian Economy Continues to Show Resilience Amid Global Uncertainties (n.d.) World Bank. https://www.worldbank.org/en/news/press-release/2023/04/04/indian-economy-continues-to-show-resilience-amid-global-uncertainties

Indian Grassland and Fodder Research Institute (Indian Council of Agricultural Research) Gwalior Road, Jhansi -284 003 (2013) Vision 2050. https://www.yumpu.com/en/document/view/3 8899124/igfri-vision-2050-indian-grassland-and-fodder-research-institute-. Accessed 30 May 2023

Jha SN, Vishwakarma RK, Ahmad T, Rai A and Dixit AK (2015) Report on assessment of quantative harvest and post-harvest losses of major crops and commodities in India. ICAR-All India Coordinated Research Project on Post-Harvest Technology, ICAR-CIPHET, P.O.-PAU, Ludhiana. https://www.researchgate.net/profile/Rajesh-Vishwakarma-3/publication/2 89637983_Assessment_of_Quantitative_Harvest_and_Post-Harvest_Losses_of_Major_ CropsCommodities_in_India/links/5690c83b08aed0aed811bae6/Assessment-of-Quantitative-Harvest-and-Post-Harvest-Losses-of-Major-Crops-Commodities-in-India.pdf. Accessed 26 Jun 2023

Kumar A, Singh DK (2008) Livestock production systems in India: an appraisal across agro-ecological regions. Indian J Agric Econom 63:902-2016-67979; https://www.researchgate.net/publication/239940030_Kumar_Anjani_and_Dhiraj_K_Singh_2008_Livestock_Production_ Systems_in_India_An_Appraisal_across_Agro-Ecological_Regions_Indian_Journal_of_Agri cultural_Economics_634577-97

Mellor JW (2004) Agricultural growth and poverty reduction: the rapidly increasing role of smallholder livestock. In: Ahuja V (ed) Livestock and livelihoods: challenges and opportunities for Asia in the emerging market environment. National Dairy Development Board of India and FAO of the United Nations, p 372; https://cir.nii.ac.jp/crid/1572261551121284992

Nanda Kumar T, Das S, Gulati A (2022a) Dairy value chain. In: Gulati A, Ganguly K, Wardhan H (eds) Agricultural value chains in India. India studies in business and economics. Springer, Singapore. https://doi.org/10.1007/978-981-33-4268-2_6; Accessed 26 Jun 2023

Nanda Kumar T, Samantara A, Gulati A (2022b) Poultry value chain. In: Gulati A, Ganguly K, Wardhan H (eds) Agricultural value chains in India. India studies in business and economics. Springer, Singapore. https://doi.org/10.1007/978-981-33-4268-2_7; Accessed 26 Jun 2023

NDDB (2020) Annual report 2018–19. National Dairy Development Board. https://www.nddb. coop/sites/default/files/NDDB-AR-2019-ENGLISH-24022020.pdf

NITI Ayog (2018) Demand & supply projections towards 2033 crops, livestock, fisheries and agricultural inputs, the working group report. https://www.niti.gov.in/sites/default/files/2019-0 7/WG-Report-issued-for-printing.pdf. Accessed 30 May 2023

PASTER (n.d.) Enhancing biodiversity through livestock keeping. https://pastres.files.wordpress. com/2022/09/en-infosheet-3of6.pdf. Accessed 18 Jul 2023

Phand J, Gummagolmath (2021) Livestock extension services: time to think beyond treatment and breed improvement. In MANAGE knowledge series, no.2/2021. National Institute of Agricul-tural Extension Management (MANAGE). https://www.manage.gov.in/publications/ knowledgeseries/livestock.pdf. Accessed 26 Jun 2023

PIB (2021) Productivity of dairy animals. https://pib.gov.in/PressReleaseIframePage.aspx? PRID=1707187. Accessed 26 Jun 2023

PIB (2022) Cabinet approves India's updated nationally determined contribution to be communicated to the United Nations framework convention on climate change. https://pib. gov.in/PressReleaseIframePage.aspx?PRID=1847812. Accessed 26 Jun 2023

Quisumbing AR, Maluccio JA (2003) Resources at marriage and intrahousehold allocation: evi-dence from Bangladesh, Ethiopia, Indonesia, and South Africa. Oxford Bull Econ Stat 65:283–327. https://doi.org/10.1111/1468-0084.t01-1-00052

Saxena R, Khan M, Choudhary B, Kanwal V (2020) The trajectory of livestock performance in India: a review. J Dairy Sci 72:1. https://doi.org/10.33785/IJDS.2019.v72i06.001

Sen M, Muthukumar (2022) Sustainable meat value chain and enhanced farmers' income. Depart-ment of Economic Analysis and Research, National Bank for Agriculture and Rural Develop-ment, Mumbai; https://www.nabard.org/auth/writereaddata/tender/2501230635paper-11-sustainable-meat-value-chain.pdf

Sharma UC (2020) Methane and nitrous oxide emissions from livestock in India: impact of land use change. J Agric Aquac 2(1):1–9; https://escientificpublishers.com/assets/data1/images/JAA-02-0014.pdf

Singh KN, Meena M, Singh RP (2012) Livestock value chains: prospects, challenges and policy implications for Eastern India. Social Science Research Network. https://doi.org/10.2139/ssrn.2020916

Singh B, Bardhan D, Verma MR, Prasad S, Sinha DK (2014a) Estimation of economic losses due to Peste de Petits ruminants in small ruminants in India. Vet World 7(4):194–199. https://doi.org/10.14202/vetworld.2014.194-199

Singh B, Prasad S, Verma MR, Sinha D (2014b) Estimation of economic losses due to Haemorrhagic Septicaemia in cattle and buffaloes in India. Agric Econ Res Rev 27(2):271. https://doi.org/10.5958/0974-0279.2014.00030.5

Singh B, Dhand N, Gill J (2015) Economic losses occurring due to brucellosis in Indian livestock populations. Prev Vet Med 119(3–4):211–215. https://doi.org/10.1016/j.prevetmed.2015.03.013

Singh B, Bardhan D, Verma MR, Shiv P, Sinha DK, Sharma VB (2016) Incidence of classical swine fever (CSF) in pigs in India and its economic valuation with a simple mathematical model. Anim Sci Rep 10(1):3–9

Smith LC, Ramakrishnan U, Ndiaye A, Haddad L, Martorell R (2003) The importance of Women's status for child nutrition in developing countries: International Food Policy Research Institute (IFPRI) research report abstract 131. Food Nutr Bull 24:287–288; https://ebrary.ifpri.org/utils/getfile/collection/p15738coll2/id/48032/filename/43490.pdf

The World Bank (2020) Implementation completion and results report. Agriculture and Food Global Practice, Sustainable Development, South Asia Region. https://www.nddb.coop/sites/default/files/pdfs/ndpi/NDPI_World_Bank_Evaluation_Report_(ICRR).pdf

Venkataramanan R, Hemadri D, Bandyopadhyay SK, Taneja VK (2006) Foot-and mouth disease in India: present status. Paper presented at a workshop on global roadmap for improving the tools to control foot-and-mouth disease in endemic settings, Agra

World Bank (2010) Demand led transformation of the livestock sector in India. Achievements, opportunities, and challenges, South Asia agriculture and rural development. Report No. 48412 IN. The World Bank, Washington, DC; http://documents.worldbank.org/curated/en/668321468041641776/Demand-led-transformation-of-the-livestock-sector-in-India-achievements-challenges-and-opportunitiesrnmentofIndia

World Development Indicators (2023) World Bank. https://databank.worldbank.org/source/world-development-indicators. Accessed 18 Jul 2023

World Popul Rev (2023). https://worldpopulationreview.com/countries/india-population. Accessed 18 Jul 2023

Aquatic Food Systems for Blue Transformation: A Vision for FAO

15

Rishi Sharma, Diana Fernandez Reguaera, Carlos Fuentevilla, Vera Agostini, and Manuel Barange

Abstract

This paper outlines a roadmap for the transformation of aquatic food systems (While applying to all aquatic food systems, the roadmap considers the critical role and potential of aquatic food systems in Low Income Countries, Food Deficit Countries and Small Island Developing States, making them a particular focus of the Blue Transformation vision outlined in this document)—'Blue Transformation' (BT), providing a compass for the FAO's work on aquatic food systems for the period 2022–2030. This roadmap for Blue Transformation aligns with the 2021 Declaration for Sustainable Fisheries and Aquaculture of the Committee on Fisheries (COFI) of the Food and Agriculture Organization of the United Nations (FAO) and FAO's Strategic Framework 2022–2031. It focuses on the elements that would maximize the contribution of aquatic food systems to the Sustainable Development Goals (SDGs). Finally, we look at approaches how this maybe applied to India.

In this paper, the BT roadmap prioritizes food systems as drivers of employment, economic growth, social development and environmental recovery, which all underpin the SDGs. The 2030 Agenda is supported through the transformation to more efficient, inclusive, resilient and sustainable aquatic food systems for better production, better nutrition, a better environment, and a better life, leaving no one behind. The overall framework provided by BT is to support planning, implementation, monitoring and communication of FAO Fisheries and Aquaculture Division (NFI)'s work. Under FAO's Strategic Framework, this BT will be used to prioritize decisions globally so as to achieve a harmonized and coherent objectives globally.

R. Sharma (✉) · D. F. Reguaera · C. Fuentevilla · V. Agostini · M. Barange
Fisheries and Aquaculture Division, Food and Agriculture Organization of the United Nations Viale delle Terme di Caracalla, Rome, Italy
e-mail: Rishi.Sharma@fao.org

K. C. Bansal et al. (eds.), *Transformation of Agri-Food Systems*,
https://doi.org/10.1007/978-981-99-8014-7_15

We discuss how best to apply this approach to systems in India, and propose an integrated framework that would meet the domestic and export demands of seafood in the Indian context. Integral in this approach is the necessity to build capacity locally so as to achieve the three pillars of the transformation; primarily to invest in Universities and training centers to make sure India can implement the three pillars, increased aquaculture production through increased and better facilities, improved management of India's marine living resources and finally adequate value-chains to increase the value and quality of the product from both supply chains, aquaculture and fisheries and improve the distribution of benefits.

Keywords

COFI · Blue transformation · Sustainable fisheries · Value chains

15.1 Introduction

Animal proteins and micronutrients found in aquatic foods play a crucial role in ensuring food and nutrition security for many coastal populations, particularly the most vulnerable. Their crucial role as suppliers of highly nutritious food, essential for physical and cognitive development, has been growing (UN Nutrition 2021), even though less than half of public health nutrition policies currently identify their consumption as key objectives (Koehn et al. 2021). In addition, fisheries and aquaculture already support over 58 million jobs in the primary sector and 600 million livelihoods, including subsistence and secondary sector and their dependents (FAO 2022), and the trade in fish products provides an important source of hard currency and income for exporting countries and regions (Figs. 15.1 and 15.2)

Given the increase in severe food insecurity post COVID-19 pandemic, the issue of food security is even more important. About 735 million (SOFI 2023) people now suffer from hunger and 2.4 billion people have severely limited access to adequate food. The challenge to feed a growing population without exhausting our natural resources continues to increase, and thus, aquatic food systems are increasingly becoming more important for their potential to provide a larger proportion of

Aquatic food systems encompass the entire range of actors and their interlinked value-adding activities involved in the production, aggregation, processing, distribution, consumption and disposal of aquatic food products that originate from fisheries and aquaculture and parts of the broader economic, societal and natural environments in which they are embedded (e.g., open oceans, coastal waters, wetlands, lakes, rivers, ponds, raceways, fields and tanks).

Fig. 15.1 Aquatic food systems definition adapted from FAO, 2018. (FAO, 2018. Sustainable food systems: concept and framework. Brief. FAO, Rome, Italy. (also available at https://www.fao.org/3/ca2079en/CA2079EN.pdf)) *(from FAO BT Roadmap 2022–2030)*

Fig. 15.2 Schematic representation of the Blue Transformation roadmap to address the strategic priorities of FAO and maximize the contribution of aquatic food systems to the Sustainable Development Goals (*from FAO BT Roadmap 2022–2030*)

humanity's nutritious food requirements, in particular in the context of climate change.

Strategies to deliver healthy, sustainable, and equitable food systems do not adequately include the critical long-term impacts of overfishing, habitat degradation and unequal access to resources and markets. In 2021, Committee on Fisheries (COFI) unanimously endorsed the Declaration for Sustainable Fisheries and Aquaculture (FAO 2021). This Declaration recognized the contributions of the sector in combating poverty and hunger since the endorsement of the 1995 Code of Conduct for Responsible Fisheries.

Blue Transformation is the vision and the process by which FAO, its Members and partners can use existing and emerging knowledge, tools and practices to secure and maximize the contribution of aquatic (both marine and inland) food systems to food security, nutrition and affordable healthy diets for all.

15.1.1 Why Do We Need Blue Transformation?

Capture fisheries production remained stable in the 25 years following the Code of Conduct for Responsible Fisheries, but aquaculture production soared 250%, enabling the sector to meet the increase in fish consumption and demand, which rose to 20.5 kg per person per year in 2019 (a growth rate double the world population). The net trade value of fish food products in developing countries (exports minus imports) is greater than the total value of all food products due to the integration of aquatic foods into global and regional supply chains.

Blue Transformation is a targeted effort to promote innovative approaches that expand the contribution of aquatic food systems to food security and nutrition and affordable healthy diets. Achieving the objectives of Blue Transformation requires holistic and adaptive approaches that consider the complex interaction between global and local components in food systems and support multi-stakeholder interventions to secure and enhance livelihoods, foster equitable distribution of benefits and provide for an adequate use and conservation of biodiversity and ecosystems.

Through Blue Transformation, aquatic food systems can:

- support the provision of sufficient aquatic food for a growing population in an environmentally, socially and economically sustainable manner;
- ensure the availability and accessibility of safe and nutritious aquatic food for all, especially vulnerable populations, and reduce food loss and waste;
- ensure that aquatic food systems contribute to improving rights and incomes of dependent communities to achieve equitable livelihoods; and,
- support resilience in aquatic food systems, which are highly influenced by dynamic human and environmental processes, including from climate change.

15.1.2 Objectives of Blue Transformation

Blue Transformation has three core objectives:

1. Sustainable aquaculture expansion—to support satisfaction of global demand for aquatic food and equitable distribution of the benefits.
2. Effective management of all fisheries—to deliver healthy stocks and secure livelihoods.

3. Upgraded value chains—to ensure the social, economic and environmental viability of aquatic food systems.

In order to meet the gap in demand and supply for aquatic foods, aquaculture must expand sustainably to satisfy the gap in global demand for aquatic foods while generating new or securing existing sources of income and employment. This requires updating aquaculture governance by fostering improved planning, legal and institutional frameworks and policies. FAO and its partners must focus on the urgent demand for the development and transfer of innovative technologies and best practices to generate efficient, resilient and sustainable operations. The continued transformation of aquaculture applies to most regions but is particularly critical in food-insecure regions; the aim is to increase global production by between 35% and 40% by 2030, according to national and regional contexts.

To rebuild capture fisheries, effective management of all fisheries is a key step. Where effective management exists, fish resources have been rebuilt and are increasingly sustainable (Hilborn et al. 2020). To achieve this objective, FAO and its partners must apply and share effective fisheries management systems that restore ecosystems to a healthy and productive state, while managing exploited resources within ecosystem boundaries. Actions to achieve this objective include building global capacity to regularly collect, analyse and evaluate data that support decision-making and consider trade-offs, particularly in regions with limited data and poor capacity.

Finally, these have successive multiplier effects enhancing value chains, public and private sector actors, including consumers, reducing food loss and waste, enhancing transparency, and improving access to lucrative markets while adopting emerging digital tools. As a result of the COVID-19 pandemic, aquatic food value chain actors have increasingly adopted these practices. Upgraded value chains also add and create value to extract more wealth and food from the sector's productive capacity and distribute it more equitably. Promoting healthy diets in an inclusive manner requires programmes and initiatives that increase consumer awareness and make healthy, safe, and nutritious aquatic foods available, including in areas with low food and nutrition security.

15.1.3 Towards Blue Transformation Globally and Within the Indian Context

FAO is committed to working with FAO Members, partners and stakeholders to implement Blue Transformation in support of the food security and nutrition of a world population expected to reach 10 billion in 2050. India being the most populous country on earth would be a large part of this vision and hence enhancing these 3 pillars are discussed below.

15.2 FAO's Vision and Strategic Framework[1]

15.2.1 FAO's Vision

FAO is a specialized agency of the United Nations (UN) that leads efforts across the globe to defeat hunger. FAO's vision is of a *world free from hunger and malnutrition where food and agriculture contribute to improving the living standards of all, especially the poorest, in an economically, socially and environmentally sustainable manner,* and carries three Global Goals:

- **eradication of hunger, food insecurity and malnutrition**, progressively ensuring a world in which people at all times have sufficient safe and nutritious food that meets their dietary needs and food preferences for an active and healthy life;
- **elimination of poverty and the driving forward of economic and social progress for all,** with increased food production, enhanced rural development and sustainable livelihoods; and,
- **sustainable management and utilization of natural resources**, including land, water, air, climate and genetic resources for the benefit of present and future generations.

15.2.2 FAO's Strategic Narrative and Key Aspirations

The strategic narrative guiding FAO's Strategic Framework is the transformation to MORE efficient, inclusive, resilient and sustainable agri-food systems for *better production, better nutrition, a better environment, and a better life, leaving no one behind.*

These four betters[2] represent an organizing principle for how FAO intends to contribute directly to its three guiding SDGs, SDG 1 (No poverty), SDG 2 (Zero hunger), and SDG 10 (Reduced inequalities) as well as support the broader SDG agenda, crucial for attaining FAO's overall vision.

---

[1] FAO's Vision, Strategic Narrative and Key Aspirations were evaluated and endorsed by the 42nd Session of the FAO Conference held from 14–18 June 2021, and provide guidance to the Blue Transformation roadmap. More information is available in the Conference Report and the Medium Term Plan Report

[2] See https://www.fao.org/strategic-framework/enfora short description of FAO's updated Strategic Framework, including the Four Aspirations (Better Production, Better Nutrition, Better Environment, Better Life).

Fig. 15.3 Objectives and targets of the Blue Transformation roadmap *(from FAO BT Roadmap 2022–2030)*

15.3 Core Components of Blue Transformation[3] and It Application Within Indian Context

This section outlines the three core global objectives of Blue Transformation with related targets and priority actions for each (Fig. 15.3). The objectives and targets require contributions from FAO, as well as Members, and international, regional and national aquatic food systems stakeholders. While FAO is not solely responsible for these, its leadership and work will help drive partners towards their achievement and align their narratives.

India accounts for approximately 8% of the annual global production in the capture and aquaculture sectors combined (see Fig. 15.4 below, 14.1 MMt of total production vs 180 MMt respectively in 2020). In 2020, India was the fourth top capture fisheries producer with 4.7% of the production at global level (5.5 million tonnes), ranking the seventh in marine capture production with 5% of the production

---

[3]The current formulation of targets and priority actions reflected in this document and described in this section builds on inputs provided by NFI Team leaders, NFI staff and FAO regional officers provided through a series of meetings and consultations between July and October 2021 (see Annex 4).

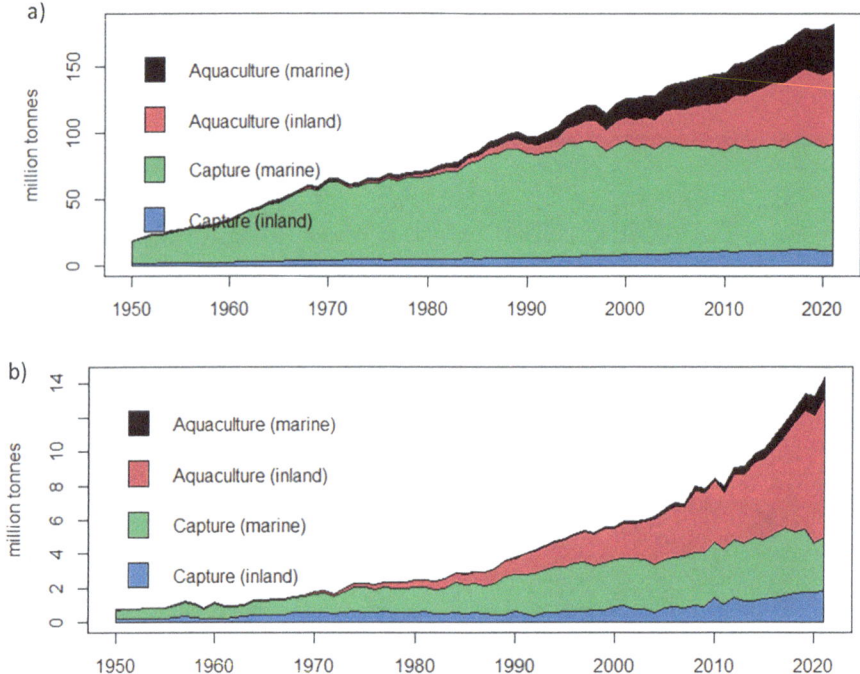

Fig. 15.4 (**a**) Total global production by marine, and aquaculture in marine and inland waters; Total production by India (as reported through FAO FishtstatJ statistics) in the same sectors as those in inset (**b**). (FAO. 2023. FishStat data. Fisheries and Aquaculture Division. Rome. https://www. fao.org/fishery/en/fishstat)

(3.71 million tonnes) and the top inland fisheries producer with 16% of the total (1.8 million tonnes). In 2020, India was the second top aquaculture producer with 9.9% of the world's share (8.6 million tonnes). Aquqculture has two components, marine and freshwater; in aquaculture in inland waters, India's production accounts for 14.3% globally but in each of the other sectors (marine and inland capture production and marine aquaculture) it is less than 5% of the global production. Given, the large area of India's EEZ (2,305,143 sq. kms), it appears that either the marine resources have not been fully optimized or overfishing is occurring. If we look at the global production trends and projections made by FAO (see Fig. 15.5 below) to a total production of 202 MMt globally by 2030. This is an increase of about 14% from the current production levels. If India were to achieve another 14% increase in production, the total value would be approximately 16.5 MMt. However, this target could probably be increased even more in each of the sectors if India did the following in marine fisheries and aquaculture areas.

In 2019, India was the third largest consuming countries after China and Indonesia, with a per capita annual consumption of 8.1%, and ranked among the five top five exporting countries of aquatic products, reaching a value of USD 5.8

a)   Global capture fisheries and aquaculture production, 1990–2030

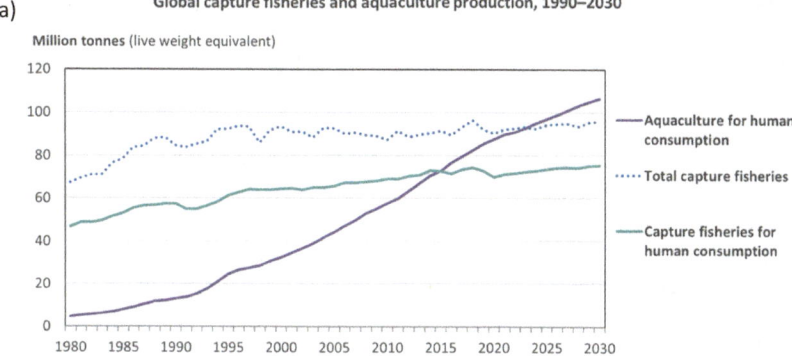

b)

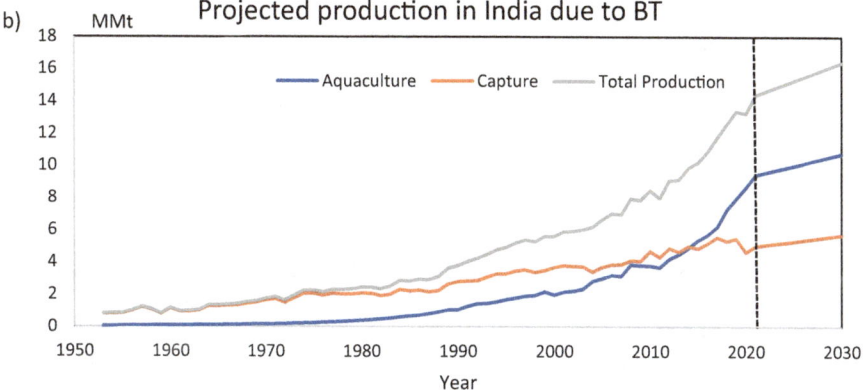

Fig. 15.5 (**a**) Projected Production globally by 2030 if Blue Transformation takes place globally and (**b**) in India

billion in exports in 2020, highlighting the importance of aquatic foods for the country.

15.3.1 Aquaculture in Inland and Marine Waters

The following targets should be focused on India in the aquaculture sector:

1. Maintain the production growth with limited resources in environmentally sustainable manner.
2. Further integration with broader agricultural systems and the maintenance of biodiversity and ecosystem integrity.
3. Sustain a predominantly smallholder base, retaining competitiveness, inclusivity, and resilience in changing economies.

4. Empower small and medium enterprises to realize technological advantages of improved aquaculture systems.
5. Promote greater organizational capacities of small and medium scale farmers, including women and youth, & facilitate access to financial services.
6. Capture potential opportunities presented by IT to improve the efficiency of farm management, business and trading of aquaculture products.
7. Ensure aquaculture products are produced safely and contribute to improved nutrition and livelihoods.
8. Provide an effective and responsible policy and regulatory environment for aquaculture development.

15.3.2 Marine and Inland Fisheries

In the marine and inland sectors in India the following needs to occur:

1. Facilitate the development of innovative data and information systems to support fisheries policy formulation, assessment and management advice.
2. Facilitate regular monitoring and reporting on the state of fisheries and the impacts of management interventions across ecological, social and economic dimensions.
3. Support the development and implementation of fisheries management plans, strategies and measures that consider tradeoffs and address ecological, social and economic objectives, particularly in capacity-limited and data-limited systems.
4. Promote the implementation of adaptive fisheries management measures that support biodiversity, facilitate ecosystem restoration, strengthen climate change adaptation and build resilience to stressors.
5. Facilitate adoption and effective implementation of new and existing international instruments, regional coordination mechanisms, plans of action and guidelines.
6. Support responsible governance of tenure to ensure inclusive, sustainable, secure and equitable access to fisheries, land and water resources.
7. Facilitate the development of organizational, technical and business skills of small-scale food producers, including women and youth groups, their organizations and institutions for equitable and effective participation in decision-making processes, resource management and value chain development.
8. Support the integration of fisheries related policies in global, regional and national development agendas (cross-sectoral, multi-risk, food systems).

Finally, the resulting benefits of the BT movement in India should be realized as an increase in value added resources in India that can be realized primarily by improved value chains using the methods adopted below:

1. Promote and support the development of diverse practices and processes to reduce fish loss and waste.

2. Add value and increase consumption through promotion of fish processing, co-products, and bycatch use.
3. Facilitate compliance with instruments addressing post-harvest issues or trade at a global, regional or national level.
4. Promote resilient, efficient and inclusive processes for value addition of fisheries and aquaculture products.
5. Ensure a growing participation of women in value chain related decision-making processes, and accelerate their access to leadership, technologies, practices, resources, and infrastructure to enhance equal distribution of benefits.
6. Promote governance frameworks to increase preparedness to, reduce impacts from and support rehabilitation of the aquatic food value chains after natural disasters, crisis and emergencies.
7. Facilitate uptake of novel technologies to improve post-harvest processes and reduce environmental impacts.
8. Improve data and information on trade and markets to reduce information asymmetries.
9. Collaborate with international organizations and Conventions on aquatic species issues involving trade (e.g., WTO, UNCTAD, CITES, among others).
10. Increase capacity to access markets, with a special emphasis on equitable participation and outcomes.
11. Support the development of market insertion possibilities for businesses, including in collaboration with other international organizations, focusing on equitable market access for small-scale actors.

In order to achieve the above transformation, India needs to invest in human resource and infrastructural capacity. A key to success in improving fisheries and aquaculture sustainability in marine and inland waters, is having the required human resources to assess and improve processes in these sectors. While in the marine and inland capture fisheries sectors these are primarily activities related to stock assessments, and overall management with monitoring and compliance measures, in aquaculture this would require improved processes and production capacity primarily in the marine, but also in the inland sector.

While the aquaculture sector is strong in terms of knowledge and processes, the capture fisheries sector could use improved techniques for assessment, management and compliance. Improved capacity would be needed in these sectors, and this could come from external forces like FAO, as well as building internal capacity through university programs targeted to achieve improved knowledge in these processes. Currently, universities target a lot of instruction and practical experience to the aquaculture sector and little is done with respect to the capture fishery sectors. Marine aquaculture could also be a place to target opportunities so this sector could expand as well. Given, the huge GDP of India ($US 3.75 Tr) and annual growth rate of 7.2%, a large investment should be put in the oceanic and fisheries sector. Building this capacity in universities should be the foremost concern of the country, so as to produce top professionals in these fields. With regard to the approaches used in instruction at the universities, integrated system thinking in terms of overall ecosystem complexities could be the focus of instruction, as this considers the impacts of fisheries and aquaculture on the ecosystem and the community as a whole.

With respect to financing programs, India per capita income is quite low so measures to increase capacity, income, social protection schemes of fishers and fish farmers in an inclusive and equitable manner are also needed. The availability and capacity of human capital employed in the aquatic food sector is a key determinant of production growth. Nevertheless, the lack of access to finance and markets for youth, women or smallholders, or the insufficient attention to the challenges faced by women in aquatic food systems are impediments to productivity gains.

Finally, using the large human capital that India has in the technology sectors could improve processes in aquaculture, capture fisheries and value chains. In the long run, India could be at the forefront of the overall realized production change through Blue Transformation and could possibly increase it by more than 14% global projections, to 30–40% from their current estimate of production within a short time frame.

Acknowledgements A lot of the material here is from FAO BT Roadmap documents, SOFIA Reports and FISHSTATJ database.

References

FAO (2021) 2021 COFI declaration for sustainable fisheries and aquaculture. Rome. p 16. https://doi.org/10.4060/cb3767en

FAO (2022) The state of world fisheries and aquaculture 2022. Toward blue transformation. FAO, Rome. https://doi.org/10.4060/cc0461en

Hilborn R et al (2020) Effective fisheries management instrumental in improving fish stock status. PNAS 117(4):2218–2224

Koehn JZ, Allison EH, Villeda K, Chen Z, Nixon M, Crigler E, Zhao et al (2021) Fishing for health: do the world's national policies for fisheries and aquaculture align with those for nutrition? Fish Fish 1:1–18; [Cited 24 November 2021]

SOFI (2023) https://data.unicef.org/resources/sofi-2023/

UN Nutrition (2021) The role of aquatic foods in sustainable healthy diets. Discussion paper. [cited 27 October 2021]. https://www.unnutrition.org/wp-content/uploads/FINAL-UN-NutritionAquatic-foods-Paper_EN_.pdf

Innovations in Fish Processing Technology

16

C. N. Ravishankar and K. Elavarasan

Abstract

Globally, fish has been realized as a nature's super-food with the global per capita availability for consumption of 20.2 kg/year/person. Aquatic system has provided 157.4 million tons of aquatic animals as food and served as a source for 25 million tonnes of easily digestible proteins packed with essential amino acids and health beneficial peptides. Marine fish is well-recognised for their richness in therapeutic fatty acids including eicosapentaenoic acid (EPA) and docosahexaenoic acid (DHA). The richness in terms of essential micronutrients like vitamins and minerals has made the fish as a better and sustainable choice to address the nutritional issues at global scale. India has contributed 8% (162.48 lakhs tonnes) to the global fish production with the very below global per capita average annual consumption (6.31 kg/year/person). Global as well as Indian fish production has significant quantity as wastage due to post-harvest loss and process related waste generation, and nearly accounts to 27–39%. In the contest of Indian fisheries, minimizing post-harvest loss and improving the waste utilization through innovative solutions could greatly transform the aquatic food system to support the food and nutritional security of India. Innovations in technologies through research and development could strengthen the strategies to minimize the post-harvest fish loss. Innovations are demanded in fish processing methods, product development, processing machineries, handling tools, and supply chain. Implementing the developments happening in allied sectors, including information technology and artificial intelligence to strengthen the forward and backward

C. N. Ravishankar (✉)
ICAR- Central Institute of Fisheries Education, Mumbai, India
e-mail: ravishankar@cife.edu.in

K. Elavarasan
Fish Processing Division, ICAR-Central Institute of Fisheries Technology, Cochin, India

© National Academy of Agricultural Sciences, under exclusive license to Springer
Nature Singapore Pte Ltd. 2023
K. C. Bansal et al. (eds.), *Transformation of Agri-Food Systems*,
https://doi.org/10.1007/978-981-99-8014-7_16

linkages will result in a well-managed supply chain and cold chain. Bridging the information, knowledge and technological gaps between producers and consumers would be a key for reducing the post-harvest fisheries loss. The innovations in technologies to be friendly with local culture and should satisfy the peculiar local needs for better adoptions by many numbers so as to achieve the transformation envisaged in aquatic food system for building a healthy and wealthy nation.

Keywords

Blue transformation · Innovations in fish processing · Food and nutritional security · Fish processing technology · Fish loss and waste

16.1 Introduction

For the people and for the planet is the globalized idea that almost all the countries have accepted for establishing the peace and prosperity irrespective of their economic development. As a result of years of work by different countries, and intergovernmental deliberations, UN Department of economic and affairs have set 'Sustainable development goals (17 goals)' with the ultimate aim to improve the human life and protect the environment. India has progressed significantly across the SDGs, through various missions like empowered and resilient India, Clean and Healthy India, Inclusive and Entrepreneurial India, sustainable India, Prosperous and Vibrant India. India continues to work collaboratively with all domestic and global stakeholders to accelerate efforts for a sustainable planet for future generations.

It is one of the global agenda to transform the aquatic food system sustainably to feed millions of people with highly nutritious food. Aquatic food system has been identified as a potential solution to address the food and nutritional security, and well-being of environment and society. The transformation has to be achieved carefully without damaging the aquatic environment by preserving the aquatic life, protecting the biodiversity, reducing aquatic pollution and promoting equality among the people who depend on the aquatic system for their livelihood. This is possible through a well-balanced inclusive approach. Three important core components of blue transformation include Aquaculture, Fisheries and Value chains. Among them, for transformation of aquatic food system sustainably, upgrading the fish value chain plays a central role. Fish value chain begins from the harvest (capture fisheries or harvest from aquaculture system) and extends up to the seller. However, the loss or waste generation is possible even at consumer's plate. Achieving the targets like increased availability of aquatic foods, enhanced consumer awareness of health benefits acquired through fish consumption, reduction in achieving food loss and waste (FLW), and improved access to lucrative markets, will help to strengthen the fish value chain.

According to FAO's road map (FAO 2022a, b) for blue transformation, the following are the targets set under value chain component

- Efficient value chains that increase profitability and reduce food loss and waste.
- Transparent, inclusive and gender-equitable value chains support sustainable livelihoods and fisheries and aquaculture products access in international markets more effectively.
- Increased sustainable consumption of sustainable aquatic food, particularly in areas with low food and nutrition security.
- Increased access to healthy, safe and high quality aquatic food.

Using appropriate technologies for processing and preservation is the heart of fish value chain. Hence innovation in fish processing aims in reducing the post-harvest loss and waste generation should be given due importance. Fish processing sector requires holistic and adaptive approaches that support multi-stakeholder interventions using existing and emerging knowledge, tools and practices to secure and maximize the contribution of aquatic food systems to global food security and nutrition. Efforts should be made to improve the capacities at all levels throughout the value chain to develop and adopt innovative technology and appropriate management practices for a more efficient and resilient fish processing industry.

16.2   Fish Processing-Global Scenario

Globally 89% of the total fish production has been used for direct human consumption. **The major form of processing and its contribution to the fish consumed directly by human is presented in** Table 16.1. Still the greater portion of fish consumed falls under the category live, fresh and chilled. In many countries, the domestic consumption mainly includes fish and fish products undergone icing and chilling. Salted and dried fish products and fermented products forms considerable portion of domestic consumption in inland areas. In Asia and Africa, the share of aquatic food production preserved by salting, smoking, fermentation or drying is higher than the world average. Freezing, curing and prepared and preserved fish products contributes significantly in export sector. In recent years, especially non-food products productions use the fish waste/by-products as raw materials. For example, over 27% of the global production of fishmeal and 48% of the total production of fish oil were obtained from by-products. Global trade of fish and fish products expands across the continents and the trading of high value fish products and value added products increased in the recent past. The largest exporter is China and the importer is European Union. In terms of volume, China is the top importing country of species not only for domestic consumption but also as raw material to be processed in China and then re-exported (FAO 2022a, b). The seafood trade significantly influences the nutritional availability for the population residing in developing countries. The macro and micronutrients from seafood is available for the people in developing nations at relatively lower price compared to the population belong to developed nations (Liu et al. 2023). Understanding the seafood trade flow and its impact on local food and nutritional insecurity would help to design futuristic strategic plans and development policies. As on today, in general, the lower income

Table 16.1 Major form of processing and its contribution to the fish consumed directly by human

S. No	Major form of processing	Contribution (%) to fish consumed directly by human
Global status		
1.	Live, fresh or chilled	44%
2.	Frozen	35%
3.	Prepared and preserved	11%
4.	Cured	10%
High in-come countries		
1.	Frozen	>50%
2.	Prepared and preserved	26%
3.	Cured	13%
Upper-middle-income countries		
1.	Live, fresh or chilled	>60%
2.	Frozen	20%
3.	Canned	11%
Low-income countries		
1.	Live, fresh or chilled	70%
2.	Frozen	7%
3.	Cured	20%

(Note: The data provided in "FAO 2022b. The State of World Fisheries and Aquaculture 2022" is used to prepare this Table)

countries catch more and export the high value fish products to the developed nations while retaining the low value fishes for food security of locals. Exporting high value seafood also generate the income through which the reduction in poverty could be achieved. However, care must be taken to ensure the availability of nutrient dense aquatic food to meet the need of locals through policy level interventions (Watson et al. 2017). The analysis on vulnerability of seafood trade to shocks has concluded that Central and West Africa are the most vulnerable. Identifying potential risks and vulnerability is more important to build a stronger food system, especially in seafood trade as it is growing significantly in current years (Gephart et al. 2016).

16.3 Scenario of Fish Processing in India

Indian fish production for the year 2021–2022 reached 16.25 million tons. In India, fish processing includes different preparations like marinating, pickling, smoking, salting and drying. However, chilled or fresh fish disposition is the highest in domestic market whereas the freezing is the major form of processing for export market. To a lesser extent, the value added products are manufactured and distributed in domestic as well as for international trading. India is the fifth largest seafood exporter and major contributor of Indian economy and foreign exchange. India mainly exports frozen shrimps, fish, cuttlefish, squids, dried items and live and chilled items Handbook on Fisheries statistics, 2022. Fish in India serve as the cheap and best source of animal proteins and helps in ensuring the nutritional security of people. Under Blue revolution scheme in India, the aquatic food system has undergone tremendous transformation through fishermen relief assistance, capacity building through trainings, development and awareness campaigns. In addition, various developmental projects particularly fisheries and aquaculture infrastructure development, etc. under PMMSY has enriched the fisheries sector. There is a continuous strive for enhancing the fisheries value chain in India through various interventions. In the context of Indian fish processing industry, biomass fishing for fish meal and oil has been raised as a concern in the recent past in the view of food security and livelihood of small scale fishers (Jyotishi 2023). On the other hand, it is also pointed out that the rural population in India is deficient in lipids which are rich in omega-3 fatty acids. In India, traditional fish processing like drying is still in practice and contribute significantly to the nutritional security as it is rich in high quality protein in the concentrated form (Siddhnath et al. 2022).

India has well equipped fish processing units with art of technology meeting the requirements and regulations of EU countries, USA etc. Similarly, the export of Indian seafood to non-EU countries also is growing tremendously. In spite of the enormous efforts taken to promote fisheries sector of India, still the major concerns are post-harvest losses and waste generation. Some strategies to minimize the post-harvest losses are outlined below, while the waste generation and its utilization is dealt in details in subsequent sections.

Strategies to reduce the post-harvest losses in fisheries sector

1. Reassessing the post-harvest losses in fisheries sector of India.
   (a) In order to identify the critical points in the supply chain where the loss is significant and needs immediate interventions.
2. Reassessing the functional and dysfunctional infrastructures available along the established cold chain facilities through fisheries departments in inland and marine sector.
3. Strengthening the cold chain facilities.
4. Improving the landing center facilities.
5. Strict implementation of cold chain through regulatory bodies at state and central level in domestic market chain.
6. Making the Time-temperature history of catch/harvest available for stakeholders/consumers.
7. Integration of processing technologies for waste utilization.
8. Implementing "waste-must be used" concept in seafood export industries as a societal-responsible component through promotional subsidies.
9. Nationwide awareness campaign on "ice-the fish" through state and central level organizations for strengthening the cold chain in domestic fish market in turn for reducing the post-harvest loss.
10. Ensuring well informed fish market network.
11. Strengthening the availability of long term storage facilities like cold storages.
12. Insisting of adoption of improved processing, storage and distribution chain for traditional fish products like dried fish products.

13. Strict monitoring of local/retail fish markets through food safety regulation bodies.
14. Strengthening basic requisites like potable water and well-connected road facilities wherever is needed.
15. Promoting newer products/processing techniques for minimizing the loss of material.

16.4 Waste Generation in Fish Processing Sector

Waste generation in fish processing sector of India remains as a challenge. Approximately more than 3.0 million tons of total fish produced is wasted. The fish waste include head, scale, skin, fin, bone, visceral mass, shell waste, shuck water, back bone, trimmings etc. Almost 0.5 million tonnes of high quality proteins are wasted during processing in to various forms. Establishing fish waste value chain models specific to various industrial applications could help to recover the nutrients, and biologically important molecules from fishery waste. This strategy would help to build a clean environment, reduction in aquatic pollution, income flow for the stakeholders, improved resource utilization and food and nutritional security. The fish waste utilization should be given national importance and industrial development in this sector should be promoted through policy level interventions for enhancing contribution of aquatic food system for food and nutritional security and nation's economic development.

16.4.1 Factors Influencing the Waste Generation

The amount of waste generated from fish processing operation depends on certain inherent aspects and processing related parameters.

16.4.1.1 Fish Related Parameters
- Species.
- Size/Age group.
- Biological nature (size of head, length of intestine, shorter fins etc.,)
- Body shape (Cylindrical, flat etc).

16.4.1.2 Process Related Parameters
- Style of dressing.
- Style of product.
- Skill of handling person/Machine operator.
- Intended use.
- Quality of raw material.

Obtaining the information on waste generation is quite difficult with reference to above parameters. Hence, generating a data base for the commercially important processed fish is essential and highly useful for any nation for industrial development, and responsible utilization.

16.4.2 Categorization of Fish Waste

Waste generated in fish processing sector can be categorized based on the site, type of aquatic animal, richness in biochemical constituent, complexity of handling and nature of waste. Categorization or classification would help to address the challenges like minimization of waste, management of waste in an environment friendly manner and responsible utilization (Elavarasan and Sathish Kumar 2017b). The details of categorization are presented below.

16.4.2.1 Based on the Site/Point of Waste Generation
- On board waste.
- Industrial waste.
- Landing center waste.
- Retail waste.
- Waste from domestic preparation.

16.4.2.2 Based on the Aquatic Animal
- Fin fish waste.
- Shellfish waste.
- Crustacean waste.
- Cephalopods waste.
- Molluscan waste.

16.4.2.3 Based on the Chemical Constituent
- Waste rich in protein.
- Waste rich in lipid.
- Waste rich in minerals.
- Waste with specialized molecules of interest.

16.4.2.4 Based on the Complexity
- Simple waste (Scale, skin, shrimp shell).
- Complex waste (Head waste, visceral waste).

16.4.2.5 Nature of Waste
- Solid waste.
- Liquid waste (effluents).

16.4.2.6 Solid Waste
Fish processing generates solid wastes that can be as high as 50–80% of the original raw material.

- Dark meat.
- Head.
- Skin.
- Scale.
- Fins.
- Frames.
- Visceral mass (including Air bladder and liver).
- Gills.
- Crab shells.
- Shrimp head and shells.
- Cuttle fish bone.
- Squid pen.
- Ink sac.
- Cuttle fish skin.
- Shells from oyster, mussels and clams.

16.4.2.7 Liquid Waste
- Effluents consist of blood, slime, mucus, wash off.
- Surimi wash water.
- Shrimp cook water.
- Exudate from phosphate treatment/tumbling operation.

16.4.3 Handling of Secondary Raw Material

Considering the importance of secondary raw material generated in seafood processing industry, the hygienic handling of raw material has to be given due importance. Without proper utilization of secondary raw material, sustainability in fish processing sector will be impossible.

The following points may be followed to maintain the quality based on the intended use and the same is represented in Fig. 16.1.

- Proper collection of waste.
- Sorting of waste parts wise and based on quality.
- Washing in chilled water.
- Appropriate pre-treatment.
- Packing in a suitable packaging material.
- Preservation based on the intended use (chilling, freezing, salting and drying or any other chemical treatment).

16.4.4 Issues Associated in Utilization of Processing Waste

The Indian fish processing industries are primarily producing fish and shellfish products. They are often eviscerated and exported in the frozen form. The

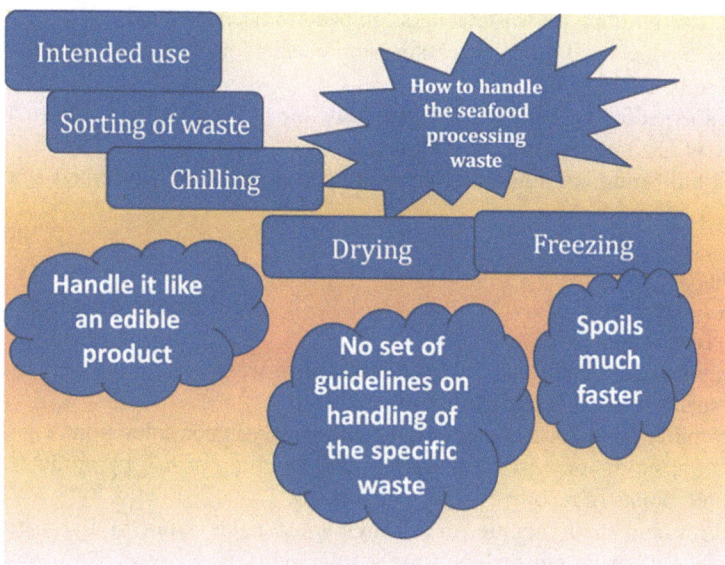

Fig. 16.1 Handling of fish waste (adopted from Elavarasan and Sathish Kumar 2017a, b)

by-products generated during processing are processed into meal, which is utilized as protein ingredient for the production of animal and poultry feed, or converted into manure for agriculture application and more often thrown away as discards. The industries producing high value products from fish processing by-products, such as pure proteins (collagen, gelatin, collagen peptide, protein isolate, hydrolysate), carbohydrate (chitin and chitosan derivatives), PUFA rich oil, enzymes, bio-chemicals, nano-materials, health promoting compounds, active pharmaceutical ingredients etc. are in the infant stage, and these products are possessing domestic consumption as well as export potential. Appropriate indigenous technologies for producing aforementioned value added products are either available or under development. However, following are some important constraints restricting the growth of secondary fish processing industry in India (Elavarasan et al. 2022).

16.4.4.1 Issues, Challenges, Opportunities and Avenues in Promoting Fish Waste Processing Industries for High Value Products

Issues
- Quality issues need to be addressed for secondary raw material with reference to various processing and preservation methods.
- Safety issues need to be identified.
- Guidelines for manufacturing high value products from secondary raw material need to be formulated.
- Feasibility and marketing related studies with reference to end products from secondary raw material need to be carried out.

- Data base building on waste generation need to be carried out.
- Novel products and green processing approaches for processing and technology need to be developed.
- Machineries for establishing simultaneous line in the existing infrastructure for waste utilization need be to developed.
- Multi-utilization and multi end product approaches to be developed among the Industrialists.

Challenges

- Lack of baseline data on availability and quality of secondary raw materials from fish processing activities.
- Highly scattered nature of domestic market fish waste.
- Non-availability of prime quality raw material.
- Availability of seafood processing industries only in coastal regions.
- Huge transportation cost for collection and handling of fish by-products due to multiple points of resources.
- Need for cold chain facilities due to highly perishable nature of raw materials.
- Lack of cold chain hubs which supports the logistics.
- Lack of quality indigenous plants and machineries.
- Requirement of high initial investment and operational cost to establish new processing plants.
- Less demand in domestic market.
- Dependency on developed countries for marketing the product.
- Trade policies and barriers.
- Lack of industries-research institutions participations in research and technology development on secondary fish processing industry.
- Lack of private-public participations in promoting secondary fish processing industry.
- Absence of focused policies to promote domestic marketing of products developed from secondary fish processing.
- Absence of focused policies to promote secondary fish processing and their export.
- Lack of awareness on possible utilization of aquatic food processing waste.

Opportunities

Unlike primary seafood processing, the secondary processing activities are entirely different. This is mainly because the outputs of the secondary fish processing may not directly be used for human consumption and animal feed. The products obtained are mainly used by specialized industries, such as pharmaceuticals, nutraceuticals, nutritional supplements, functional and formulated food and feed industries (including aquaculture), cosmetics, packaging material industries, agro-biochemical (for agriculture uses), etc. Therefore, the Secondary fish processing industry needs holistic approach since from beginning and the activities should be initiated in value-chain mode coupled with complete backward and forward linkages.

Fish Waste Processing May Be Categorized into Three Types

1. **Secondary aquatic raw material processing units in seafood processing areas/retails markets (Feeder units):** Establishing processing units for segregation, processing and preservation of by-products for making it available in the usable/processable forms has the huge economical potential. The unit operations include sorting, washing, pretreatment, chilling/freezing/drying, packaging and storage. These units can be established in small capacities, designed for handling multiple by-products (scale, skin, viscera, bones, belly flaps, fins, shell waste etc) to ensure round the year availability of secondary raw material for high value products processing industries. Such units should be established nearby seafood processing industrial areas and major fish markets. Establishing call center or Ola/Uber type interface communication platform models linked with refrigerated vehicles may be tried for collection, transportation of raw materials from major fish sellers/fish cutters/online fish traders to processing centers.

2. **Secondary fish processing units for industrial products:** The possible industrial products which can be produced from bulk quantity of secondary aquatic raw materials include fish meal, fish silage, hydrolyzed fish protein, bone meal, crustacean meal, whole fish oil, fish feed etc. For producing such products the raw material need not be sorted out into various parts/categories. Unit operations include grinding, proteolysis (autolysis/exogenous enzymes assisted proteolysis), cooking, oil separation, evaporation, drying (steam drying, spray drying etc), packing and storage.

3. **Secondary fish processing units for the production of high value compounds and specialty products/formulated products:** Relatively large capacity extraction units to separate high value components from the biomass/by-products. The raw material for these enterprises can be obtained from Secondary aquatic raw material processing units (Feeder units) and have facility for processing different biomass such as fish scale, fish skin, fish bone, fish head, shrimp shell, etc. The unit operations may include pre-treatments (with acid/alkali/enzymes/any other chemicals), extraction (enzyme/thermal/ultrasound/high pressure/microwave/pulsed electric field/combination), concentration, drying, formulation, packing and storage. These units aim to produce products like chitin, chitosan, chitooligomers, nanochitin/nanochitosan, collagen peptide, bioactive fish muscle peptides (with specific claims), fish bone calcium, PUFA rich fish oil, shrimp oil, pigments, gelatin, fish visceral enzymes, functional feed ingredients, functional agro-ingredients etc.

Main element of success in fish waste processing is flow of raw materials from seafood processing units/retails markets to enterprises units for making use of wastes of various fish and shellfish. For example: Fish scale in the retail markets are collected, washed in water or mild alkali, dried, packed and stored by local individuals. This dried scale is directly supplied to the collagen peptide processors (located within the country and outside the country) through agents, used for extraction of collagen peptide in SFP units, and supplied to specialty products processing SFP. Fish scale collagen peptide has applications as nutraceuticals, functional food ingredient, cosmetic ingredient, and pharmaceutical ingredient (Elavarasan et al. 2022).

16.4.5 Promoting Fish Waste Processing

Followings are some immediate but not limited suggestions for promoting Secondary Fish Processing in India.

1. Conducting nationwide awareness campaigns on fish waste utilization.
2. Building a nationwide database on availability of secondary raw material and quality at various hot spots (Processing industries/local markets). This has to be carried out in collaboration with state fisheries departments. For this the NFDB's fish market price information system (FMPIS) model can be adopted.
3. Identification of potential technologies on fish waste utilization available with ICAR institutes like CIFT for establishing technology demonstration centers at various location.
4. Establishing the technology demonstration units (pilot scale) for handling of secondary aquatic raw material and processing into high value products.
5. Promoting fish waste utilization though separate government supported schemes.
6. Certification programme to the primary fish processing units for responsible fish waste utilization and encouraging waste utilization by providing incentives.
7. Promoting small scale enterprises in fish waste utilization.
8. Customization of technologies available for high value product production from secondary aquatic raw material.
9. Development of linkages between research organization and nutraceuticals/pharmaceutical industries in consortia mode.
10. Attracting MSMEs and industry for manufacturing high value products at designated sites.
11. Encouraging process innovation and product innovation through providing financial support to target oriented focused group projects.
12. Establishment of fish waste processing units involving SHG (men/women)/fisheries societies to own and operate specific industrial products of high values in collaboration with research organizations and private companies having technical support.
13. Formation of dedicated networks and teams of scientists in ICAR research institutes and researchers from state fisheries universities and colleges in a multidisciplinary framework for research in the field of Secondary Fish Processing of unexplored primary fish processing industries' byproducts.

16.5   Innovations in Process/Products Development

Innovative energy efficient eco-friendly processing techniques need to be developed and promoted. In recent years processing technologies like pulsed electric field, NIR assisted drying, microwave drying process, atmospheric freeze drying, cold plasma, high hydrostatic pressure, irradiation, and retort pouch processing has to be explored in India for shelf-life extension and post-harvest loss reduction Kontominas et al. 2021. The area like bio-preservations using extracts from essential oils, algal, plants etc., and using alternative natural preservatives like chitosan yet to be explored in commercial firms. Similarly, the major form of fish consumption in India is in fresh form/chilled and main method of preparation is cooking, frying and baking. The regional recipes can be explored in commercial processing as India's domestic market for ready to eat food products is growing tremendously. The product innovations have to target the healthy fish products liked by children as they are the next generation consumers. Similarly the nutraceutical, and functional food properties of aquatic food system has to be transformed in the form of product in such a way to reach the aged population, pregnant women, lactating mother and school going children. Population targeted fish products has be developed. In another 25 years, the consumer expectation in India will be on varieties in fish products, and convenience. The establishments involved in fish processing should aim in integrating the processing techniques for complete utilization of raw materials available with them. The waste stream should be converted into high value products. Multiple fish products processing strategies would help in a better value realization and also to compete in international market.

16.6   Innovations in Machineries/Tools

Fish processing machineries are much in use in seafood processing establishments mainly produces fish products for export markets. The involvement of machineries in fish processing intended for domestic market is very limited. As the online fish selling platforms are increasing in India, there is a demand for machineries and tools mainly useful in pre-processing operations, and packing. In majority of the cases, machineries used in washing, sorting, descaling, de-skinning, filleting, prawn peeling, fish meal and oil processing, canning and retorting, surimi processing, coated shrimp and crab analogue products production, shrimp cooker, plate freezer, IQFs, Spiral freezers, accelerated freeze dryers, and tumbling machines, are all imported items. Rarely any indigenous efficient alternative machineries and tools are available in India. These machineries are highly capital intensive which prevents the expansion of seafood processing industries, heavily taxed for importing and often demands and depends on subsidiary schemes. In addition, the analytical equipment used in the in-house laboratories of seafood processors for quality management are also imported. Considering the continuous growth in seafood industry as well as demand in fish processing for domestic market and retail distribution chain, innovations in aquatic food processing machineries and tools would be much helpful in transforming the aquatic food system sustainably. The machineries design and efficiency plays a major role in achieving the reduction in fish processing waste generation.

16.7 Innovations in Supply Chain Management

Simplified fish supply chain model in India is as follows:
   Wild/Farming ---- > Harvesting ---- > Landing ---- > Wholesale ---- > Transport ---- > Retail---- > Consumption.
   This model mainly has 7 major stages. At each stage there is a loss of fish in terms of quantity and economy. Such loss is a threat to the food and nutrition security in the near future. The players involved in each stage should be familiarized about the impact of time temperature delay on the final quality. Supply chain management refers to the efficient coordination, utilization and management of the components of such a chain for maximizing value or/and minimizing the costs. There should be proper linkage in terms of information, services, and communication. A well informed value chain/supply chain from producer to consumer can guarantee a reduction in post-harvest loss and fish waste. The developments happened in the information technology especially the revolution in mobile phone technology should be better exploited for an efficient supply chain management.
   Innovations in supply chain management should address the following aspects.

- Strengthening the distribution networks.
- Efficient procurement of high quality raw material, production, packaging and distribution of safe fish products.
- Logistics decisions.
- Making well informed decisions in advance throughout the supply chain.
- Regular business based on the demand in the local market as well as export market.
- Maintaining efficiency in Cash flows.

16.8 Innovations in Seafood Logistics

Cold chain logistics refers to that the products are transferred from origin to destination under temperature-controlled conditions including refrigerated production, distribution, and storage activities. Seafood logistics involves a complex cold chain especially in international trade. The break in cold chain during logistics cause irreparable quality loss which leads to rejection of containers. Therefore, it is necessary to develop reliable techniques which can be applied in cold chain logistics for quality control and safety supervision. Non-destructive evaluation technologies in aquatic food quality assessment are coming up in a great way. Non-destructive technologies (gas-sensitive freshness indicator and Radio Frequency Identification (RFID), help in cold chain trackability and assessment of real time changes in quality and safety during shipment. Innovations in seafood logistics minimize loss of quality, ensure consumer safety, reduce the tremendous waste of resources and related ethical implications for distributors, sellers, and consumers.

16.9 Internet of Things (IOT) in Fish Processing

Providing valuable and timely information to the stakeholders of value chain to trace, track and verify the authentic information related to the fish product from harvest to retail end would greatly minimize quality and safety related issues which are ultimately leading to loss and waste in fish supply chain. Advanced technologies like internet of things, Block Chain technology, Machine learning, and Distributed ledger technology, will transform the traditional fish value chain into smart -fish value chain. India is one of the fastest growing economy in the world which implies that tomorrow's consumer would be very much advanced and aware of such technologies and obviously todays innovations with emphasis on specific local needs is very much essential. Not only the seafood safety, IoT aid in comprehensively assessing the parameters related to sustainable development like energy, water consumption, and carbon foot printing in the seafood industries.

16.10 Innovation in Machine Vision and AI in Fish Processing

Mapping the business processes, understanding the system-level interactions with the governing material, understanding the data and the information flows could facilitate the implementation of AI in the digitalization of fish supply chain. The impact on data monetisation related to sustainability depends on the parameters and objectives set by stakeholders involved in supply chain. For enhancing the value delivery in fish value chain, building AI enabled food supply chains is of essential in future. Innovation in AI enabled practices in fish value chain of India should be given priority especially in fish processing industries to understand the stand of our industries in sustainability map.

Innovations in AI assisted machineries design and automation in various fish processing operations would minimize the amount of fish waste generated due to processing activities such as cutting, and filleting. With the advancements in imaging and computing technologies, machine vision and AI have been applied to the fish industry for improved precision and automation levels of fish cutting. The accuracy in blade cutting has been improved through involving AI methods especially in automated process lines. They could be further implemented in other operations like fish beheading, trimming, and portioning operations. The question in front of us which bothers is how good the machine vision and AI-driven fish cutting automated machines will work in commercial fish processing conditions particularly in countries like India. Machine vision and AI has the potential application in the whole fish processing industries. Hence, it is worth investigating and can be potential.

16.11 Need for New Generation Human Resource Development in Fish Processing Sector

Currently the human resource in fish processing sector and in fish value chain mainly comprise of skilled labour force, trained technologists and machine operators, engineers, food technologists, business managers etc. In domestic market, the major human resource is again skilled labour force especially in fish dressing, cutting and portioning operations. In order to maintain the better quality of fresh fish, the baseline human resource should be well equipped and trained with recent advancements in machineries usage, cold chain management, quality and safety management. For the seafood processors involved in export industry, the human resource development should take the smart value chain techniques into account. The capacity building programs should focus on developing human resources in line with the advancements happenings in machineries development, employing artificial intelligence in value chain, block chain technology, machine vision, big data analysis, machine learning and IoT, and imaging technology for transforming the aquatic food system in an efficient way for ensuring the food and nutritional security of future generation.

16.12 Challenges and Opportunities

Transforming the aquatic food system through interventions and innovations in fish value chain specifically in processing technologies for reducing the wastage and loss is highly challenging in countries like India. However, it comes with the plenty of opportunities. In India, we find the extreme situations like processing of fish using automated line as well as processing fish by involving a huge number of labour forces. Introducing any advanced techniques in commercial fish processing industries should not affect the livelihood of a huge labour force mainly the women. Minimizing the gap in understanding of the fish value chain by the stakeholders and consumers would be critical and challenging in achieving the envisaged transformation. The innovations should address the gap between the science and economy for putting them into practice especially in a larger scale. Next generation scientific man power in the sector should be well equipped with the advancements taking in the allied sector. Research and educational institutions should cope-up with the technological developments and their curriculum and research funding should focus on holistic approach while being more innovative. There should be a balance and the developments should be inclusive in nature. The transformation in aquatic food system (marine fisheries, aquaculture) is inevitable and beneficial. The sustainable development strategies in the fisheries sector should not leave anyone behind or unaddressed. The transformation should aim in inclusive sustainable development in aquatic food system.

16.13 Conclusion

Transformation in aquatic food systems would help to support the need of ever growing population for a healthy and sustainable diet. Millions of people associated with the fish processing sector will be supported, socially and economically. Ensuring the reduction in fish loss and waste would improve the contribution of aquatic food system to the food and nutritional security as well as socio-economic-livelihood status of stakeholders. In the near future, as a responsible consumer, he/she may demand that as soon as the fish is purchased a message showcasing the traceability in terms of sustainability and safety has to be delivered to him. The responsibility of all the stakeholders should be realized in order to transform the aquatic food system in to more sustainable and transparent.

References

Elavarasan K, Sathish Kumar K (2017a) Waste generation profile in industrial fish and shellfish processing. In: Elavarasan K, Kumar A, Kumar KS, Rejula K (eds) Protocols for the production of high value secondary products from fish and shell fish processing. Central institute of Fisheries Technology, Kochi, pp 25–31

Elavarasan K, Stahish Kumar K (2017b) Nature and composition of fish processing waste; an introduction. In: Elavarasan K, Kumar A, Kumar KS, Rejula K (eds) Protocols for the production of high value secondary products from fish and shell fish processing. Central Institute of Fisheries Technology, Kochi, pp 32–35

Elavarasan K, Karthikeyan M, Ravishankar CN (2022) Promoting secondary fish processing: challenges, avenues and a way forward. MPEDA News Lett 10(9):42–26

FAO (2022a) Blue transformation—roadmap 2022–2030: a vision for FAO's work on aquatic food systems. FAO, Rome. https://doi.org/10.4060/cc0459en

FAO (2022b) The state of world fisheries and aquaculture 2022. Towards blue transformation. FAO, Rome. https://doi.org/10.4060/cc0461en

Gephart JA, Rovenskaya E, Dieckmann U, Pace ML, Brännström Å (2016) Vulnerability to shocks in the global seafood trade network. Environ Res Lett 11(3):035008

Handbook on Fisheries statistics (2022) Department of fisheries, ministry of fisheries, animal husbandry and dairying. Government of India, New Delhi

Jyotishi A (2023) Fish for food or fish for feed: new populism and blue economy perspective. In: Deshpande M et al (eds) Encyclopedia of new populism and responses in the 21st century, pp 1–6. https://doi.org/10.1007/978-981-16-9859-0_12-1

Kontominas MG, Badeka AV, Kosma IS, Nathanailides CI (2021) Innovative seafood preservation technologies: recent developments. Animals 11(1):92

Liu Y, Smith M, Dietz D, Leaning DC, Abbott J, Smyth A (2023) The seafood trade and nutritional access. Res Square 1:1. https://doi.org/10.21203/rs.3.rs-2675210/v1

Siddhnath RA, Mohanty BP, Saklani P, Dora KC, Chowdhury S (2022) Dry fish and its contribution towards food and nutritional security. Food Rev Intl 38(4):508–536

Watson RA, Nichols R, Lam VW, Sumaila UR (2017) Global seafood trade flows and developing economies: insights from linking trade and production. Mar Policy 82:41–49

Meeting Emerging Challenges in Aquatic Animal Health

17

Neeraj Sood, Pravata Kumar Pradhan, Anutosh Paria, Chandra Bhushan Kumar, Ravindra, and Uttam Kumar Sarkar

Abstract

Several technological innovations in the field of aquatic animal health in recent years hold immense potential to address the emerging challenges of infectious diseases. At the core of advancements is the One Health perspective which takes into account the interconnectedness between human, animal and environment health. Implementation of aquaculture biosecurity including use of specific pathogen-free, specific pathogen-resistant and specific pathogen-tolerant stocks as well as vaccines can pave way for preventing development of antimicrobial resistance in aquaculture. Studies on gut microbiome can help in increasing our understanding about host-pathogen interaction. Advanced diagnostic techniques such as quantitative PCR, digital PCR, isothermal amplification and CRISPR-mediated detection have helped in increasing the sensitivity and specificity of detection of pathogens, whereas use of rapid diagnostic tools like lateral flow kits can help in providing point of care diagnosis even by the untrained aquaculturists. The application of metagenomics and environmental DNA can help to identify novel pathogens and also early detection of pathogens. These techniques combined with environmental data will further help in predicting and preventing disease outbreaks. The new-age technology like artificial intelligence will add new dimension to the disease diagnosis and prediction modelling in aquaculture. Herein, we discuss some of these technological advancements and their role in minimizing the disease risks in aquaculture thereby ensuring sustainable aquaculture production and meeting the global food security.

N. Sood (✉) · P. K. Pradhan · A. Paria · C. B. Kumar · Ravindra · U. K. Sarkar
ICAR-National Bureau of Fish Genetic Resources, Lucknow, India

© National Academy of Agricultural Sciences, under exclusive license to Springer Nature Singapore Pte Ltd. 2023
K. C. Bansal et al. (eds.), *Transformation of Agri-Food Systems*,
https://doi.org/10.1007/978-981-99-8014-7_17

Keywords

One health · PCR · Antimicrobial resistance · Biosecurity · Lateral flow kits · CRISPR · Metagenomics · eDNA · Microbiome · Artificial intelligence

17.1 Introduction

Aquaculture is considered as one of the important sectors to meet the increasing demand for animal protein of the burgeoning population. However, diseases pose a significant threat to the growth of aquaculture, resulting in huge economic losses and serious concerns about sustainability of the sector and food security. In recent years, several technological advancements in the field of aquatic animal health provide innovative solutions to reduce the disease risks significantly, and also improve aquatic animal health. These innovative solutions include one health approach, the use of health/genetic principles to develop specific pathogen-free (SPF), specific pathogen-tolerant (SPT) and specific pathogen-resistant (SPR) stocks of commercially-important aquaculture species, advanced diagnostic techniques, namely polymerase chain reaction, isothermal amplification, lateral flow kits that enable quick detection of pathogens. Besides, metagenomics, environmental DNA and artificial intelligence-based techniques would play a key role in aquatic animal health management and sustainability of the aquaculture sector.

17.2 One Health

One Health is a holistic approach to address health issues that recognizes the interrelationship between human, aquatic animal and the environmental health. For addressing the issues, there is a need of multidisciplinary approach involving experts from human, veterinary, aquatic animal health sector and also from environmental science. Some of the important points about One Health and its connection to aquaculture include:

- Disease management: Diseases in aquaculture affect aquatic animals and the environment, leading to reduced farmers' income. One Health approach can help to manage and prevent the diseases, considering both the farmed aquatic animals and their environment.
- Preventing misuse of antimicrobials: Antimicrobials are often used in aquaculture to treat or prevent diseases. But misuse of the antimicrobials can lead to development of resistance to antimicrobials and raises human health concerns when farmed aquatic animals carrying these resistant microorganisms are either ingested or resistant aquatic microorganisms transfer their genetic element to human pathogenic microorganisms. Therefore, the One Health approach encourages judicious use of antimicrobials in aquaculture.

- Reducing adverse environmental impacts: Aquaculture, if not practised properly, can pollute water and affect the surrounding environment. Therefore, One Health approach supports sustainable aquaculture practices which are beneficial to the environment, aquatic animals and the people who depend on it.
- Zoonotic Diseases: Some diseases in aquatic animals can be transmitted to humans through contact with aquatic animals or by eating contaminated aquatic animals. One Health approach can reduce such risks.
- Research and Collaboration: One Health encourages experts from different disciplines for working together to study and find solutions to the health problems in aquaculture. This teamwork can lead to new ways of managing diseases, ensuring sustainability, and producing safe foods.

One Health recognizes how human, animal, and environmental health are interlinked, and promotes collaboration to solve health-related issues in aquaculture. It involves many professionals from different sectors and considers social and economic factors, not concentrating only the economic benefits. This approach is increasingly supported by international organizations like the Food and Agriculture Organization of the United Nations (FAO), the World Organisation for Animal Health (WOAH), World Health Organisation (WHO) etc. (Rabinowitz et al. 2013).

17.3 Antimicrobial Resistance in Aquaculture

Aquaculture is one of the fastest growing food sectors globally. It is unique due to diverse species cultured in different agro-climatic regions. This growth has led to challenges, including increased stocking density, water quality issues, and disease outbreaks, often treated with antimicrobials. Antimicrobials used for humans and livestock are also commonly used in aquaculture, although their efficacy is not understood properly. These are generally administered with feed and applied at the population level. However, fish do not metabolize antmicrobials efficiently, and monitoring feed intake by animals is quite difficult in the aquatic environment. Therefore, a large proportion can remain in the environment as uneaten and undigested feed, and also as antimicrobial metabolites. These antimicrobials can subsequently interact with an aquatic microbiome that harbours a large variety of mobile genetic elements where significant genetic exchange and recombination can spread the antimicrobial resistance (AMR). This can promote the emergence of antibiotic-resistant bacteria in aquatic ecosystems (Rathore et al. 2020). There is a need of research and capacity-building efforts to address this emerging issue, with focus on biosecurity, diagnostics, education, vaccines, alternative treatments, and legislation. Regulations regarding antimicrobial use in aquaculture vary globally. Furthermore, global AMR surveillance systems are fragmented and primarily focused on humans, making it challenging to assess the risk of AMR emergence in aquaculture.

AMR surveillance primarily tracks genetic resistance determinants in microorganisms. Culture-based methods and molecular techniques are used to identify resistance pattern, which can emerge through mutation or horizontal gene transfer mechanisms. The data on the incidence and distribution of AMR in aquaculture is limited despite the risks associated with aquaculture expansion. FAO has prepared an action plan on Antimicrobial Resistance 2021–2025 which would serve as a roadmap for focusing global efforts to address the issue of AMR in the food and agriculture sectors (FAO 2021).

17.4 Aquaculture Biosecurity

Aquaculture biosecurity refers to a set of measures and practices designed to prevent the introduction, establishment and spread of pathogens in aquaculture facilities. These practices minimize the stress on aquatic animals, thus making them less susceptible to diseases. Biosecurity is important for maintaining the aquatic animal health and improving aquaculture productivity and minimizing economic losses due to diseases for sustainable aquaculture growth. Key aspects include selecting healthy seed stock, emergency preparedness, diagnostics, microbial control, disease surveillance, trade regulations, policy frameworks, animal welfare, research, addressing antimicrobial resistance, unconventional pathogen transmission, and implementing the Progressive Management Pathway (PMP) (Subasinghe et al. 2023).

In the context of PMP for improving aquaculture biosecurity (PMP/AB), it refers to the cost-effective management of risks posed by infectious agents to aquaculture through a strategic approach at enterprise, national and international levels with shared public-private responsibilities (FAO 2020). It comprises four stages: risk assessment, initiating biosecurity systems, enhancing preparedness, and establishing sustainable biosecurity and health management systems. The PMP/AB enhances aquaculture biosecurity by building on existing resources and public-private partnerships. The goals of PMP/AB include reducing disease burdens, improving health standards, limiting global disease spread, maximizing economic benefits, attracting investments, and promoting One Health principles. The proactive and preventive biosecurity measures are cost-effective, while reactive responses to outbreaks are more expensive. Biosecurity should be integrated from the outset of aquaculture development, and reducing response time after an outbreak is a crucial step for effective biosecurity.

In shrimp farming, to address the threat of diseases, a biosecurity strategy is essential for each farm, system, and sanitary area. It involves measures to prevent, control, and manage health risks in shrimp with the aim of minimizing economic losses due to diseases. When choosing shrimp for farming and designing a biosecurity plan, it's vital to consider their health status and genetic traits (Alday-Sanz et al. 2020). Shrimp stocks can be categorized as given below;

1. **Specific Pathogen-Free** (SPF) shrimp stocks are created through rigorous screening and selection processes to ensure they are free from known shrimp diseases. These disease-free shrimps are used in breeding programs to improve genetics and produce post larvae for commercial hatcheries. However, SPF status only indicates the sanitary health of the shrimp stock and doesn't imply disease

resistance. SPF shrimp may not be free from all pathogens; they must be free from specific pathogens listed at the time of certification, which may change as new pathogens are discovered. Two main approaches to develop SPF shrimp stocks exist: capturing and screening wild shrimp in pathogen-rare areas (natural SPF stocks) or selecting shrimp from disease-prevalent areas and continuous screening (cleansed SPF stocks). Natural SPF stocks may lack genetic resistance, while cleansed SPF stocks may have higher disease resistance.

2. **Specific Pathogen-Resistant** (SPR) stocks are primarily defined by their genetic traits rather than their current health status. The SPR stocks focus on the inherent genetic resistance of the animals, allowing them to remain healthy even when exposed to lethal doses of these pathogens. In essence, SPR highlights the genetic makeup of shrimp that enables them to withstand particular pathogens without falling ill. An important distinction to make is that being labelled as SPR for one pathogen does not guarantee resistance to all diseases. Shrimp stocks labelled as SPR for a specific pathogen can still be vulnerable to other infectious agents. This means that while they may exhibit resistance to one or more known pathogens, they might be susceptible to other diseases. The development of SPR stocks represents a significant advancement in shrimp breeding. It has been possible to identify and selectively breed shrimp with inherent resistance to specific pathogens, thereby creating populations of animals that can thrive even in the presence of these disease-causing agents. Notably, there is a population of *Penaeus monodon* known as the WSSV SPR, which has been established through the collection and breeding of naturally-resistant mutants from wild populations (Alday-Sanz et al. 2020).

3. **Specific Pathogen Tolerant (SPT)** stocks is a recent concept in shrimp breeding that implies that they can tolerate infection by specific pathogens without displaying obvious disease symptoms. Their tolerance is influenced by genetics, pathogen strains, and environmental factors. SPT can be specific to certain pathogens or strains but doesn't guarantee protection against other diseases. Non-SPF SPT stocks may carry pathogens without showing symptoms or transmit them to other shrimps. SPT, like SPR stocks, focuses on genetic traits rather than overall health status. Combining health-based SPF designation with genetic traits is possible, creating SPF + SPR, SPF + SPT, or SPF + SPT + SPR stocks. However, these combined statuses must meet criteria for both health and genetic traits.

4. **Uncharacterized, selected survivor (USS)** stocks represent a new term in the world of animal stocks. These stocks are created by selecting survivors from multiple generations of shrimp breeding, primarily based on their size and overall health appearance. These selections occur in non-biosecure farming conditions, typically in regions where various known and unknown pathogens exist. In the past, these stocks have been referred to as 'all pathogen exposed' (APE) stocks. However, it's important to note that not all pathogens are present in every geographical region, and even within a specific region, not all pathogens are found in every pond. Consequently, it's practically impossible to develop a stock that is exposed to 'all' known and unknown pathogens. Therefore, the term 'APE

stock' is not scientifically valid and renamed as USS stock. It's crucial to understand that while USS stocks may outwardly appear healthy, they can be carriers of pathogens that have the potential to spread horizontally or vertically. This transmission can occur unless these stocks have undergone testing for specific pathogens and received negative results for at least two consecutive years, effectively converting them to a 'cleansed SPF status'. It's evident that USS shrimp stocks pose a higher risk compared to SPF shrimp stocks when it comes to transboundary movement.

17.5   Vaccines

The annual losses due to infectious diseases in aquaculture exceed 10 billion US $ globally, thereby posing a significant threat to the growth of aquaculture (Micuchova et al. 2022). In last few decades, vaccines have emerged as a highly promising solution to combat infectious diseases in aquaculture. The shift towards vaccines is also crucial as it avoids the use of chemotherapeutics like antibiotics, which can have detrimental effects due to residues in cultured products and increased bacterial resistance. Moreover, vaccination aligns with the holistic One Health approach, benefiting human, animal, and environmental health. Although most of the vaccines elicit humoral immunity but some vaccines can induce cell-mediated immunity through cytotoxic T-lymphocytes as well, thereby providing protection against viral diseases.

Currently, there are three main vaccination strategies for fish: injection, immersion, and oral vaccination. Injection vaccines provide the highest protection but are logistically demanding, costly, and labour-intensive. These can also lead to stress and unwanted side effects. Immersion vaccines are less effective but are more suitable for mass vaccination. However, these are often used as a booster for injectable vaccines. Oral vaccination is the most convenient and environmental-friendly option but faces challenges in antigen uptake and mucosal tolerance. Protecting orally-delivered antigens from the harsh gastrointestinal environment is very crucial. Importantly, understanding fish immune responses and optimizing oral vaccine delivery methods are essential for advancing aquaculture disease control. Development of effective oral vaccines has huge potential by offering a convenient and sustainable solution for immunizing fish against infectious diseases. The improvements in antigen delivery to the gut and the inclusion of molecular adjuvants can enhance efficacy of vaccines. Various encapsulation methods, including biopolymers, bacterial cultures, and exploiting fish appetite for crustaceans, have shown promise in this regard (Micuchova et al. 2022).

Biopolymers like alginate, chitosan, and Poly Lactic-co-Glycolic Acid (PLGA) are commonly used to encapsulate antigens in oral vaccines. Alginate comes from brown algae, chitosan from crustaceans and fungi, and PLGA is synthetic. These biopolymers protect antigens from digestion and effectively deliver them to the intestine. Moreover, biopolymers are biodegradable and FDA-approved for human drug delivery. Besides, they have mucoadhesive properties, ensuring controlled antigen release in the gastrointestinal tract and mimicking the prime-boost vaccination approach. This makes biopolymers promising for improving efficacy of oral vaccines. The fish orally vaccinated with biopolymer encapsulated antigens exhibit improved immunity and enhanced immune responses.

Bacterial expression systems, particularly non-pathogenic probiotic bacteria like *Lactococcus* are used to encapsulate subunit vaccines. These systems allow for precise localization and high-level production of antigens, leading to specific immune responses and fish protection. Additionally, biofilms formed by some pathogenic bacteria can enhance vaccine effectiveness, as demonstrated in Asian seabass experiments with *Vibrio anguillarum* (Ram et al. 2019).

Crustaceans like *Artemia*, *Moina*, and *Daphnia* often used as fish food for young fish, provide another option for vaccine delivery. They can naturally consume bacteria, yeasts, or microalgae, making them carriers for nutritional or antibacterial agents. *Artemia*, in particular, is effective in delivering specific antigens. For instance, feeding of orange-spotted grouper with *Artemia*-encapsulated inactivated recombinant bacteria expressing envelope proteins from nervous necrosis virus resulted in increased survival and antibody levels (Chen et al. 2011). A recent study used *Artemia* to deliver an oral vaccine with recombinant yeast expressing envelope proteins from cyprinid herpes virus 3 to common carp larvae, leading to high antibody levels and improved immune-related gene expression (Ma et al. 2020).

17.5.1 Molecular Farming of Fish Vaccines

Molecular farming involves using plants to produce various molecules, including proteins, for pharmaceutical, veterinary, or agricultural applications. Plant-based production offers advantages in terms of yield, cost, and proper protein folding. The process of creating transgenic plants involves designing a transformation construct, selecting suitable promoters, and integrating the antigen-coding gene into the plant genome. Various methods like *Agrobacterium*-mediated transformation, biolistic methods, and direct delivery into protoplasts can be used to introduce these genes into plants. Plants' natural encapsulation properties due to composition of their cell walls can protect recombinant antigens from digestion in the gastrointestinal tract. Once these reach the intestine, cellulolytic bacteria release the intracellular content of plant cells allowing for antigen absorption. Besides, the mucoadhesive properties of plant cell wall components enhance antigen attachment to the mucin layer and prolong its retention in the intestine. Plants can also positively affect fish health and immune competence, indirectly improving vaccination efficacy. In addition to terrestrial plants, the aquatic plants and microalgae, like duckweed, offer promising avenues for molecular farming due to their nutritional content and encapsulation properties (Sirirustananun 2018). Microalgae in particular, have a favourable composition of essential nutrients and a rigid cell wall for encapsulation (Sarker et al. 2020).

17.6 Importance of Gut Microbiome

Microbiome, the microorganisms living in an organism's exposed surfaces like the gastrointestinal tract, plays a vital role in nutrient absorption, immune system stimulation, and disease prevention. These live microorganisms offer benefits to the host and are a valuable source of potential probiotics, which are widely used in aquaculture to stimulate the innate immune response and protect against pathogens. Probiotics for aquaculture can originate from mammals (allobiotic) or fish's own gut microbiota (autochthonous probiotics). Autochthonous probiotics adapt well to water conditions and compete effectively with gut organisms. The probiotics reduce the use of antibiotics, improve appetite and growth, and have been demonstrated to have antagonistic activity against aquatic animal pathogens. Moreover, use of probiotics also aligns well with the One Health approach.

Recently, the importance of the gut microbiome in the development and function of the intestinal mucosa, disease protection, and digestive processes has been emphasized. In addition to probiotics, the use of prebiotics, synbiotics, and biofloc systems can modulate the gut microbes of aquatic animals thereby controlling the diseases and promoting healthy growth. However, there is a need to understand the mechanisms behind improving immune status and fish quality through the use of prebiotics, synbiotics, and biofloc technology. Importantly, a more diverse microbial community in the gut has higher probability of preventing the infection with a pathogen as evidenced by higher species diversity in gut of healthy fish compared with diseased fish (Diwan et al. 2023).

17.7 Technological Advancements in Diagnostics

17.7.1 Polymerase Chain Reaction

17.7.1.1 Conventional Polymerase Chain Reaction

Polymerase chain reaction (PCR) is one of the most widely used basic molecular biology techniques by which a target region of a DNA can be specifically amplified from a small amount of DNA template in a controlled thermal cycling conditions (denaturation of DNA, annealing of primers and synthesis of new DNA strands) using a thermostable polymerase, deoxynucleotide triphosphate (dNTP) bases and small synthetic oligonucleotide sequences known as primers. Typically, a PCR thermal cycle ranges from 30 to 35 cycles and after the completion of the PCR reaction, a large number of DNA copies identical to the target region of the template DNA is produced. However, when the amount of initial target DNA is very less, single-step PCR cannot amplify sufficient amount of target DNA copies, hence cannot be detected during post-PCR processing. Therefore, a second round of PCR known as nested PCR is generally undertaken using a new set of primers internal to the PCR product generated in the first-step. Nested PCR has increased sensitivity and specificity compared to the single-step PCR reaction and is very useful in detection of low level of infection or latent infection.

For performing a conventional PCR, DNA or RNA from the tissues of animals need to be isolated based on the target pathogen or the purpose of use. If the starting material for the PCR is RNA, then it must be reverse-transcribed to complimentary DNA (cDNA) before conducting the PCR as Taq polymerase cannot synthesis new DNA strands from RNA template. The type of PCR in which RNA is used as starting material, known as reverse transcription-polymerase chain reaction (RT-PCR).

PCR has revolutionized the research and diagnostic field due to its high sensitivity, specificity, and rapidity. Specifically, the method is very useful in disease diagnosis in both apparent and inapparent cases, understanding genetic diversity among different pathogenic isolates, and forensics. Although, the method is comparatively cheaper than many other disease diagnostic tools, it requires trained manpower and well-equipped laboratory facility for generating efficient results. At present, both conventional and nested PCR methods are available for most of the World Organization for Animal Health (WOAH)-listed and other economically-important fish/shellfish pathogens.

17.7.1.2 Quantitative PCR

Quantitative PCR (qPCR) or real-time PCR is a widely used molecular diagnostics technique for measuring the level of infection in real-time in terms of quantitative load of microorganisms. The method eliminates the post-PCR procedures such as gel electrophoresis, thereby reducing the processing time and cross-contamination. Further, the limitation of end-point detection in conventional PCR is also overcome in qPCR as the detection or quantification happens in log phase of amplification and denoted as cycle threshold (Ct) value. The basic of the methodology behind qPCR is detection and measurement of target nucleic acid depending on the principle of fluorescence resonance energy transfer (FRET). There are many fluorescent-based chemistries available for qPCR, however the two most commonly used chemistries are sequence-specific fluorescent-labelled oligonucleotide probes (e.g., TaqMan) and sequence-independent double-stranded DNA intercalating dyes such as SYBR Green. The presence of initial copies of target nucleic acid corresponds to the Ct value, which in turn is inversely proportional to the initial target DNA. This method of molecular detection has proven to be one of the most important techniques in routine diagnosis of many human and animal pathogens as its highly sensitive, rapid, and can even detect a single copy of target DNA in a reaction under a wide dynamic range (8–9 log10 scale). Moreover, qPCR can be employed to understand disease progression dynamics on temporal scale and detection of the infection in apparently healthy individuals. As, development and application of therapeutics is extremely difficult in managing fish/shellfish disease outbreaks especially those due to shrimp viruses, this technique holds enormous promise in early detection of the pathogen before onset of the outbreak and screening of brooders or seeds for identifying the carriers of the targeted pathogens, thereby taking necessary management measures for prevention of the aquatic animal diseases and their spread in the culture system. qPCR can be also efficiently used in screening multiple shrimp/fish pathogens using probes labelled with different fluorescent dyes in a single reaction.

17.7.1.3 Digital PCR

Digital PCR (dPCR) is one of the new entries in the list of molecular detection techniques and can be employed for determining the exact number of the target nucleic acid (Hou et al. 2023). Hence, this particular method is very important in specific detection of rare target DNA. dPCR also based on the principle of FRET similar to qPCR, but in dPCR, the reaction happens in a water oil emulsion droplet or in nano wells on a microfluidics chip. When dPCR performed in a droplet-based module, it is termed as droplet dPCR (ddPCR), and it's one of the most commonly used modes of digital PCR presently. In dPCR, every partition or droplet acts as a single micro-PCR reaction chamber, and proportion of the positive fluorescence signal from each micro chamber to the complete set of micro chambers follows Poisson statistical distribution. As only positive or negative result can be obtained from each micro reaction chamber, hence the sample should be properly diluted to obtain a dilution in which each micro partition should contain either one or no target DNA molecule. Although the dPCR works on a narrow dynamic range compared to qPCR, but the most important advantage over qPCR is that dPCR doesn't require any standard of the target DNA for quantification. Hence, the absolute quantification in dPCR is truly absolute or digital. dPCR can be used for accurate detection of pathogens in complicated samples, early detection of diseases, detection of mutations responsible for genetic diseases etc. However, dPCR or ddPCR is comparatively new in modern diagnostic era especially for aquatic fish pathogens, and despite having a high sensitivity and specificity, the technique is yet to gain popularity in the aquaculture sector as it requires sophisticated equipments to perform the test. At present, dPCR has been standardized for a few pathogens infecting aquatic animals (Netzer et al. 2021) which includes Infectious spleen kidney necrosis virus (ISKNV), and *Streptococcus agalactiae*, where it showed better and accurate quantification ability compared to existing qPCR for the respective pathogens (Lin et al. 2020).

17.7.2 Isothermal Amplification

Isothermal amplifications are the methods for amplifying target nucleic acid at a constant temperature using strand displacing polymerases. Owing to the rapid turnaround time and working principle based on a single incubation temperature, these methods are very popular for field-level diagnosis. Some of the common isothermal amplification methods are loop-mediated isothermal amplification (LAMP), recombinase polymerase amplification (RPA), and nucleic acid sequence-based amplification (NASBA). Apart from these, other isothermal methods such as helicase-dependent amplification (HDA), rolling circle amplification (RCA) and cross-priming amplification (CPA) are being experimented for diagnosis of different pathogens infecting humans and animals.

17.7.2.1 Loop-Mediated Isothermal Amplification

Loop-mediated isothermal amplification (LAMP) is probably the most widely used isothermal amplification. The LAMP generally uses a polymerase from *Bacillus stearothermophilus* (Bst) and 4–6 primers targeting different regions of the target nucleic acid (Becherer et al. 2020). Compared to conventional PCR, LAMP requires incubation at 60–65 °C for about one hour in a simple water bath. LAMP method has been standardized for many fish and shellfish pathogens and is successfully being used in rapid detection of the target pathogens including WSSV, EHP, ISKNV, TiLV, *Flavobacterium columnare*, *S. agalactiae*, most of the viruses infecting salmonids etc. Similar to conventional PCR, the LAMP products can be visualized in a common gel electrophoresis method, but for on-site detection of the aquatic animal pathogens, dyes such as SYBR Green, calcein and hydroxynapthol blue are incorporated in the reaction mixture for easy visualization. However, most of the LAMP methods can only generate a positive or negative reaction and size-based discrimination of the amplified product is hardly possible in LAMP as compared to PCR.

17.7.2.2 Recombinase Polymerase Amplification

Recombinase polymerase amplification (RPA) is another isothermal amplification method gaining popularity due to its several advantages over LAMP. It has more rapid turnaround time than LAMP and can be completed within 20 min at a relatively lower incubation temperature i.e. 37–42 °C. Hence it is very suitable for field-level detection of target pathogens. A typical RPA method requires two enzymes namely recombinant enzyme T4 UvsX and *Bacillus subtilis* Pol I (Piepenburg et al. 2006) and two primers specific to the target nucleic acid. The recombinase enzyme unwinds the DNA double strand and DNA polymerase synthesizes the new DNA strands. In the RPA reaction, single-stranded DNA-binding proteins are also incorporated which prevent the unwinded DNA strands from reannealing. The RPA method can be coupled with many end-point detection techniques such as agarose gel electrophoresis and lateral flow or detection of positive signals in real-time using a fluorescent dye. Presently, this particular technique is standardized for a few pathogens infecting finfish and shellfish, however it is one of the tools which holds enormous potential in rapid and on-site detection of the aquatic animal pathogens.

17.7.3 CRISPR-Mediated Detection

One of the major drawbacks in isothermal amplifications is non-specific amplification as the reaction temperature is low. Integrating CRISPR with isothermal amplifications increases both specificity and sensitivity of any isothermal amplification method. Among the several CRISPR proteins, Cas12a or Cas13a are ideal for detecting the target nucleic acid. However, the most widely used Cas protein i.e. Cas9 cannot be used for this purpose as it lacks the nonspecific secondary cleavage activity. Following recognition of the target molecule, Cas proteins can digest the strands of nucleic acid leading to emission of fluorescence signal in case of positive reaction. The fluorescent signal can be detected either with naked eye through a colorimetric reaction or can be coupled with a lateral flow dipstick. Although CRISPR can be integrated in any isothermal amplification, but the RPA is most suitable as both Cas proteins and RPA works at similar temperature (37 °C) (Kanitchinda et al. 2020). Generally, Cas proteins are extremely specific in recognizing the target DNA and CRISPR assay is designed to target a specific region of the amplicon generated by isothermal amplification, thereby increasing the specificity of the reaction. Due to this advantage, CRISPR-mediated assays may be an easier option in identifying different genotypes and geographical isolates of a particular pathogens. There are very few CRISPR mediated detection assays for aquatic animal pathogens, however in years to come, this technique may be extensively used for finfish and shellfish pathogens as well in line with human pathogens.

17.7.4 Lateral Flow Kits

Disease diagnosis in aquaculture has traditionally relied on conventional and advanced molecular techniques. However, these methods have limitations due to need for laboratory, specialized equipment, trained personnel, and are time-consuming. Meanwhile, diseases can rapidly spread among fish farms, and pathogens can also move through the water. Therefore, swift and accurate disease diagnosis is critical to prevent further transmission. The lateral flow assays (LFAs) offer a simple, rapid, cost-effective, and portable point-of-care detection method for the pathogens. LFAs have been extensively employed in human medicine and hold potential for detection of pathogens in aquaculture. In LFAs, reactant molecules move through a series of capillary membranes within a device, driven by capillary forces. These molecules bind to specific bio-recognition elements strategically positioned along the capillary bed, producing visible signals. LFAs employing antibody as the bio-recognition element are categorized as LF immunoassays (LFIAs). In LFIA reactions, two antibodies, primary and secondary, are typically employed together. The primary antibody attaches to the specific antigen on the analyte, and the secondary antibody detects the resulting antigen-antibody complex. LFIAs support both qualitative and semi-quantitative analyses (Koczula and Gallotta 2016). Qualitative results are indicated by the presence or absence of a coloured band, while semi-quantitative results are determined by the intensity of the coloured band when using a portable strip reader. LFIAs can be designed in two formats: sandwich and competitive, with the sandwich model generally considered superior.

LFIAs are favoured for their ease of use, speed, and affordability. They have been developed for detecting pathogens in finfish and shrimp, enabling unskilled personnel to conduct on-site testing without requiring a laboratory. Importantly, LFIAs can detect fish and shrimp pathogens qualitatively and semi-quantitatively, providing results within 10–20 min. It is anticipated that in the coming years, LFIAs will find extensive use in screening for pathogens on aquaculture farms, offering a practical and efficient disease detection solution.

17.7.5 Metagenomics

Metagenomics, a recent field, explores complex microbial genomes in various environments. It can be divided into single-gene surveys targeting specific genes and random shotgun studies sequencing all genes. Metagenomics offers insights into microbial diversity through the analysis of taxonomically informative gene segments, namely 16S gene, thereby enabling the study of intracellular pathogens that are difficult to culture. Shotgun metagenomics is promising for identifying unknown pathogens without prior genomic information, and is crucial for precise clinical diagnostics, detecting coinfections, and preventing outbreaks. Metagenomics can help in identifying novel bacterial and viral pathogens before they become problematic in aquaculture systems (Martınez-Porchas and Vargas-Albores 2017), and offers insights into role of microbial diversity in pathogen proliferation.

17.7.6 Environmental DNA

Environmental DNA (eDNA) refers to genetic material from environmental samples like water or soil. This has revolutionized the detection of organisms in various environments and has a wide range of applications in disease surveillance as well as conservation. The technique enables the early detection of aquatic animal pathogens, thereby helping to predict and prevent disease outbreaks by providing early and critical information. It also reduces the need to acquire and analyze diseased hosts, making disease monitoring more efficient and cost-effective.

Traditional methods for detecting pathogens in aquaculture are often time-consuming and lack sensitivity until disease outbreaks are advanced. eDNA methods, on the other hand can detect pathogens in water or sediments before animals show clinical signs, offering early pathogen detection and risk assessment (Bohara et al. 2022). Regular eDNA sampling, combined with quantitative PCR and environmental data can provide insights into the environmental factors driving disease outbreaks and predict likelihood of mortality in aquatic animals. Frequent sampling allows stronger correlations between pathogen abundance and mortality, enabling timely preventive measures such as water exchange and adjustment of water quality parameters. Therefore, eDNA-based methods can be designated as a powerful approach to monitor and manage disease risks in aquaculture.

17.7.7 Use of Artificial Intelligence

Artificial intelligence (AI) holds immense potential for diagnosing diseases in aquatic animals, and enhances the accuracy and efficiency of disease diagnosis. By utilizing AI, algorithms can be trained to recognize diseases by analyzing images of infected aquatic animals, particularly focusing on external lesions and behavioural changes (Li et al. 2022). The analysis of the images includes three main features i.e. segmentation to separate the lesions from the rest of the image, features extraction to obtain information from region of interest, and classification to combine the information from the features into reliable identification of disease (Barbedo 2014). AI-based systems offer the advantage of rapid aquatic animal disease detection compared to traditional methods, enabling prompt decision-making by healthcare experts. The machine learning algorithms have the capability to process and analyze extensive datasets containing information about water quality, environmental conditions, and fish health records. They can identify patterns and correlations within the data that might indicate impending disease outbreaks. AI can also be employed for developing disease prediction models based on historical data and real-time monitoring. This approach can help fish farmers in taking preventative measures well in advance. Furthermore, AI can assist in determining most effective strategies for disease management, by analyzing data from different sources including sensors and other records. In salmonid aquaculture, for instance, sensors for monitoring water quality parameters, automated feeding systems, and in-tank/cage camera systems are widely utilized. Integrating AI with these technologies can offer automated Level I diagnostic alerts, thereby enhancing disease management in aquaculture (Dong et al. 2023).

17.8 Way Forward

The recent technological advances in the field of aquatic animal health management would pave way for minimizing the disease risks in aquaculture. The application of One Health principles would facilitate in sustainable aquaculture and safe food production. The development and use of SPF/SPR/SPT stocks of commercially-important fishes holds potential for enhancing production particularly in areas where the pathogens affecting aquaculture species are endemic. The development of edible vaccines and methods for enhancing uptake of antigen in the gut hold promise for combating infectious diseases in aquaculture in the coming years. The advanced diagnostic techniques such as quantitative PCR, digital PCR, isothermal amplifications can help in detection of carriers thereby preventing the spread of the pathogens, whereas lateral flow kits would empower the aquaculturists with point-of-care diagnostics for detection of pathogens on the farms. Besides, application of metagenomics and eDNA offer early detection of pathogens thereby helping to take preventive actions well in advance. Further, the application of AI would help in enhancing early disease diagnosis and holds promise for disease prediction in the farms. Finally, to harness the potential of technological advancements for sustainability of the aquaculture sector, there is a need of close collaboration between all the stakeholders including researchers, industry and policy makers.

References

Alday-Sanz V, Brock J, Flegel TW, McIntosh R, Bondad-Reantaso MG, Salazar M, Subasinghe R (2020) Facts, truths and myths about SPF shrimp in aquaculture. Rev Aquac 12:76–84

Barbedo JGA (2014) Computer-aided disease diagnosis in aquaculture: current state and perspectives for the future. Rev Innover 1:19–32

Becherer L, Borst N, Bakheit M, Frischmann S, Zengerle R, von Stetten F (2020) Loop-mediated isothermal amplification (LAMP) – review and classification of methods for sequence-specific detection. Anal Methods 2:717–746

Bohara K, Yadav AK, Joshi P (2022) Detection of fish pathogens in freshwater aquaculture using eDNA methods. Diversity 14:1015

Chen YM, Shih CH, Liu HC, Wu CL, Lin CC, Wang HC, Chen TY, Yang HL, Lin JHY (2011) An oral nervous necrosis virus vaccine using *Vibrio anguillarum* as an expression host provides early protection. Aquaculture 321(1-2):26–33

Diwan AD, Harke SN, Panche AN (2023) Host-microbiome interaction in fish and shellfish: an overview. Fish Shellfish Immun Rep 4:100091

Dong HT, Chaijarasphong T, Barnes AC, Delamare-Deboutteville J, Lee PA, Senapin S, Mohan CV, Tang KFJ, McGladdery SE, Bondad-Reantaso MG (2023) From the basics to emerging diagnostic technologies: what is on the horizon for tilapia disease diagnostics? Rev Aquac 15 (S1):186–212

FAO (2020) Progress towards development of the progressive management pathway for improving aquaculture biosecurity (PMP/AB): highlights of 2019 activities. FAO Fisheries and Aquaculture Circular No. 1211. FAO, Rome

FAO (2021) The FAO action plan on antimicrobial resistance 2021–2025. FAO, Rome. https://doi. org/10.4060/cb5545en

Hou Y, Chen S, Zheng Y, Zheng X, Lin J-M (2023) Droplet-based digital PCR (ddPCR) and its applications. TrAC Trends Anal Chem 158:116897

Kanitchinda S, Srisala J, Suebsing R, Prachumwat A, Chaijarasphong T (2020) CRISPR-Cas fluorescent cleavage assay coupled with recombinase polymerase amplification for sensitive and specific detection of *Enterocytozoon hepatopenaei*. Biotechnol Rep 27:e00485

Koczula KM, Gallotta A (2016) Lateral flow assays. Essays Biochem 60(1):111–120

Li D, Li X, Wang Q, Hao Y (2022) Advanced techniques for the intelligent diagnosis of fish diseases: a review. Animals (Basel) 12(21):2938

Lin Q, Fu X, Liu L, Liang H, Niu Y, Wen Y, Huang Z, Li N (2020) Development and application of a sensitive droplet digital PCR (ddPCR) for the detection of infectious spleen and kidney necrosis virus. Aquaculture 529:735697

Ma Y, Liu Z, Hao L, Wu J, Qin B, Liang Z, Ma J, Ke H, Yang H, Li Y, Cao J (2020) Oral vaccination using *Artemia* coated with recombinant *Saccharomyces cerevisiae* expressing cyprinid herpesvirus-3 envelope antigen induces protective immunity in common carp (*Cyprinus carpio* var. Jian) larvae. Res Vet Sci 130:184–192

Martinez-Porchas M, Vargas-Albores F (2017) Microbial metagenomics in aquaculture: a potential tool for a deeper insight into the activity. Rev Aquac 9:42–56

Micuchova A, Piackova V, Frebort I, Korytar T (2022) Molecular farming: expanding the field of edible vaccines for sustainable fish aquaculture. Rev Aquac 4:1978–2001

Netzer R, Ribicic D, Aas M, Cave L, Dhawan T (2021) Absolute quantification of priority bacteria in aquaculture using digital PCR. J Microbiol Methods 183:106171

Piepenburg O, Williams CH, Stemple DL, Armes NA (2006) DNA detection using recombination proteins. PLoS Biol 4(7):e204

Rabinowitz PM, Kock R, Kachani M, Kunkel R, Thomas J, Gilbert J, Wallace R, Blackmore C, Wong D, Karesh W, Natterson B, Dugas R, Rubin C, Stone Mountain One Health Proof of Concept Working Group (2013) Toward proof of concept of a one health approach to disease prediction and control. Emerg Infect Dis 19(12):130265

Ram MK, Naveen Kumar BT, Poojary SR, Abhiman PB, Patil P, Ramesh KS, Shankar KM (2019) Evaluation of biofilm of *Vibrio anguillarum* for oral vaccination of Asian seabass, *Lates calcarifer* (Bloch, 1790). Fish Shellfish Immunol 94:746–751

Rathore G, Lal KK, Bhatia R, Jena JK (2020) INFAAR – a research platform for accelerating laboratory-based surveillance of antimicrobial resistance in fisheries and aquaculture in India. Curr Sci 119(12):1884–1885

Sarker PK, Kapuscinski AR, McKuin B, Fitzgerald DS, Nash HM, Greenwood C (2020) Microalgae-blend tilapia feed eliminates fishmeal and fish oil, improves growth, and is cost viable. Sci Rep 10(1):19328

Sirirustananun N (2018) Appropriate proportion of water meal (*Wolffia arrhiza* [L.]) and commercial diet in combined feeding for tilapia fingerlings rearing. Int J Agric Technol 14(2):249–258

Subasinghe R, Alday-Sanz V, Bondad-Reantaso MG, Jie H, Shinn AP, Sorgeloos P (2023) Biosecurity: reducing the burden of disease. J World Aquacult Soc 54:397–426

Post-Harvest Management of Horticultural Crops: Use of Sensors and New Molecules

18

Ram Krishna Pal

Abstract

Fruits and vegetables are important components of the daily diet for good health. These are a powerhouse of essential vitamins, minerals, fibres and antioxidants, making them protective food that builds immunity in the human body. It is more relevant now since the recent COVID-19 pandemic compromised the human immune system of a large section of the affected population. The perishability of these products after harvest is a primary concern, particularly in tropical and subtropical agro-climatic regions worldwide. The living nature of harvested horticultural produce makes it more challenging for their quality assurance during subsequent distribution and marketing. Numerous pre- and post-harvest management techniques can increase the shelf life of horticulture produce. However, treatments concerning food safety issues such as PHI (postharvest interval for consumption after pesticide application) and MRL (Minimum Residue Level) of chemicals are the foremost concern. In the new artificial intelligence arena, sensors, new molecules, and intelligent packaging play a vital role in the postharvest management of harvested perishable horticultural crops. This paper highlights some critical issues in using these relevant technologies in the postharvest management of horticultural produce.

Keywords

Biosensor · ClO_2 · Harvest maturity · DPA · Magnetoresistive sensors · Melatonin · Polyamines · 1-MCP

R. K. Pal (✉)
ICAR-National Research Centre on Pomegranate, Solapur, Maharashtra, India

18.1   Introduction

The horticultural crops comprise various fruits, vegetables, spices, flowers, mushrooms, bamboo, plantation crops and non-food crops like medicinal and aromatic plants. The worldwide market had predicted that the horticulture industry would be worth USD 20.77 billion by 2021. By 2026, the market's projected value of horticultural produce is expected to be USD 40.24 billion (Anon 2023). Substantial development in the agricultural sector and the growing perception of sustainable horticultural practices are crucial considerations influencing the market. In developed countries, many growers and other horticulture professionals are using sensor technologies. Solutions based on artificial intelligence (AI) and the Internet of Things (IoT) for the production and post-harvest handling of horticultural crops are the major way forward for achieving sustainable development goals. In addition, augmented urban demand for organic food by the demographic of health-conscious individuals also influences the marketing opportunities of horticultural produce.

India first recognised the value of horticulture through crop diversification as a lucrative, feasible, and long-term alternative to the rice-wheat cropping system and its benefit in ensuring livelihood security three decades ago. As a result, horticulture today provides 30% of India's overall agricultural GDP and 15% of India's total cultivated land, contributing 52% of the country's agricultural exports. Next to China, India is now the second-largest producer of fruits and vegetables.

18.2   Postharvest Losses

Although India has touched new heights in producing fruits and vegetables comparable to Brazil and China, the production still needs to be improved for the basic nutrition security of its increasing population. Part of this problem could be overcome by saving a fair proportion of huge postharvest losses of these perishables. The annual losses of harvested fruits and vegetables in India are approximately comparable to the annual production of these commodities by the United Kingdom. These losses occur at several stages in the postharvest management chain, during harvesting, grading, packaging, transport, storage, distribution and marketing. Further, in economic terms, saving one unit of loss of harvested produce is relatively less expensive than increasing the equivalent quantity of production, particularly in developing countries, as the farmers face challenges like agricultural input crunch and difficulty in managing biotic and abiotic stresses. Hence, more attention is needed to manage postharvest losses besides increasing the production and productivity of horticultural crops using scientific knowledge and judicious utilization of land, water, forest, animals and environment.

18.3 Factors Affecting Postharvest Life

Critical abiotic factors influencing the postharvest life of horticultural produce are temperature, relative humidity, gaseous combination of the storage environment, light, type of packaging material, mode of transport, etc. In addition, bruising due to mechanical injury is a significant cause of concern as it serves as the primary entry port for microorganisms. Therefore, hygiene maintenance remains a focal point for HACCP (Hazard Analysis and Critical Control Points) evaluation during postharvest operations for quality assurance. Critical points need to be identified using standard problem solving techniques and PDCA (Plan-Do-Check-Act) cycle of quality assurance. Appropriate postharvest management is essential for controlling physiological and other metabolic functions to prolong market life, check market gluts, and increase financial gains. The postharvest behaviour of perishable horticultural crops depends further on species, cultivar, harvest maturity, postharvest treatments and storage conditions. Critical gaps observed in postharvest management in India and many developing countries are poor farm practices, improper harvesting, improper ripening practices, poor postharvest handling, improper packaging, poor infrastructure and logistic facilities and an ill-equipped marketing system (Chadha and Pal 2023). In recent years, several developments have occurred in Western countries and China on the judicious use of innovative sensors, new molecules and intelligent packaging for wholesale and retail marketing of fresh horticultural produce.

18.4 Application of Sensors in PHM of Horticultural Produce

18.4.1 Sensors for Sorting and Grading Based on Fruit Maturity

Maturation, ripening, and senescence are the three important stages of the harvested produce. The maturity of the horticultural crops can be defined as their indicator for harvest. Although the produce is fully developed concerning size, these may not be fit for immediate consumption. Further, the harvest maturity dictates the subsequent marketability and quality of horticultural produce. Premature harvesting of horticultural produce leads to excessive shrivelling, bruising, susceptibility to physiological disorders, short storage life and poor quality after ripening. Similarly, overmatured harvest exhibits off-flavour and insipid taste. In general, non-climacteric fruits (e.g. citrus, pineapple, pomegranate, grapes, etc.) are harvested when they are ripened on the tree. However, climacteric fruits like mango and banana are usually harvested at optimum maturity but at an unripe stage to facilitate postharvest handling system and distant marketing.

Consumer preference, competitive price realization and early catching up on the seasonal demand are driving factors that govern the harvest maturity of some fruits and vegetables in India. However, for export purposes, mandatory quality standards for most fresh horticultural produce include objective maturity indices to ensure the consumers' acceptance of their desired flavour and visual quality. Ripening makes the commodity edible with characteristic colour, texture and flavour.

In most cases, the maturity of horticultural produce is determined by subjective and objective evaluation. Subjective evaluation of maturity refers to using human sensory organs of visible colour, flavour, feel (texture), size, appearance, etc. However, total soluble solids content (cherry, grape, grapefruit, kiwifruit, mandarin and pear, palmyrah, cocoa), titratable acidity (pomegranate, lime, lemon), TSS: acid ratio (citrus fruits and grapes), flesh firmness (apple, pear, banana), total fat (avocado), oil content (olive), juice % (oranges) are some of the objective parameters used as maturity indices. In addition, surface colour development (apricot, nectarine, peach, persimmon, plum, raspberry, and strawberry fruits); tannin content (tea bud and leaves); aroma (cashew apple and cocoa); colour change and oil content (Fresh fruit bunches of oil palm); caffeine content (coffee) etc. are predominantly used as maturity indices.

Sensors could transform the subjective measurement of maturity indices into a quantifiable objective measure. For example, the visual colour of a fruit or vegetable can be distinguished by the colour difference meter with 'L', 'a' and 'b' values and their combination. The L value denotes the lightness or darkness of the surface colour. Similarly, the -ve value of 'a' in the colour coordinate indicates the greenness. As the commodity matures, the 'a' value increases from –ve to +ve, where +a value represents the redness. Again, a negative 'b' value in the colour coordinate represents blue and a positive 'b' corresponds with yellow. Therefore, measuring the 'b' value using a colour sensor is most important in fruits and vegetables, which turn yellow at their optimum maturity. The sensors can differentiate the intensity of the colour, i.e., rednesss, yellowness and greenness as a quantifiable objective measurement otherwise impossible by human eyes.

Non-destructive techniques for measuring quality attributes of fruit have been reported by several workers (Nicola et al. 2014). Humans' sense of touch and smell seems more reliable in grading fruits according to their maturity and the degree of ripeness, which traders and consumers mostly follow in India. A sensory gripper with a capacitive tactile sensor array correctly classified 88% of the mangoes according to their stage of ripeness (Carvalho et al. 2021; Scimeca et al. 2019). Cilia sensors were also reported to assess the fruit maturation stage, reaching a classification accuracy of 96% for apples (single cilia with 400 μm diameter and 3 mm height) and 83% for strawberries (nine cilia in a 3 × 3 array, each with 360 μm diameter and 1.6 mm height) (Ribeiro et al. 2020). Tactile sensors consisting of artificial cilia provide exquisite sensing performances due to their high aspect ratio and surface-to-volume ratio (Alfadhel and Kosel 2015).

Piezoelectric films are crystals like quartz that generate a proportional voltage on compressive, mechanical or tensile stress. Various tactile sensors have recently been reported for fruit quality assessment (Zhang et al. 2020) using piezoelectric films and strain gauges. These tactile sensors were used to evaluate and classify the surface roughness of cucumbers, cantaloupes and apples, achieving a 94% accuracy (Zhou et al. 2017) that directly correlates to maturity.

Commercial sorting lines for determination of external features such as colour, size, and absence of external defects are mostly machine vision-based tools. However, different fruit skin colours, blemishes, and lighting conditions at the sorting area affect monitoring reliability (Blasco et al. 2017). Furthermore, slow speed, high cost and requirement of specific laboratories for assessing internal quality attributes are major impediments and thus non-applicable for industrial sorting and grading lines of fruits and vegetables (Zhang et al. 2016). Therefore, the development of sensors should preferably be based on biomimetic principles, which can mimic the output close to human perceptions.

Researchers in Portugal and other European countries developed a touch-based texture sensor using simple surface scanning that could differentiate fruits (strawberries and blueberries) according to their skin, directly related to the degree of fruit ripeness (Bourne 1979). Compared to other techniques, advanced magnetic sensor chips commonly accepted in biocompatible applications in industrial sorting line was developed by Carvalho et al. (2021). This texture sensor comprises magnetized nanocomposite cilia attached to a chip with magnetoresistive sensors. They demonstrated using ciliary sensors in scanning fruits (blueberries and strawberries) in different maturation stages. The contact of the cilia with the fruit skin provided qualitative information about its texture regarding the ripeness stage. On average, less mature fruits exhibited a peak voltage of 0.14 mV for blueberries and 0.12 mV for strawberries, while overripe fruits showed 0.58 mV and 0.56 mV, respectively. The results were confirmed by sensory assessment of the fruit freshness, therefore attesting to its potentiality for fruit quality assurance application.

18.4.2 Sensors for Packaging of Fresh Horticultural Produce

Tracking and traceability are the growing needs of the food industry for long-term storage of perishable horticultural produce with quality assurance and food safety. The status of the product can be monitored using intelligent sensors and labels on packaging. Widespread application can be made for identification, testing of authenticity of raw material, GM food, presence of allergens, pathogens and chemical contaminants using suitable sensors.

Ethylene enhances ripening even at an extremely low concentration due to its low threshold requirement for initiating autocatalytic action. The ripe fruits and vegetables continuously produce ethylene. Mechanical injury, environment, and physiological nature of the crop (climacteric or non-climacteric type) are responsible for production of stress induced ethylene. With a very low threshold requirement of exogenous ethylene for triggering endogenous ethylene production, the unripe fruits undergo a rapid ripening process when kept close to ripe fruits. Ethylene scavengers (absorbers or oxidizers) can arrest rapid ripening and senescence. Potassium permanganate ($KMnO_4$) is the most commonly available ethylene scavenger. It oxidizes ethylene to CO_2 and H_2O. The ethylene chemisorption by $KMnO_4$ has led to the development of a range of commercial scavengers as granules over clays or activated carbon (Vermeiren et al. 1999). A technique for removing ethylene by oxidation at low temperatures over a platinum catalyst on mesoporous silica was proposed by Jiang et al. (2013). Even at 0 °C, the catalyst could remove 50 ppm of ethylene.

Ethylene sensors could help detect its concentration, which causes rapid ripening and fruit degradation. Many researchers have reported several sensors for ethylene detection. One such sensor was created by Esser et al. (2012) using single-walled carbon nanotubes (SWNTs) combined with a copper complex and sandwiched between gold electrodes. Resistance change occurs when the electrodes come in contact with hydrocarbon compounds, viz. ethylene.

The recognition capabilities of biomolecules like enzymes, antibodies and aptamers (short sequences of synthetic DNA, RNA, and Xeno Nucleic Acid) are used by biological sensors, also known as biosensors (Turner et al. 1987). Numerous techniques are used to monitor the binding process, such as colorimetric (Yang et al. 2011), electrochemical (Viswanathan et al. 2009), optical (Narsaiah et al. 2012) etc. While electrochemical biosensors for glucose monitoring, which account for 85% of all biosensors on the market (Wang 2008), have made a significant impact on the field of medical diagnosis, the use of biosensors in postharvest management and the food industry is still in its infancy with only a few examples currently available. A biosensor (sensor cell line chip—AIRCHO-SEAP) for detecting ethylene and acetaldehyde generated by plants was created by Weber et al. in 2009. The sensor detects acetaldehyde, which is produced when ethylene is oxidised.

A number of volatile organic compounds (VOC) might accumulate in closed containers or packages of fresh fruits and vegetables. Terpenes, carboxylic acids, alcohols, aldehydes, sulphur compounds, ammonia, and jasmonates are only a few examples of substances detected using these indicators (Toivonen 1997). The measurement of VOC evolution in the headspace of guava packaging has been made possible by Kuswandi et al.'s (2013) development of a colour-based pH indicator employing bromophenol blue immobilised on a cellulose membrane. The indicator's hue will switch from blue to green to indicate over ripeness, which is visible to the human eye. The results demonstrated that at ambient temperatures (28–30 °C), the colour indicator may be used to assess the guava's level of freshness. The pH of the headspace of the guava packing is reflected in the colour shift of the indicators. Additionally, comparable changing trends can be observed in many characteristics (soluble solids concentration, texture, and sensory evaluation) that traditionally define the freshness of guava. As a result, the indicator can be used to visually check the freshness status of packaged guavas in real-time. The freshness of the guava is indicated on this label, but further research is required to determine the safety of utilizing such chemical colours and the likelihood that any chemicals would migrate into the food. Increased use of sensors in intelligent or smart packaging has led to some commercial success for food-freshness assessment (Kuswandi et al. 2011; Bahadır and Sezgintürk 2015), with prominent instances for chemical rather than rather than biological sensors.

18.5 New Molecules for Postharvest Management of Horticultural Produce

18.5.1 1-Methylcyclopropene (1-MCP)

A new option for improving the shelf life and quality of fruits is the use of 1-methylcyclopropene (1-MCP), a potent ethylene activity inhibitor. Because 1-MCP has an approximately ten-fold higher affinity for the ethylene receptor than ethylene does, it is active at significantly lower concentrations (Sisler and Serek 1997). The USA's Environmental Protection Agency (EPA) has given its approval, and Agro Fresh, Inc., a Rohm and Hass subsidiary, has taken on the commercial production of 1-MCP. Several nations are benefiting from this chemical, which is currently sold under the brand name Smart Fresh™. The magic gas, l-MCP, has been utilised to extend the shelf life of perishables because it irreversibly attaches to ethylene receptors, causing a conformational shift in the receptors and blocking the action of ethylene in plant tissues. Since its 1996 patent, various scientific studies have been published demonstrating the effect of 1-MCP in slowing the ripening and senescence processes in fresh horticultural crops.

Fruit from guavas (*Psidium guajava* L. cv. 'Allahabad Safeda') collected at the mature light-green stage was treated to 300 and 600 nL L^{-1} 1-methylcyclopropene (1-MCP) for 6, 12, and 24 h at 20 1 °C, and then stored at 10 °C for 25 days and for 9 days at ambient temperatures (25–29 °C). 1-MCP influenced most of the physiological and biochemical changes that occurred during storage and ripening in a dose-dependent way. Depending on the 1-MCP concentration and exposure time, ethylene production and respiration rates were dramatically reduced throughout storage and ripening under both storage conditions. Fruit firmness changes during storage under both circumstances were strongly influenced by 1-MCP treatment. The efficiency of 1-MCP in delaying fruit ripening was demonstrated by the decreased deviations in the soluble solids contents (SSC), titratable acidity (TA), and vitamin C content. Fruit treated with 1-MCP had a considerably increased vitamin C content than fruit not treated with 1-MCP or with 300 nL L^{-1} 1-MCP for 6 h. During cold storage and ripening, 1-MCP-treated fruit showed a higher reduction in chilling injury symptoms. Fruit treated with 1-MCP had a significantly lower incidence of fruit degradation in both storage scenarios. To increase the shelf life of guava cv, 'Allahabad Safeda', 1-MCP treatment at 600 nL L^1 for 12 h in combination with cold storage (10 °C) appears promising. According to Singh and Pal (2008), 1-MCP at 300 nL L^{-1} for 12 and 24 h or 600 nL L^{-1} for 6 h may extend the marketability of fruit by 4–5 days in ambient environments. It was challenging to obtain quantitative measurements for 1-MCP because of its small molecular weight and low boiling point, especially at the residual level. The analytical method developed by Dong et al. (2021) was utilised to monitor the 1-MCP residue levels in commercially available samples, and all values were below 5 g/kg, satisfying the EU or Japan MRLs for 1-MCP. Table 18.1 (Watkins 2006; Schotsmans et al. 2009) shows the effects of 1-methylcyclopropene on postharvest storage behaviour and quality of horticultural produce.

Table 18.1 Effects of 1-methylcyclopropene on horticultural produce

Crop	1-MCP concentration	Altered postharvest physiology and biochemistry
(a) Fruits crops		
Apple	1 µLper L	Reduced softening, delayed ripening, inhibited superficial scald and decay
	2 µL per L	Less softening, and reduced superficial scald
Apricot	0.5 µL per L	Reduced respiration rate and tissue softening
Banana	0.01–1.0 µL per L	Fruit softening and peel colour change delays, longer shelf life
Ber	1 µL per L	Delayed senescence, maintained quality, reduced decay
Cherry	1 µL L^{-1}	Preserved firmness and prolonged shelf life
Grape	0.5–1.0 µL per L	Increased shelf life, and improved quality
Guava	0.6 µL per L	Improved shelf life, ameliorated chilling injury, and reduced decay
Kiwifruit	0.5 µL per L	Reduced ethylene production and softness of fruit
Lime	1 µL per L	Maintenance of the green colour of the fruit
Mandarin	5 µL per L	Inhibited chlorophyll degradation and senescence
Mango	2 mg kg^{-1}	Delayed ripening, increased antioxidant enzymes
Peach	10 µL per L	Inhibited degradation of starch, prolonged shelf life
Pineapple	0.1 µL per L	Effectively managed internal browning or black heart
Plum	0.5 µL per L	Slowed fruit ripening, reduced fruit softening
(b) Vegetable crops		
Bell pepper	1 µL per L	Deferred senescence through decreased chlorophyll breakdown
Brinjal	1 µL per L	Prevents browning, maintains quality, and delays senescence
Broccoli	0.6 µL^1 per L	Minimised loss of chlorophyll pigments and green colour
Carrot	1 µL per L	Decreased respiratory rate, minimized loss of sucrose
Coriander	0.5 g m^{-1}	Increased shelf life of leaves, retarded chlorophyll degradation
Muskmelon	1 µL per L	Delayed mesocarp softening and reduced water soaking
Okra	5 µL per L	Reduced development of chilling injury and extended storage life
Onion	1 µL per L^1	Sprout growth was reduced during storage
Parsley	10 µL per L	Retard senescence of leaves
Spinach	1 µL per L	Extended postharvest life, maintained higher chlorophyll
Tomato	5–20 µL per L	Delayed ripening and increased shelf life
Watermelon	5 µL per L	Protected against ethylene-induced water-soaking
(c) Flower crops		
Carnation	1–20 nL per L	Inhibited wilting, ethylene effects, and extended vase life four times
Freesia	4 nL per L	Vase life extension to 9 days

(continued)

Table 18.1 (continued)

Crop	1-MCP concentration	Altered postharvest physiology and biochemistry
Lilium	500 nL per L	Superior quality, inhibition of ethylene response
Phalaenopsis	500 nL per L	Reduced ethylene-induced floral bud drop
Tulip bulbs	200 nL per L	Inhibited ethylene-induced splitting

Source: Watkins 2006; Schotsmans et al. 2009

18.5.2 Diphenylamine (DPA)

Diphenylamine (DPA) is a colourless, aromatic amine possessing antioxidant properties. Postharvest drenching of fruits with DPA provides reasonable control against scald, a severe physiological disorder of apples and pears during storage manifested as brown discolouration that makes the fruit unmarketable and limits its storage potentialThe fruits stay rigid and green throughout storage because postharvest DPA treatment of pear suppresses normal rates of respiration and ethylene production during ripening. DPA reduces the chilling-induced pitting of green bell pepper. DPA is commercially used on apple and pear fruit in the USA and Europe to improve their storage life and maintain quality. The Codex Alimentarius norms describe the MRL of DPA as 5 mg/Kg of food.

18.5.3 Chlorine Dioxide (ClO_2)

Washing fruit and vegetables in chlorinated water produces harmful by-products (chlorinated derivatives) that can taint taste and colour. Instead, using ClO_2 is recommended to effectively disinfect with a low dose for high effectiveness without any residue of chlorinated derivatives. The powerful and effective action of chlorine dioxide is not affected by organic compounds, and it can successfully inactivate dangerous bacteria. In addition, ClO_2 can penetrate biofilms and bacteria embedded in fresh produce's cuticle, eliminating the spoilage source before the product leaves the facility.

18.5.4 Melatonin

Plants use the multifunctional chemical melatonin (MLT) to take part in many physiological functions. Reactive oxygen species (ROS) are typically produced by fruits and vegetables during their postharvest lives. Excessive ROS promote tissue ageing, protein structure disintegration, and cell damage, which lowers quality. Several enzymatic and non-enzymatic systems have been found in numerous investigations to influence ROS homeostasis when exogenous melatonin is applied. Melatonin frequently interacts with hormones and other signalling molecules. The postharvest preservation of fruits and vegetables benefits from the interaction between melatonin and ROS, particularly the H_2O_2 produced by Respiratory Burst Oxidase Homolog (RBOHs) (Li et al. 2023). The idea that melatonin is essential for preserving the quality of fruits and vegetables throughout the postharvest period has been put forth in numerous papers for several years (Fan et al. 2022a, b). For instance, under cold storage settings, exogenous melatonin administration preserved fruit firmness, maintained a higher colour of the fruit's rind, and preserved more lightness of papayas than the control group. Reductions in weight loss, deterioration, and titratable acid (TA), as well as the maintenance of fruit firmness, total soluble solids (TSS), and the TSS/TA ratio during storage in cherry tomatoes MLT treatment was favourable to maintaining fruit quality (Yan et al. 2022). The fruit colour index (a*/b*) was also raised after melatonin treatment in sweet cherry and guava fruits (Carrión-Antolí et al. 2022). Exogenous melatonin administration also slowed down the breakdown of chlorophyll in pepper, broccoli, and Chinese cabbage (Tan et al. 2020; Cano et al. 2022; Wang et al. 2022). Exogenous melatonin treatment significantly boosted the levels of other essential compounds found in fruits and vegetables, such as total soluble solids, sugar, protein, ascorbic acid, carotenoids, total flavonoids, and phenol (Cano et al. 2022; Sun et al. 2021; Fan et al. 2022c; Verde et al. 2022). Additionally, melatonin regulated pear fruit's volatile substances (propyl acetate and hexyl acetate) during the postharvest period (Liu et al. 2019; Wei et al. 2022).

18.5.5 Methyl Jasmonate, Methyl Salicylate, Salicylic Acid and Polyamines

Methyl Salicylic Acid (MeSA), a plant molecule generated from salicylic acid, is crucial for plants' growth and development as well as their defence mechanisms, ability to adapt to abiotic stressors, and fruit ripening (Kumar 2014). With their pre- and postharvest applications, the signalling molecules methyl jasmonate (MeJA), methyl salicylate (MeSA), and salicylic acid (SA) increase the levels of bioactive chemicals and preserve the postharvest life of several horticultural crops. These essentially function as plant immune modulators. MeJA is a volatile substance that was first identified in *Jasminum grandiflorum* blossoms and has since been found in plants worldwide. MeJA is essential for plant cellular responses, defence mechanisms against infections and environmental stresses such as drought, salinity, low temperatures, plant-herbivore interactions, and plant-plant interactions, according to Cheong and Choi (2003). Additionally, it has been demonstrated that the exogenous administration of MeJA can prevent stress-related lesions in whole and freshly cut fruits and vegetables (Gonzalez-Aguilar et al. 2007). According to the Joint FAO/WHO Expert Committee on Food Additives (JECFA), methyl jasmonate is non-toxic as a food additive with the highest predicted pesticide residues of 373 g ai/kg of seed.

According to Klessig and Malamy (1994), SA is an endogenous signal molecule that controls plant developmental processes such as heat production, photosynthesis, stomatal conductance, transpiration, ion uptake and transport, disease resistance, seed germination, sex polarisation, and crop yield in addition to stress responses. Additionally, SA is a promising substance that delays the ripening and spoiling of fruits and vegetables that have been stored after harvest (Asghari and Aghdam 2010). Plants naturally contain salicylic acid as a natural compound. Hence, Salicylic acid application does not appear to have any effect on endogenous levels, according to studies on kiwifruit, grapes, and pome fruit conducted by the New Zealand Food Safety Authority.

In living things, polyamines (PAs) are low-molecular-weight tiny aliphatic amines linked to various biological activities, including plant growth, development, and stress response (Smith 1985). Putrescine (PUT), spermidine (SPD), and spermine (SPM), which are present in every plant cell, are the three most prevalent polyamines. PAs frequently exist in nature as free molecular bases, but they can also be attached to other macromolecules and small molecules, including phenolic acids (conjugated forms). According to Kumar et al. (1997), PAs are thought to have an antisenescence effect, yet their levels typically drop off as most fruits ripen. The shelf life and textural qualities are also impacted by this general reduction. Thus, it has been observed that the exogenous application of PAs can improve the texture and shelf life of various fruits, including the strawberry (Khosroshahi et al. 2007), plum (Pérez-Vicente et al. 2002), apricot (Martínez-Romero et al. 2002), and mango (Malik and Singh 2005).

The shelf life of kiwifruits (*Actinidia deliciosa*) cv. Allison, when stored in ambient settings [(22 ± 4 °C) and RH: 65 5%], was best extended by postharvest treatment with spermine at 1.5 mM and spermidine at 2.0 mM, according to Jhalegar et al. (2011). It has been suggested that PAs may have a role in lowering chilling injury due to the accumulation of putrescine when exposed to cold stress in apples (Kramer et al. 1989). A uniform and significant increase in PUT and a decrease in SPD and SPM were the results of investigations on citrus and zucchini squash during storage at low temperatures (McDonald and Kushad 1986; Wang and Ji 1989). Pomegranate (*Punica granatum* L., cv. Mridula) fruits were treated with putrescine, carnauba wax, and putrescine+carnauba wax before being placed in cold storage at 2 °C to lessen chilling damage and maintain quality. Fruits were exposed to post-cold storage exposure at 20 °C for 3 days before analysis of physical, physiological, and biochemical characteristics. During storage, untreated fruits quickly developed chilling damage, showing brown skin discoloration, surface pitting, weight loss, and loss of firmness. The application of putrescine and carnauba wax greatly suppressed all these unfavourable alterations. Additionally, respiration and the rate of ethylene evolution were decreased by the putrescine and carnauba wax treatment. Due to the synergistic advantages of PUT's antisenescence activity and carnauba wax's barrier qualities, the combination therapy was superior to untreated pomegranate fruits (Barman et al. 2011). A collection of the broad effects of a few novel molecules on the postharvest handling of horticultural produce is shown in Table 18.2.

Table 18.2 Effect of a few new molecules on PHM of horticultural produce

Molecule	Commodity	Concentration	Desired effect
Putrescine	Peach	5–20 mM	Reduced ethylene production, delayed softening, and retained titratable acidity
Salicylic acid	Sweet cherry	2 mM	Induced disease resistance and reduced disease incidence
	Banana (*Musa acuminate* cv 'Hari chhal')	500 (μM) and 1000 (μM) 6 h at 20 °C	Delayed fruit ripening
	Mango (*Mangifera indica*L. cv. 'Matisu')	1 (mmol L^{-1}) 2 min at 20 °C	Enhanced disease resistance
Chorine dioxide	Fruits and vegetables	2–3 ppm	Surface disinfection for bacteria and fungi
Diphenylamine (DPA)	Apples & Pears	Max 1000 ppm with MRL of 10 ppm	Control against scald of apples and pears
Methyl Jasmonate	Kiwifruit (*Actinidia deliciosa* cv. 'Jinkui')	0.1 (mmol L^{-1}) 24 h at 20 °C	Reduced postharvest softening during storage
Methyl salicylic acid	Mango (*Mangifera indica* L. cv. 'Zill')	0.1 (mM) for 12 h at 20 °C	Enhanced resistance to low -temperature stress
	Banana (*Musa acuminate* cv. 'Sucrier')	2 (mM) for 30 min at 25 ± 1 °C	Delayed the senescence

References

Alfadhel A, Kosel J (2015) Magnetic nanocomposite cilia tactile sensor. Adv Mater 27(47): 7888–7892. https://doi.org/10.1002/adma.201504015

Anonymous (2023). https://www.globalmarketestimates.com/market-report/horticulture-market-3414. Accessed 29 May 2023

Asghari M, Aghdam M (2010) Impact of salicylic acid on postharvest physiology of horticultural crops. Trends Food Sci Technol 21:502–509. https://doi.org/10.1016/j.tifs.2010.07.009

Bahadır EB, Sezgintürk MK (2015) Applications of commercial biosensors in clinical, food, environmental, and bio threat/biowarfare analyses. Anal Biochem 478:107–120

Barman K, Asrey R, Pal RK (2011) Putrescine and carnauba wax pretreatments alleviate chilling injury, enhance shelf life and preserve pomegranate fruit quality during cold storage. Sci Hortic 130(4):795–800

Blasco J, Munera S, Aleixos N, Cubero S, Molto E (2017) Machine vision-based measurement systems for fruit and vegetable quality control in postharvest. In: Measurement, modeling and automation in advanced food processing. Springer, pp 71–91

Bourne MC (1979) Texture of temperate fruits. J Texture Stud 10(1):25–44. https://doi.org/10.1111/j.1745-4603.1979.tb01306.x

Cano A, Giraldo-Acosta M, García-Sánchez S, Hernández-Ruiz J, Arnao M (2022) Effect of melatonin in broccoli postharvest and possible melatonin ingestion level. Plan Theory 11: 2000. https://doi.org/10.3390/plants11152000

Carrión-Antolí A, Martínez-Romero D, Guillén F, Zapata PJ, Serrano M, Valero D (2022) Melatonin pre-harvest treatments leads to maintenance of sweet cherry quality during storage by increasing antioxidant systems. Front Plant Sci 13:863467. https://doi.org/10.3389/fpls.2022.863467

Carvalho M, Ribeiro P, Romão V, Cardoso S (2021) Smart fingertip sensor for food quality control: fruit maturity assessment with a magnetic device. J Magn Magn Mater 536:168116

Chadha KL, Pal RK (2023) Managing post harvest quality and losses in horticultural crops, 2nd revised and enlarged edition. Daya Publishing House® a Division of Astral International Pvt. ltd, New Delhi; ISBN 978-93-5461-960-1 (HB)

Cheong J, Choi Y (2003) Methyl jasmonate as a vital substance in plants. Trends Genet 19:409–413. https://doi.org/10.1016/S0168-9525(03)00138-0

Dong M, Wen G, Li J, Wang T, Huang J, Li Y, Tang H, Sun Q, Wang W (2021) Determination of 1-methylcyclopropene residues in vegetables and fruits based on iodine derivatives. Food Chem 358:129854

Esser B, Schnorr JM, Swager TM (2012) Selective detection of ethylene gas using carbon nanotube-based devices: utility in determination of fruit ripeness. Angew Chem Int Ed 51: 5752–5756

Fan S, Li Q, Feng S, Lei Q, Abbas F, Yao Y et al (2022a) Melatonin maintains fruit quality and reduces anthracnose in postharvest papaya via enhancement of antioxidants and inhibition of pathogen development. Antioxidants 11:804. https://doi.org/10.3390/antiox11050804

Fan S, Xiong T, Lei Q, Tan Q, Cai J, Song Z et al (2022b) Melatonin treatment improves postharvest preservation and resistance of guava fruit (*Psidium guajava* L.). Foods 11:262. https://doi.org/10.3390/foods11030262

Fan Y, Li C, Li Y, Huang R, Guo M, Liu J et al (2022c) Postharvest melatonin dipping maintains quality of apples by mediating sucrose metabolism. Plant Physiol Biochem 174:43–50. https://doi.org/10.1016/j.plaphy.2022.01.034

Gonzalez-Aguilar GA, Villegas-Ochoa MA, Martinez-Tellez M, Gardea A, Ayala-Zavala JF (2007) Improving antioxidant capacity of fresh-cut mangoes treated with UV-C. J Food Sci 72:197–202. https://doi.org/10.1111/j.1750-3841.2007.00295.x

Jhalegar MJ, Roshan SR, Krishna PR, Vishal R (2011) Effect of postharvest treatments with polyamines on physiological and biochemical attributes of kiwifruit (*Actinidia deliciosa*) cv. Allison Fruits 67(1):13–22

Jiang C, Hara K, Fukuoka A (2013) Low-temperature oxidation of ethylene over platinum nanoparticles supported on mesoporous silica. Angew Chem Int Ed 52:6265–6268

Khosroshahi MRZ, Esna-Ashari M, Ershadi A (2007) Effect of exogenous putrescine on postharvest life of strawberry (*Fragaria ananassa* Duch.) fruit cultivar Selva. Sci Hortic 114:27–32

Klessig D, Malamy J (1994) The salicylic acid signal in plants. Plant Mol Biol 26:1439–1458. https://doi.org/10.1007/BF00016484

Kramer GF, Wang CY, Conway WS (1989) Correlation of reduced softening and increased polyamine levels during low-oxygen storage of McIntosh apples. J Am Soc Hort Sci 114: 942–947

Kumar D (2014) Salicylic acid signalling in disease resistance. Plant Sci 228:127–134. https://doi.org/10.1016/j.plantsci.2014.04.014

Kumar A, Altabella T, Taylor MA, Tiburcio AF (1997) Recent advances in polyamine research. Trends Plant Sci 2:124–130

Kuswandi B, Wicaksono Y, Abdullah A, Heng LY, Ahmad M (2011) Smart packaging: sensors for monitoring of food quality and safety. Sens Instrumen Food Qual 5:137–146

Kuswandi B, Maryska C, Abdullah A, Heng LY (2013) Real time on-package freshness indicator for guavas packaging. J Food Meas Charact 7:29–39

Li N, Zhai K, Yin Q, Gu Q, Zhang X, Melencion MG, Chen Z (2023) Cross talk between melatonin and reactive oxygen species in fruits and vegetables postharvest preservation: an update. Front Nutr 10:1143511. https://doi.org/10.3389/fnut.2023.1143511

Liu J, Liu H, Wu T, Zhai R, Yang C, Wang Z et al (2019) Effects of melatonin treatment of postharvest pear fruit on aromatic volatile biosynthesis. Molecules 24:4233. https://doi.org/10.3390/molecules24234233

Malik AU, Singh Z (2005) Pre-storage application of polyamines improves shelf-life and fruit quality of mango. J Hortic Sci Biotechnol 80(3):363–369

Martínez-Romero D, Serrano M, Carbonell L, Burgos F, Valero D (2002) Effects ofpostharvest putrescine treatment on extending shelf life and reducing mechanical damage in apricot. J Food Sci 67(5):1706–1711

McDonald RE, Kushad MM (1986) Accumulation of putrescine during chilling injury of fruits. Plant Physiol 82:324–326

Narsaiah K, Jha SN, Bhardwaj R, Sharma R, Kumar R (2012) Optical biosensors for food quality and safety assurance—a review. J Food Sci Technol 49:383–406

Nicola BM et al (2014) Nondestructive measurement of fruit and vegetable quality. Annu Rev Food Sci Technol 5:285–312. https://doi.org/10.1146/annurev-food-030713-092410

Pérez-Vicente A, Martínez-Romero D, Carbonell A, Serrano M, Riquelme F, Guillén F, Valero D (2002) Role of polyamines in extending shelf life and the reduction of mechanical damage during plum (*Prunus salicina* Lindl.) storage. Postharvest Biol Technol 25:25–32

Ribeiro P, Cardoso S, Bernardino A, Jamone L (2020) Fruit quality control by surface analysis using a bio-inspired soft tactile sensor. In: International conference on intelligent robots and systems, pp 1094–1100

Schotsmans WC, Prange RK, Binder BM (2009) 1-Methylcyclopropene: mode of action and relevance in postharvest horticulture research. Hortic Rev 35:263–313

Scimeca L, Maiolino P, Cardin-Catalan D, del Pobil AP, Morales A, Iida F (2019) Nondestructive robotic assessment of mango ripeness via multi-point soft haptics. In: International Conference on Robotics and Automation (ICRA), pp 1821–1826. https://doi.org/10.1109/ICRA.2019.8793956

Singh SP, Pal RK (2008) Response of climacteric-type guava (*Psidium guajava* L.) to postharvest treatment with 1-MCP. Postharvest Biol Technol 47(3):307–314

Sisler EC, Serek M (1997) Inhibitors of ethylene responses in plants at the receptor level: recent developments. Physiol Plant 100:577–582

Smith TA (1985) Polyamines. Annu Rev Plant Physiol 36:117–143

Sun H, Cao X, Wang X, Zhang W, Li W, Wang X et al (2021) RBOH-dependent hydrogen peroxide signaling mediates melatonin-induced anthocyanin biosynthesis in red pear fruit. Plant Sci 313:111093. https://doi.org/10.1016/j.plantsci.2021.111093

Tan XL, Zhao YT, Shan W, Kuang JF, Lu WJ, Su XG et al (2020) Melatonin delays leaf senescence of postharvest Chinese flowering cabbage through ROS homeostasis. Food Res Int 138:109790. https://doi.org/10.1016/j.foodres.2020.109790

Toivonen PM (1997) Non-ethylene, non-respiratory volatiles in harvested fruits and vegetables: their occurrence, biological activity and control. Postharvest Biol Technol 12:109–125

Turner A, Karube I, Wilson GS (1987) Biosensors: fundamentals applications, vol 201. Oxford University Press, New York, NY, p 363

Verde A, Míguez JM, Gallardo M (2022) Role of melatonin in apple fruit during growth and ripening: possible interaction with ethylene. Plan Theory 11:688. https://doi.org/10.3390/plants11050688

Vermeiren L, Devlieghere F, van Beest M, de Kruijf N, Debevere J (1999) Developments in the active packaging of foods. Trends Food Sci Technol 10:77–86

Viswanathan S, Radecka H, Radecki J (2009) Electrochemical biosensors for food analysis. Monatsh Chem Chem Mon 140:891

Wang J (2008) Electrochemical glucose biosensors. Chem Rev 108:814–825

Wang CY, Ji ZL (1989) Effect of low-oxygen storage on chilling injury and polyamines in zucchini squash. Sci Hortic 39(1):1–7

Wang L, Shen X, Chen X, Ouyang Q, Tan X, Tao N (2022) Exogenous application of melatonin to green horn pepper fruit reduces chilling injury during postharvest cold storage by regulating enzymatic activities in the antioxidant system. Plan Theory 11:2367. https://doi.org/10.3390/plants11182367

Watkins CB (2006) The use of 1-methylcyclopropene (1-MCP) on fruits and vegetables. Biotechnol Adv 24:389–409

Weber W, Luzi S, Karlsson M, Fussenegger M (2009) A novel hybrid dual-channel catalytic-biological sensor system for assessment of fruit quality. J Biotechnol 139:314–317

Wei S, Jiao H, Wang H, Ran K, Dong R, Dong X et al (2022) The mechanism analysis of exogenous melatonin in limiting pear fruit aroma decrease under low temperature storage. PeerJ 10:e14166. https://doi.org/10.7717/peerj.14166

Yan R, Li S, Cheng Y, Kebbeh M, Huan C, Zheng X (2022) Melatonin treatment maintains the quality of cherry tomato by regulating endogenous melatonin and ascorbate-glutathione cycle during room temperature. J Food Biochem 46:e14285. https://doi.org/10.1111/jfbc.14285

Yang C, Wang Y, Marty JL, Yang X (2011) Aptamer-based colorimetric biosensing of Ochratoxin a using unmodified gold nanoparticles indicator. Biosens Bioelectron 26:2724–2727

Zhang C, Guo C, Liu F, Kong W, He Y, Lou B (2016) Hyperspectral imaging analysis for ripeness evaluation of strawberry with support vector machine. J Food Eng 179:11–18. https://doi.org/10.1016/j.jfoodeng.2016.01.002

Zhang B, Xie Y, Zhou J, Wang K, Zhang Z (2020) State-of-the-art robotic grippers, grasping and control strategies, as well as their applications in agricultural robots: a review. Comput Electron Agric 177:105694. https://doi.org/10.1016/j.compag.2020.105694

Zhou J, Meng Y, Wang M, Memon MS, Yang X (2017) Surface roughness estimation by optimal tactile features for fruits and vegetables. Int J Adv Robot Syst 14(4):1–8. https://doi.org/10.1177/172988141772

Total Quality Management in High-value Seed Spice Production

19

M. K. Mahatma, S. N. Saxena, and Vinay Bhardwaj

Abstract

Seed spices are an important ingredient of Asian cuisine from ancient times due to their characteristic aroma and flavour. In spite of being used in very small quantities, they have significant effects in treating a variety of ailments at home. Quality is very important to trade at international and national markets. Quality standardizing bodies at the international and national level has fixed some chemical and physical specification for seed spices. Seed spices are mostly used for adding flavour and peculiar taste in prepared dishes. Fixed oil or total oil also consists trace amount of essential oil constitutes and unique fatty acids (petroselinic acid and linolenic acid). Amount of essential oil (EO) and composition are largely affected by sowing and harvesting time and agronomic management. Immature seeds have a low content of EO and major constituent of respective seed spices (Linalool in coriander, anethole, trans-anethole in fennel, cuminaldehyde in cumin and thymol in ajwain). The sowing time has a significant impact on the accumulation of essential oil and its constituents. Late sowing may shorten crop duration and cause higher temperatures during crop maturation which may reduce EO amount and alter EO composition. Application of FYM increased EO amount and improved major constituents in dill seeds, similarly organic cultivation increased EO amount in fennel compared to conventional practices. Constant care is required from pre-harvest operations until the product reaches the customers. In order to prevent contamination and degradation of seed spices, scientific and recommended procedures and practices must be followed at every stage of cultivation, harvesting, post-harvest handling, processing, packing, storage, and transportation to ensure consumer satisfaction.

M. K. Mahatma · S. N. Saxena · V. Bhardwaj (✉)
ICAR-National Research Centre on Seed Spices, Ajmer, Rajasthan, India
e-mail: director.nrcss@icar.gov.in

© National Academy of Agricultural Sciences, under exclusive license to Springer Nature Singapore Pte Ltd. 2023
K. C. Bansal et al. (eds.), *Transformation of Agri-Food Systems*,
https://doi.org/10.1007/978-981-99-8014-7_19

Keywords

AGMARK · Seed spices · Quality · ISO · Essential oil · Petroselinic acid ·
Bioactive compounds

19.1   Introduction

Seed spices are primarily produced and exported by India. Coriander (*Coriandrum sativum* L.), cumin (*Cuminum cyminum* L.), fennel (*Foeniculum vulgare* Mill.) and fenugreek (*Trigonella foenum-graecum* L.) occupy large areas thus considered as major seed spices while ajwain, dill (*Anethum graveolens* L.), celery (*Apium graveolens* L.), anise (*Pimpinella anisum* L.) and nigella (*Nigella sativa* L.) are grown in small areas and considered as minor seed spices. During 2022–23, the export of seed spices contributed about 17.2% of export earnings of a total export value of spices (Rs. 31,761 crore). About 16% of India's total seed spice production (1.8 mt) was exported, with cumin occupying the highest share of the market (29.7%), followed by fennel and fenugreek (15.4% each) and least (6.4%) with coriander (Spice board estimate 2022–23). It is estimated that only a small part of the total production is exported, which can be increased if quality management of seed spices is implemented.

Seed spices have been used as a taste enhancer, flavouring agent, aromatic, and appetiser since ancient times to season or garnish food items and drinks. Also, they are industrially important and are used in pharmaceuticals and cosmetics due to their health-benefiting compounds. Phytochemicals present in seed spices are classified based on their functional groups. Each seed spice has a particular aroma that is attributed to specific/predominant compounds present in the essential oil. Aldehydes, saponins, phenolic acids, flavonoids, tannins, glycosides, steroids, and terpenoids are among the bioactive compounds present in seed spices. Phenolics possess a variety of biological functions such as anti-inflammatory, anti-cancer, antiallergenic, antiproliferative, and antimicrobial properties. Looking at the importance of chemical and physical characteristics for trade, quality standardising bodies have fixed the chemical and physical specifications of each seed spice for the International market.

19.2   What Is Quality?

19.2.1   Quality Has Been Defined in Several Ways

According to Juran's definition "Quality is fitness for use". Quality is the totality of features and characteristics of a product or service that bear on its ability to satisfy stated or implied needs (ISO 8402-1994). The official ISO definition is more informative than Juran's definition. The most important quality features expected by consumers include food safety ('good for health'), and no uncertainty regarding spoiling, composition and weight or volume. (Van den Berg 1993). Thus, it is imperative to consider professional quality criteria such as sensory characteristics, composition, etc. when determining quality.

19.2.2 Why Food Quality Is So Important?

Food regulations are meant to protect customers from poor quality food or food containing contaminants or harmful substances. Developed countries place a high value on the health of their citizens. Thus, any food item that we export meets the importing country's quality standards.

19.2.3 Quality Standardising Bodies

The evolution of standardizing authority has been the most spectacular historical development for standardization. Standardizing authority becoming an economic force in both national and international life which makes voluntary standards for guiding trade and industry. Standards are mostly developed by adhering to the consensus principle. Standards exist at several levels, including company, trade, national, and international. Figure 19.1 depicts the quality standardising bodies at the international and national levels.

Indian standards for the quality of spices have been prepared by the Bureau of Indian Standards (BIS) and the Food Safety and Standards Authority of India (FSSAI). The International Organisation for Standardisation (ISO), a global network of national standards institutes working together, develops voluntary technical standards for a wide range of internationally traded items. The technical committees

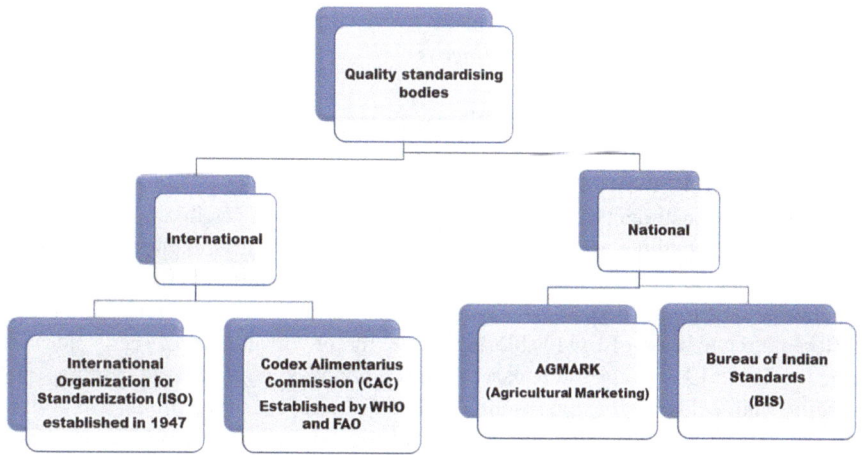

Fig. 19.1 Quality standardising bodies at the International and National Levels

representing the producer nations developed and adopted ISO standards for individual spices and essential oils, which are constantly updated. The ISO standards contribute to higher levels of quality by ensuring minimum standards and specifying standardised analytical methodologies.

Various codes were prepared by Codex Alimentarius Commission for spices. In 1995, the Codex Alimentarius Commission (CAC/RCP 42–1995) approved a Code of Hygienic Practice for Spice and Dried Aromatic Plants. This Code dealt with Good Agricultural Practices (GAPs) and Good Manufacturing Practices (GMPs). This code describes sanitary criteria in the manufacturing/harvesting area and facilities, staff hygiene, hygienic processing requirements, and end-product specifications. The code of practice for the prevention and reduction of mycotoxins in spices (CXC 78–2017) and the code of hygienic practice for low-moisture foods (CXC 75–2015) are available at https://www.fao.org/fao-who-codexalimentarius/codex-texts/codes-of-practice/en/, and the standard for cumin is available at https://www.fao.org/fao CXS 327–2017. The standard, which was adopted in 2017, can be found at https://www.fssai.gov.in/upload/uploadfiles/files/Standard_Cumin_16_0 8_2018.pdf.

19.3   Quality Attributes in Seed Spices

19.3.1 Extrinsic Quality Parameters

The physical characteristics and colour of seed spice primarily determine its marketability. These parameters are generally measured through visualisation and are given in Table 19.1. Both small or large-seeded fenugreek and green or yellow seeds are considered for commercial purposes. The European Spice Association (ESA) has set a 1% limit for extraneous matter in spices and a 2% limit for foreign matter.

19.3.2 Intrinsic Quality Parameters

These parameters include essential oil and its constituents, total oil and its profile, bioactive compounds (phenol, flavonoids, antioxidants and alkaloids). The chemical specification decided by the Spice board is given in the Table 19.2. Each seed spice has a specific constituent in its EO *viz.* cuminaldehyde (cumin), linalool (coriander), estragole and anethole (fennel), diosgenin (fenugreek), thymoquinone (nigella), trans-anethole (anise), Carvone and apiol (dill) and thymol (ajwain). The sweetness or bitterness of fennel depends on its EO composition. Sweet and bitter fennel have different composition of EO (Table 19.3). The major constituents of seed spices are listed in Table 19.4. Some seed spices are also good sources of oils and contain some specific fatty acids such as petroselinic acid (C18:1n12c) is predominant in coriander, fennel, cumin, dill anise and celery oleoresin (fat). Linoleic acid is the major fatty acid of nigella and fenugreek. Fenugreek oil also has high content of linolenic acid generally known as omega-3 fatty acid (Table 19.5). Seed spices are also good

Table 19.1 Physical specification for seed spices (ASTA[a] cleanliness specification)

Specification	Seed spices				
	Coriander	Cumin	Fennel	Dill	Anise
Whole dead insects (count)	4.0	4.0	–	4.0	4.0
Mammalian excreta (mg/lb)	3.0	3.0	3.0	3.0	3.0
Other excreta (mg/lb)	10.0	5.0	–	2.0	5.0
Mould (% weight)	1.0	1.0	1.0	1.0	1.0
Insect defiled/ infested (% weight)	1.0	1.0	1.0	1.0	1.0
Extraneous matter (% weight)	0.5	0.5	0.5	0.5	1.0
Colour	Yellow-brown or green	Light brown	Light green, light brownish green	Brown	Brownish grey

Source: http://www.indianspices.com/spices-development/properties/chemical-physical-specifications-of-spices.html
[a]American Spice Trade Association

Table 19.2 Chemical specification for seed spices

Specification	Seed spices					
	Coriander	Cumin	Fennel	Fenugreek	Dill	Anise
Volatile oil (%) min	0.3	2.5	1.5	0.25	2.0	2.5
Moisture (%) max	9.0	9.0	10.0	12.0	10.0	10.0
Total ash (%) max	6.0	8.0	9.0	3.0	8.0	6.0
Acid insoluble ash(%) max	1.0	1.0	1.0	0.25	1.0	1.0
Average bulk index (mg/100 gm)	285	240	210	120	190	230

Source: http://www.indianspices.com/spices-development/properties/chemical-physical-specifications-of-spices.html

Table 19.3 Comparative essential oil compounds of sweet and bitter fennel

Name of compounds	Compounds (%)	
	Sweet fennel	Bitter fennel
Trans-anethole	52–95	48–75
Estragole	2.5–2.9	8.3–15.5
Fenchone	0.62–3.2	5.03
γ-Terpinene	–	0.84
α-Pinene	0.0–4.0	0.57–18.10
p-Anisaldehyde	0.24	–
α-Phellandrene	–	13.0

Reference: Karlsen et al. (1969); Coşge et al. (2008)

Table 19.4 The major constituents of coriander, cumin and fennel essential oil with their chemical groups

Chemical group	Constituents					
	Coriander	Cumin	Fennel	Dill	Anise	Fenugreek
Alcohols	Linalool (60–80%), geraniol (1.2%–4.6%), terpinen-4-ol (3%), α-terpineol (0.5%)	Geraniol (0.02–2.4%), p-thymol (2.2%), Anethole + Estyragol (4–9%), Cumic alcohol (10–14%) Cymol (2.9–4.3%)				α-Terpineol (2.77%)
Hydrocarbons	γ-Terpinene (1–8%), r-cymene (3.5%), limonene (0.5%–4.0%), α-pinene (0.2%–8.5%), camphene (1.4%), myrcene (0.2%–2.0%)	γ-Terpinene (13–34%), β-Pinene (12–15%), Isoterpinolene (0.83–1.12%), Santolina triene (2–5%), Isoterpinolene (0.47–1.14%)	γ-terpinene (6–11%), 3-Thujene (0.34–.58%), Myrecene (0.14–0.54%), (β-terpinen (0.18–0.33%)	α-Phellandrene (0.19%), p-cymene (0.40%), Limonene (15%)		β-Pinene (15.5%), γ-Terpinene (2.08%), α-Pinene (2.61%)
Ketones	Camphor (0.9%–4.9%)			Dihydrocarvone (13.13%), Carvone (33.57%)		Camphor (16.32%), 3-Octen-2-one (4.32%), 6-Methyl-5-hepten-2-one (4.48%)
Aldehydes		2-caren-10-al (0.02–1.2%), Cuminaldehyde (25–40%)				Campholenal (2.63%), Geranial (4.81%)
Esters	Geranyl acetate (0.1%–4.7%), linalyl acetate (0%–2.7%)	Ethyl Mandelate (0.03–0.24%), Geranyl acetate (0.1–0.2%)	Geranyl acetate (0.07–0.15)			Neryl acetate (17.32%),

Sesquiterpens and pyrazine	β-Caryophyllene (14.63%), α-Selinene (4.04%), 2,5-Dimethylpyrazine (6.14%),	α-Himachalene (0.71%), β-Himachalene (0.44%) γ-Himachalene (7.0%), β-Elemene (0.45%), α-Zingiberene (0.77%), β-Bisabolene (0.38%)			
Phenylpropene		Estragole (0.33%), trans-anethole (82.1%)	Thymol (0.25%), Carvacrol (4.92%), Myristicin (24.21%) Dill apiole (15–19%) Apiol (17–33%)	Anethole (38–44%), estragole (22–34%)	Anethole (0.21–0.71%), estragole (0.03–0.1%)
References	Hamden et al. (2011)	Ullah et al. (2014)	Dimov et al. (2018), Kaur et al. (2021)	Coşge et al. (2008)	Dubey et al. (2017)

Additional Phenylpropene / References column:

Phenylpropene	Thymol (1.85%) Carvacrol (0.46%)
References	Nadeem et al. (2013)

Table 19.5 Fatty acid and sterols profile of total oil in seed spices

S. No.	Fatty acid (%)	Coriander	Cumin	Fennel	Dill	Anise	Fenugreek
1	Myristic acid (C14:0)	0.08–0.09	0.1–0.2	0.57	0.9	0.07	0.1–0.3
2	Hexadecanoic acid/palmitic acid (C16:0)	3.69–3.67	3.9–4.3	0.12–7.8	8.50	6.15	5.0–13.8
3	9-Hexadecenoic acid/ Palmitoleic acid (C16:1n-7)	0.22–0.23	0.2–0.3	–	0.3	0.8	0.1
4	Octadecanoic acid/Stearic Acid (C18:0)	0.77–0.88	0.3–1.0	1.0–2.5	1.5	1.3	0.8–7.6
5	9-Octadecenoic acid /Oleic acid (C18:1n9c)	5.6–5.8	11.2-12.2	0.04–6.4	15.3	0.09	2.5–18.3
6	6-octadecenoic acid/ Petroselinic acid (C18:1n12c)	75.4–75.9	47.4–51.5	62.0–69.2	50.4	65.2	–
7	Cis-vaccenic acid (C18:1n7)	–	1.2–1.5	–	–	–	–
8	9,12-Octadecadienoic acid/ Linoleic acid (C18:2n6c)	13.3–13.8	30.5–32.9	6.9–7.6	20.5	25.1	33.4–43.4
9	9,12,15-Octadecatrienoic acid/ Linolenic acid (C18:3n3)	0.14–0.15	0.2–0.5	0.01–0.03	–	–	26.8–40.5
10	Arachidic acid (C20:0)	0.9–0.11	0.1	–	0.2	–	0.7–2.0
11	Sterols (mg/100 g of oil)	518	237–356.0	517.3	361.4	551.9	1420–1883
12	Campesterol (%)	7.9–8.3	10.6–11.7	3.8	5.8	3.5	8.7–20.5
13	Stigmasterol (%)	24.1–24.8	38.3–40.2	35.3	30.2	23.5	1.8–6.0
14	Δ7 –stigmasterol (%)	4.5–5.0	4.0–5.1	–			1.1–5.7
15	Δ5 –stigmasterol (%)	15.4–16.3	–	–			–
16	β-Sitosterol (%)	28.8–34.9	43.4–46.5	25.7	15.9	22.0	31.8–49.6

#							
17	α-Spinasterol	–	–	–	17.3	19.9	–
18	Cycloartenol	–	–	–			6.1–14.8
	References	Sriti et al. (2011)	Merah et al. (2020)	Bernath et al. (1994); Agarwal et al. (2018)	Saini et al. (2021a, b)	Alfekaiki (2018)	Ciftci et al. (2011); Rathore et al. (2017)

Table 19.6 Phenolics profile of seed spices

Phenolics	Coriander mg^{-1} 100 g	Cumin	Fennel	Dill	Anise	Fenugreek
Phenolic acid	113.48	145–186	780	215–713	238	490–10,631
Gallic acid	1.58	9.0	27.7–66.0	11–68.0	60.7	1.71
Caffeic acid	1.10	7.0–22.0	29.0–83.4	285.0	44.5	
p-Coumaric acid	4.38	233.0–483.0	5.45–42.4	11.1	5.0	1.4
Ellagic acid	–		9.9		10.6	
Ferulic acid	0.17	14.0–47.0	1.3–69.7	2.9	36.6	3.4
Chlorogenic acid	1.34	6.0–22.0	9.6–232.5		11.8	16.8
Rosmarinic acid	0.76	70.0–104.0	18.3	34.7	4.9	
Cinnamic acid	0.89	27.0–94.0	0.9		24.3	
Syringic acid						6.18
Flavanoids	11.42	177–288	18–84	52–672	156	18–84
Epicatechin				276.2	11.4	
Myricetin	–		19.8	0.70		
Quercetin	–	2.0–8.0	21.5–145	48–110	5.00–13.30	4.2 (1.4[a])
Kaempferol	–	215.81	6.5	16–30.5		0.16 (2.15[a])
Luteolin	3.31	39.0-129.0	9.3			0.24
Catechin	–	23.0–52.0	2.1–10.0	91.0	71.4	
Flavone	–	12.0–61.0				
Rutin	–		10.4–69.7	84.6	1.17–11.02	
Apigenin					5.59–6.4	2.9
Saponins						5.12
Vitexin						0.86
Isorhamnetin				15–72		
Resveratrol				100.4		
References	Sriti et al. (2012)	Rebey et al. (2012), Shahidi and Hossain (2018)	Shahidi and Hossain (2018), Malin et al. (2022)	Shahidi and Hossain (2018), Bota et al. (2021)	Shahidi and Hossain (2018), Sakr et al. (2019)	Chatterjee et al. (2009); Pasha et al. (2017)

[a]Amount of bound Quercetin and Kaempferol

sources of phytosterols (Table 19.5) and bioactive phenolics and flavonoids, which are listed with quantity in Table 19.6.

19.4 Good Agriculture Practices (GAP) for Production of Quality Seed Spices

19.4.1 Date of Sowing

Timely sowing is very crucial for the yield and quality of seed spices. The accumulation of EO and its constituents is largely affected by sowing time. Late sowing may shorten crop duration and cause higher temperatures during crop maturation which may reduce EO amount and alter EO composition. Moreover, it also affects the extent to which incidence of disease and pests can take place. The earlier sowing date (24 May) accumulated higher EO in coriander compared to the second sowing date (8 June) in Canada. Linalool (64% to 84.6%) was the predominant ingredient of coriander EO (Tables 19.5 and 19.6). Camphor, alpha-pinene, phellandrene, linalyl acetate, limonene, para-cymene, and geranyl acetate were minor components in the EO. Linalool (78.1%), camphor (4.2%), and linalyl acetate (2.8%) contents were higher in seed EO from the first sowing date than the second sowing date for linalool (70.7%), camphor (0.0%), and linalyl acetate (2.5%), respectively (Zheljazkov et al. 2008).

Selim et al. (2013) conducted an experiment on fennel to study the effect of sowing dates (15th Sep., 7th Oct. and 1st Nov.), plant spacing (25, 35 and 50 cm between two plants on a hill) and bio-fertilization (three spray per season with active dry yeast at 0, 2 and 5 gm^{-1} L) on growth, seed yield and EO production. They observed a maximum EO yield $plant^{-1}$ when fennel was sown on seventh Oct. at 50 cm spacing and spraying with 5 gm^{-1} L of yeast. The lowest content of estragole (48.96%) and maximum content of anethole (15.75%) was detected in fennel seeds produced from sowing on seventh October with 35 cm spacing and 2 gm^{-1} L of dry yeast spraying. While the highest estragole (87.98%) was obtained on late sowing (first Nov.) with 35 cm spacing and 2 gm L^{-1} of dry yeast spraying. The maximum content of limonene (23.99%) was observed with 15th September +25 cm + 5 gm^{-1} L treatment.

19.4.2 Harvesting Stages

The harvesting stage of seed spices greatly affects both intrinsic and extrinsic quality parameters including seed colour, essential oil yield and its composition. The EO content reported in coriander seeds varied from 0.5% to 2.5%, and it increases with the maturity of the seeds. The EO content in coriander was varied in immature (full green), intermediate (green-brown), and mature (Brown) seeds with a yield of 0.01%, 0.12% and 0.35% (w/w), based on dry weight, respectively. In the first stage of harvesting (immature), geranyl acetate (46.27%) was a major constituent of EO followed linalool (10.96%). At the intermediate and the final stage of maturity, the chemical composition of seed EO had similar profiles and significantly differed from immature seeds. At the intermediate stage, linalool (76.33%) content was highest in EO composition followed by cis-dihydrocarvone (3.21%) and geranyl acetate (2.85%). Essential oils of the mature seeds possessed maximum content of linalool (88.51%) and minimum content of geranyl acetate (0.83%). Mature seeds also had 2.36% of cis-dihydrocarvone (Msaada et al. 2007).

The EO content of the fennel declined steadily from the first stage to the last stage. As an average of two seasons, stage 1 where the primary umbel was at the start of the waxy stage has more than twice as much essential oil (2.37%) as stage 8 where the primary umbels were dropping. However, the highest oil yield (1.82 and 1.90 ml/plant) was observed when the primary umbel was totally matured (stage 4). Major components in fennel EO were varied uring seed development. The major component was methyl chavicol (estragole) ranged from 72.34% to 88.67%. It was observed that delaying harvest till the last harvest stage led to a higher accumulation of methyl chavicol, which is not good for market demand because it increases bitterness. Methyl chavicol levels were lowest (72.34%) in stage 4 (72.34%), followed by stage 3 (72.90%) where the colour of the secondary umbels was changed. While, limonene, the second largest component, declined with maturity from 16.80% to 5.27% (El-Gamal and Ahmed 2017a). In another study, El-Gamal and Ahmed (2017b) observed an appreciably higher concentration of EO (4.82–4.95%) in anise seeds, when the primary umbel was at the beginning of waxy stage (stage 1). However, the highest essential oil yield per plant (1.57–1.68 mL/plant) with high trans-Anethole content (94.2%) and low estragole content (0.34%) was observed when the primary umbel was matured completely (in the fourth harvesting stage). (Moghaddam et al. 2015) determined cumin essential oils content and their constituents, total phenolic content and antioxidant activity of EO at four different stages of maturity (immature, intermediate, premature and fully mature). Mature seeds showed the highest EO content (4.3%), whereas immature seeds had the lowest (2.7%). GC/MS analysis of the essential oil profile revealed that the concentrations that the amounts of alpha-pinene, beta-pinene, alpha-phellandrene, alpha-terpinene, and gamma-terpinene declined, whereas the levels of p-cymene, alpha-terpineol and cumin aldehyde enhanced with maturation. Phenolic contents (Gallic acid equivalent: GAE) and antioxidant activities of EO had a positive correlation and enhanced at intermediate and premature stages. Intermediate-stage seeds had the highest antioxidant activity (132.0 μmol Fe^{2+}/mg EO) and phenolic content (40.0 mg GAE/g EO), followed by premature seeds (120.98 μmol Fe^{2+}/mg EO and 36.86 GAE/g EO), while immature (89.32 μmol Fe^{2+}/mg EO and 25.52 GAE/g EO) and mature seeds (106.0 μmol Fe^{2+}/mg EO and 30.0 GAE/g EO) had lower antioxidant activity and phenolic content. Numerous research studies have demonstrated that the growth/maturity stages of a crop are very important for obtaining higher essential oil yields and the desired level of major constituents.

19.4.3 Agronomic Management

The fennel seeds contain 1–4% of essential oil. The EO yield of fennel was observed to be higher (1.37–4.3%) when grown in organic cultivation than in conventional cultivation practice (1.37–3.24%). The highest EO yields of 4.3% and 3.24% respectively were recorded in E7 accession grown in organic and conventional cultivation modes (Ben Abdesslem et al. 2020). The essential oil content of coriander was increased from 2.20 to 7.32 kg ha^{-1} (3.3 times higher) due to the application of 100 kg of nitrogen, phosphorus and sulphur fertilizers as compared to control or zero fertilizer (Abnet et al. 2023). Application of various amounts of FYM (7.5, 10, 12.5 and 15 t ha^{-1}) and ammonium nitrate (30, 60, 90 and 120 kg ha^{-1}) as a nitrogen fertilizer significantly improved EO yield and its constituents in dill seeds (Ozliman et al. 2021). Maximum EO content (6.18%) was reported with the 15 t ha^{-1} of FYM application followed by 30 kg ha^{-1} of ammonium nitrate (5.98%) and FYM @ 12.5 t ha^{-1} (5.91%) application. Moreover, FYM @ 15 t ha^{-1} application also improved the amount of carvantonacetone (21.76%), alpha-phellandrene (10.34%), limonene (12.53%) and myrcene (1.56%) in the EO compared to control (12.28, 2.66.11.26 and 1.25%, respectively).

19.5 Pre and Post-Harvest Operations for Quality Seed Spice Production

The raw materials and processes utilised in processing, packing, storage, and shipping all have an impact on the quality of any food product. Continuous attention is essential for agricultural products from pre-harvest operations till the product reaches its consumer.

19.5.1 Important Points during Harvesting

Coriander and fennel should be harvested before full maturity stage to retain their green colour while other seed spices may be harvested at the full maturity stage. It is critical to avoid injuring the seeds while harvesting. The importing countries strongly oppose chemical colourings, so colouring substances should not be used to improve the appearance of goods.

19.5.2 Storage Precautions

Harvested materials should be stored in a dry place. To prevent moisture entry from the floor and wall, the packed bags should be stacked 50–60 cm away from the wall on dunnage. Under no circumstances should insecticide be put directly on dried material. Insecticide should not be used directly on the dried material under any circumstances. Stored materials should be subjected to periodic fumigation and exposure to sunlight. Entry of insects, rodents, and other animals into the storage building should be prevented.

19.5.3 Microbial Load

Salmonella and *E coli* contamination limits are set by both ASTA and ESA. Seed spices may also contain other pathogens such as *Bacillus cereus, Clostridium perfringens*, and *Shigella* spp. Most products have strict limits for mould in their import regulations. It is possible to reduce or eliminate microbial contamination in spices in a number of ways. Steam heat sterilization results in the loss of volatile oils and flavours in seed spices. Irradiation is another option that is both economical and useful. It is reported that about 40 countries have accepted sterilization by irradiation. Mostly microbial contamination results from careless handling and storage after harvest. Microbial contamination issues can be significantly reduced with proper drying and storage and following good agricultural and hygienic practices.

19.5.4 Aflatoxins

There are several types of aflatoxins produced by *Aspergillus flavus* and *Aspergillus parasiticus*, including B1, B2, G1, G2, M1 and M2. Contamination with aflatoxins in food products has become a major health hazard thus importing countries are enforcing strict standards. Most of the developed countries set a maximum limit for total aflatoxin content between 1–2 ppb. It has been observed that improper storage practices, as well as careless handling before, during, and after harvest, contribute to *Aspergillus* fungal infestation. This will greatly lessen the issue if modern dryers are used during post-harvest procedures, and the moisture content is kept below 6%.

19.5.5 Trace Metal Contamination

In developed nations, trace metal contamination is also seen as a health concern. Mercury, cadmium, arsenic, chromium, and lead are among the metals that are currently regarded as harmful. To comply with importer requirements, certain developing nations that previously used lead as a solder for tin cans were able to switch to another soldering substance. It should be mentioned that copper is a component of commonly used fungicides in the agricultural sector, such as the Bordeaux mixture. Thus care should be taken to reduce trace metal contamination.

19.5.6 Pesticide Residues

Pesticide residue is the most significant non-tariff restriction on the trade of spices. It is impossible to entirely remove pesticide residues during processing because they are acquired during the pre-harvest period. Therefore, it is essential to develop workable and affordable pre-harvesting solutions to solve this issue. Spice seeds should not be treated with pesticides that are banned in the countries where they are imported. Use fungicides or insecticides in cases of disease or pest infestation only

Table 19.7 Maximum residue Limit (mg/kg) for seed spices decided as per international food standards

Pesticide	Maximum residue limit (MRL) mg/Kg			
	Cumin	Coriander	Fennel	Dill
Dithiocarbamates	10.0	0.1	0.1	–
Phorate	–	0.1	0.1	–
Profenofos	5.0	0.1	0.1	–
Trizophos	–	0.1	0.1	–
Fluopyram	–	0.1	0.1	70.0

NB: There are currently no MRLs for the Anise and Fenugreek
From: https://www.fao.org/fao-who-codexalimentarius/codex-texts/dbs/pestres/commodities/en/

after consulting professionals and adhering to their dosage and schedule. Seed spices should be devoid of pesticide residue or have a maximum residue level (MRL) of pesticides according to the importing country. A MRL is the highest level of pesticide residue that is legally tolerated in or on food or feed when pesticides are applied correctly in accordance with GAP (Table 19.7).

19.5.7 Grading

Seeds should be graded using a grading machine or gravity separator after they have been properly dried. Coriander is categorised in the Indian market into different categories based on seed colour. Brown colour seeds are called *Badami*, while medium green and green colour seeds are called *Eagle/ Scooter* and *Single/Double Parrot*, respectively. Graded produce should be packaged in smaller packs of high quality polythene or aluminium pouches in a variety of sizes (from 100 g to 5.0 kg).

19.6 Value Addition

19.6.1 Powder Purpose

Seeds are mainly processed into powder by crushing them and the powder is used as a food ingredient for specific aroma. Seed spices powder can be prepared using cryogenic grinding to retain maximum aroma and EO constituents. ICAR-NRCSS has developed a cryogenic grinding technique for seed spices. It has been reported that cryogenic grinded powder of seed spices has higher EO and Oleic acid (Sharma et al. 2017). To ensure cleanliness, seeds should be washed with water to remove dust and any other adhering materials and then dried. Sun drying is widely used being cheaper. However, cabinet drying at 40 °C may be also used to retain a green colour and more aroma.

19.6.2 Essential Oil, Oleoresins and Bioactive Compounds

Each seed spice is packed with a specific aroma due to its unique EO compound. The essential oil is extracted by distilling mature dried seeds. The hydro or steam distillation process is commonly employed for essential oil extraction. EO of all the seed spices have been reported to have antimicrobial and antioxidant activity along with various nutraceutical properties. Essential oil and Oleoresins can be promoted other value-added products in the market. Coriander, cumin, fennel, dill and anise seeds oleoresins have rare predominant specific fatty acid (45–75%), petroselinic acid, which is used as moisturising and anti-ageing agent in cosmetic industry.

Fenugreek seed has about 25% gum or mucilage content. Fenugreek seed has galactomannan as a major component (80%) followed by protein (5%). When galactomannan is dissolved in water, it increases viscosity. These qualities distinguish it from other natural hydrocolloids as a useful component for a variety of food applications (Khorshidian et al. 2016). Due to its high fibre, protein, and gum content, fenugreek is utilised as a food stabiliser, adhesive, and emulsifying agent. Sapogenins in fenugreek control cholesterol levels, while 4-Hydroxyisoleucine controls blood sugar levels.

Fenugreek contains steroidal sapogenins (diosgenin), and 4-hydroxyl isoleucine (free unnatural amino acids) which control cholesterol and blood sugar levels, respectively.

19.6.3 Dehydrated Green Leaves

The dehydrated green leaves and stem of coriander, fennel, dill, fenugreek, and celery are also used for drying. The leaves are generally sun-dried, but to retain maximum color, aroma and nutritive value these can be dehydrated in a suitable dehydrator or lyophilized for further use in off-season.

19.7 Conclusions

Aroma, colour and cleanliness are considered the most vital parameters when evaluating the quality of spices. According to the hygiene standards, seed spices should neither contain pathogenic microbes nor must insects rather be practically free from extraneous substances and defective material. Seed spices should contain minimum content of EO and follow other chemical and physical specifications described by ASTA and ESA. Various factors affect the chemical composition of seed spices EO, including geography, climate conditions, varieties, cultural practices, and extraction methods. However, the impact of these conditions on fatty acids, sterols, phenolic acids and flavonoids is still largely unknown. Since seed spices possess important chemical constituents, EO and bioactive compounds, used in food and pharmaceutical manufacturing, it is important to focus research on improving seed spices' quality. To get better quality of the harvested produce GAP protocols for cultivation of spices are important steps to be followed by growers.

References

Abnet A, Abrham S, Abraham B (2023) NPS fertilizer and spacing effects on yield and quality of coriander (*Coriandrum sativum* L.). J Plant Sci 18:12–21

Agarwal D, Saxena SN, Sharma LK et al (2018) Prevalence of essential and fatty oil constituents in fennel (*Foeniculum vulgare* Mill) genotypes grown in semi-arid regions of Indian. J Essent Oil Bear Plants 21:40–51

Alfekaiki DF (2018) Chemical and physical characteristics and fatty acid profile of some oil seeds of Apiaceae family in Iraq. Chem Proc Eng Res 58:17–27

Ben Abdesslem S, Boulares M, Elbaz M, Ben Moussa O, St-Gelais A, Hassouna M, Aider M (2020) Chemical composition and biological activities of fennel (*Foeniculum vulgare* Mill.) essential oils and ethanolic extracts of conventional and organic seeds. J Food Process Preserv 45:1. https://doi.org/10.1111/jfpp.15034

Bernath J, Kattaa A, Nemeth E, Franke R (1994) Production-biological investigation of fennel (*Foeniculum vulgare*) populations of different genotypes. Atti Convegno Intern:287–292

Bota SR, Stanasel OD, Blidar CF, Serban G (2021) Phenolic constituents of *Anethum graveolens* seed extracts: chemical profile and antioxidant effect studies. J Pharm Res Int 33:68–179

Chatterjee S, Variyar PS, Sharma A (2009) Stability of lipid constituents in the radiation processed fenugreek seeds and turmeric: role of phenolic antioxidants. J Agric Food Chem 57:9226–9233

Ciftci ON, Przybylski R, Rudzinska M, Acharya S (2011) Characterization of fenugreek (*Trigonella foenum-graecum*) seed lipids. J Am Oil Chem Soc 88:1603–1610

Coşge B, Kiralan M, Gürbüz B (2008) Characteristics of fatty acids and essential oil from sweet fennel (*Foeniculum vulgare* Mill. Var. dulce) and bitter fennel fruits (*F. Vulgare* Mill. Var. vulgare) growing in Turkey. Nat Prod Res 22(12):1011–1016

Dimov M, Georgieva K, Denev Y, Dobreva K, Stoyanova A (2018) Analysis of the chemical composition of dill essential oils (*Anethum graveolens* L.) by the method of infrared spectroscopy. Scientific works of University of Food Technologies 65:55–58

Dubey PN, Saxena SN, Mishra BK et al (2017) Preponderance of cumin (*Cuminum cyminum* L.) essential oil constituents across cumin growing Agro-ecological sub regions, India. Ind Crop Prod 95:50–59

El-Gamal S, Ahmed H (2017a) Influence of different maturity stages on fruit yield and essential oil content of some apiaceae family plants B: fennel (*Foeniculum vulgare* Mill.). J Plant Prod 8: 127–133

El-Gamal S, Ahmed H (2017b) Influence of different maturity stages on fruit yield and essential oil content of some Apiaceae Family plants A: Anise (*Pimpinella anisum*, L.). J Plant Prod 8:119–125

Hamden K, Henda K, Belhaj S (2011) Inhibitory potential of omega-3 fatty and fenugreek essential oil on key enzymes of carbohydrate-digestion and hypertension in diabetes rats. Lipids Health Dis 10:226

Karlsen J, Baerheim SA, Chingova B et al (1969) Fruits of *Foeniculum* species and their essential oils. Planta Med 17(3):281–293

Kaur V, Kaur R, Bhardwaj U (2021) A review on dill essential oil and its chief compounds as natural biocide. Flavour Fragr J 36:412–431

Khorshidian N, Asli MY, Arab M, Mortazavian AM, Mirzaie AA (2016) Fenugreek: potential applications as a functional food and nutraceutical. Nutr Food Sci Res 3:5–16

Malin V, Elez Garofulic I, Repajic M et al (2022) Phenolic characterization and bioactivity of fennel seed (*Foeniculum vulgare* Mill.) extract isolated by microwave assisted and conventional extraction. PRO 10:510

Merah O, Sayed-Ahmad B, Talou T, Saad Z, Cerny M, Grivot S, Evon P, Hijazi A (2020) Biochemical composition of cumin seeds, and biorefining study. Biomol Ther 10:1054. https://doi.org/10.3390/biom10071054

Moghaddam M, Khaleghi Miran SN, Pirbalouti AG, Mehdizadeh L, Ghaderi Y (2015) Variation in essential oil composition and antioxidant activity of cumin (*Cuminum cyminum* L.) fruits during stages of maturity. Ind Crop Prod 70:163–169

Msaada K, Hosni K, Ben Taarit M, Chahed T, Kchouk ME, Marzouk B (2007) Changes on essential oil composition of coriander (*Coriandrum sativum* L.) fruits during three stages of maturity. Food Chem 102:1131–1134

Nadeem M, Anjum FM, Khan MI, Tehseen S, El-Ghorab A, Sultan J (2013) Nutritional and medicinal aspects of coriander (*Coriandrum sativum* L.)—a review. Br Food J 115:743–755

Ozliman S, Yaldiz G, Camlica M, Ozsoy N (2021) Chemical components of essential oils and biological activities of the aqueous extract of *Anethum graveolens* L. grown under inorganic and organic conditions. Chem Biol Technol Agric 8:20

Pasha I, Shabbir MA, Haider MA, Afzal Ab, Chughtai MF, Ahmad S, Shabbir, Manzoor, MS (2017) Biochemical evaluation of Trigonella foenum graecum (Fenugreek) with special reference to phenolic acids. Pakistan J Scient Ind Res Series B: Biol Sci 60:154–161

Rathore SS, Saxena SN, Kakani RK, Sharma LK, Agrawal D, Singh B (2017) Genetic variation in fatty acid composition of fenugreek (*Trigonella foenum-graecum* L.) seed oil. Legum Res 40:609–617

Rebey IB, Zakhama N, Karoui IJ, Marzouk B (2012) Polyphenol composition and antioxidant activity of cumin (*Cuminum cyminum* L.) seed extract under drought. J Food Sci 77:C734–C739

Saini RK, Assefa AD, Keum YS (2021a) Spices in the Apiaceae Family represent the healthiest fatty acid profile: a systematic comparison of 34 widely used spices and herbs. Foods 10:854

Saini RK, Song MH, Yu JW, Shang X, Keum YS (2021b) Phytosterol profiling of apiaceae family seeds spices using GC-MS. Foods 10:2378

Sakr AAE, Taha KM, Abozid MM, El-saed HEZ (2019) Comparative study between anise seeds and mint leaves (chemical composition, phenolic compounds and flavonoids). Menoufia J Agric Biotechnol 4:53–60

Selim SM, Abdella EMM, MSH T, Abou-Sreea AI (2013) Effect of sowing date, sow spacing and bio-ertilizer on yield and oil quality of fennel plant (*Foeniculum Vulgare*, Mill.). Australian J Basic Appl Sci 7:882–894

Shahidi F, Hossain A (2018) Bioactives in spices, and spice oleoresins: phytochemicals and their beneficial effects in food preservation and health promotion. J Food Bioact 3:8–75. https://doi.org/10.31665/JFB.2018.3149

Sharma LK, Agarwal D, Rathore SS, Malhotra SK, Saxena SN (2017) Effect of cryogenic grinding on volatile and fatty oil constituents of cumin (Cuminum cyminum L.) genotypes. J Food Sci Technol 53:2827–2834

Sriti J, Neffati M, Msaada K, Talou T, Marzouk B (2012, 2013) Biochemical characterization of coriander cakes obtained by extrusion. J Chem. Article ID 871631

Sriti J, Talou T, Faye M, Vilarem G, Marzouk B (2011) Oil extraction from coriander fruits by extrusion and comparison with solvent extraction processes. Ind Crop Prod 33:659–664

Ullah H, Mahmood A, Honermeier B (2014) Essential oil and composition of anise (*Pimpinella anisum* L.) with varying seed rates and row spacing. Pak J Bot 46:1859–1864

Van den Berg MG (1993) Kwaliteit van levensmiddelen (quality of food). Khrwer, Deventer

Zheljazkov VD, Pickett KM, Caldwell CD, Pincock JA, Roberts JC, Mapplebeck L (2008) Cultivar and sowing date effects on seed yield and oil composition of coriander in Atlantic Canada. Ind Crop Prod 28:88–94

Genomic Innovations for Improving Crops: The CRISPR Way

20

Rutwik Barmukh and Rajeev K. Varshney

Abstract

The rapid growth in the human population and increasing environmental fluctuations have created a pressing necessity to develop crops that exhibit greater yields and heightened resilience to climate change. Genome editing mediated by Clustered Regularly Interspaced Short Palindromic Repeat (CRISPR)/CRISPR-associated protein (Cas), hold the potential in addressing these challenges by enabling targeted modifications in the crop genomes, thereby creating novel variations and expediting breeding efforts. The use of genome editing for improving crops is not restricted by narrow genomic diversity or the necessity for multiple breeding generations to select desired alleles. However, the deployment of this technology for editing crop genomes face limitations due to the absence of complete, high-quality reference genomes, limited understanding of possible editing targets, and a lack of functional assays to assess the effect of specific gene edits. To overcome these obstacles, advancements in next-generation sequencing are being utilized at different stages of the genome editing process. These techniques enable the analysis of CRISPR off-target effects, confirmation of gene knockouts, and validation of additional edits. Various high-throughput sequencing methods are presently employed to evaluate the influence of CRISPR/Cas-mediated genome edits on the structure and functionality of genes. When integrated with precise phenotyping and functional genomic studies, genomics is providing novel foundations for designing future crops using genome editing.

R. Barmukh · R. K. Varshney (✉)
WA State Agricultural Biotechnology Centre, Centre for Crop and Food Innovation, Murdoch University, Murdoch, WA, Australia
e-mail: Rajeev.Varshney@murdoch.edu.au

© National Academy of Agricultural Sciences, under exclusive license to Springer Nature Singapore Pte Ltd. 2023
K. C. Bansal et al. (eds.), *Transformation of Agri-Food Systems*,
https://doi.org/10.1007/978-981-99-8014-7_20

Keywords

Genome editing · CRISPR/Cas · Genomics-assisted breeding · Next-generation sequencing · Agriculture · Crop improvement

20.1 Introduction

The rising global human population and environmental fluctuations are placing significant stress on agriculture to satisfy the growing demands for food and nutritional security. Projections indicate that within the next three decades, the world's population will surge from the current 8 billion to 10 billion (https://www.worldometers.info/world-population/), while the food requirement is estimated to surge by up to 56% (van Dijk et al. 2021). However, the expected advancements in agricultural practices are predicted to yield only modest improvements for major crops (Ray et al. 2013). Considering the diminishing availability of arable land worldwide due to land degradation and urban expansion, a viable approach is development of crops that can proficiently utilize existing resources and possess resistance against various abiotic and biotic stresses. By complementing agronomic practices, this strategy of designing climate-resilient crops can considerably lessen the environmental influence of farming, resulting in reduced reliance on pesticides and fertilizers while achieving higher yields.

Despite the continuous advancements in conventional plant breeding, which result in improved traits and better varieties, this process depends on the germplasm crossing that can be time-consuming, taking about 4–6 years to develop a successful variety. As a result, traditional breeding is falling short to maintain pace with the projected food demand. While genetic markers can be utilised to enhance breeding efficiency, there are limitations in improving crop germplasm because of the non-targeted nature of recombination (Epstein et al. 2023). In contrast, genome editing technologies employing CRISPR/Cas system induce targeted edits at specific sites in the genome (Doudna and Charpentier 2014; Jinek et al. 2012). The CRISPR/Cas system's enhanced versatility and cost-efficiency, attributed to its sequence-specificity, position it as the preferred choice when compared to zinc finger nucleases and transcription activator-like effector nucleases (Molla et al. 2021). Furthermore, the CRISPR/Cas tool expedites the rapid development of improved varieties without the need for transgenes, thereby accelerating progress in both basic research and crop improvement (Pixley et al. 2022; Bansal et al. 2022; Zhu et al. 2020).

The utilization of the CRISPR/Cas technology as an innovative method for improving crops not only provides a potent breeding tool but also requires additional genomic information. Successful integration of genome editing into crop enhancement depends on high-quality reference genomes and comprehensive functional genomics data to discover suitable editing targets, involving potential genes and regulatory sequences present within the genome (Varshney et al. 2019; Zhu et al. 2020). Modern breakthroughs in next-generation sequencing (NGS) techniques offer valuable enhancements to various stages in the genome editing process. For instance, employing whole-genome sequencing enables the investigation of CRISPR off-target effects (Wang et al. 2021), while targeted sequencing facilitates the confirmation of CRISPR knockouts and other specific modifications (Fraiture et al. 2023). Subsequent investigations can then be conducted using techniques such as methylation analysis and RNA sequencing for expression profiling, enabling the evaluation of the functional impact resulting from a particular gene edit.

This article proffers an overview of the essential prerequisites for obtaining accurate genome-wide information, which plays a central role in identifying potential genes and regulatory sequences for genome editing. We shed light on the genomic techniques that hold promise to enhance editing precision, multiplexing abilities, and gene regulation via the CRISPR/Cas system. Finally, we discuss applications of genome editing and advocate for a transition in crop improvement strategies, urging the integration of genomic innovations and genome editing. By combining these state-of-the-art approaches, we aim to underscore the immense potential of genomic advancements in improving the precision of genome editing for designing future crops.

20.2 Revolutionizing Crop Improvement: The Power of Genomics-Assisted Breeding before CRISPR/Cas Took Centre Stage

Several research endeavours have been conducted thus far to expedite crop improvement by enhancing the efficiency of selection. These efforts rely on the utilization of methodologies such as marker-assisted backcrossing (MABC), marker-assisted recurrent selection (MARS), and genomic selection (GS) (Varshney et al. 2021a). MABC seeks to alter particular traits in selected lines while maintaining the remaining characteristics of the desired cultivar intact (Barmukh et al. 2022; Hasan et al. 2015). It employs the linkage disequilibrium (LD) existing within molecular markers and quantitative trait loci (QTLs) to identify crops that possess the desired traits for crop breeding programs. Despite significant advancements in marker technologies within the past few decades, MABC encounters complications while dealing with complex polygenic traits, primarily due to the intricate nature of genotype × environment interactions. When dealing with the selection of complex traits, it appears that MARS or GS may be appropriate for amalgamating beneficial alleles. For example, MARS facilitates the enhancement of polygenic characters by combining numerous favourable alleles from the most significant regions closely associated with target traits (Singh et al. 2022; Beyene et al. 2016). By contrast, GS involves the prediction of breeding values for different genotypes within a training population that has undergone both phenotyping and genotyping. This enables the selection of breeding populations via genomic-estimated breeding values (GEBVs) and not depending solely on phenotypes or gene-linked markers (Crossa et al. 2017). Implementing GS in breeding programs underscores that the precision of predictions diminishes with increasing generational gaps from the training population. This decline is partly attributed to the limitations of GS models in identifying markers that are in strong LD with the causal loci. This is where genome-wide association study (GWAS) excels, as it is utilized to unravel the genetic composition of complex quantitative characters in crop plants (Uffelmann et al. 2021).

With the growing comprehension of crop genomes, genomics-assisted breeding methods may transition towards strategies that rely on candidate genes that have been functionally validated. One important advancement in functional genomics that provides insights into the function of gene(s) through reverse genetics is the utilization of approach known as Fast Identification of Nucleotide variants by droplet DigITal PCR (FIND-IT) (Knudsen et al. 2022). The FIND-IT approach combines efficient sample pooling and splitting with extremely vulnerable droplet digital PCR-based genotyping. Within crop breeding, this approach enables the screening of large variant populations with low mutation density, allowing for the targeted detection of desired traits at the level of individual nucleotides. By specifically targeting SNPs in candidate genes that govern yield performance, grain structure, and quality traits of barley, the FIND-IT approach overcomes the limitations of comparable techniques (Knudsen et al. 2022). Nevertheless, as breeding programs increasingly emphasize potential genes for specific target traits, CRISPR/Cas emerges as a more precise technology for generating new allelic variants of these genes. This enables subsequent agronomic evaluations to be conducted with greater accuracy and control.

20.3 Harnessing High-Quality Reference Genomes as the Gateway to Genome Editing

The presence of a well-constructed and fully annotated reference genome is essential for targeted modifications in crops, as it provides crucial information about the foundational genetic material. High-quality reference genomes play a critical role in identifying functional regions associated with specific traits of interest, like candidate genes, enhancers, and promoters. For instance, the efficacy of a high-quality pigeonpea reference genome (Varshney et al. 2012a) was exemplified by the successful cloning and incorporation of a resistance gene called *CcRpp1* from pigeonpea into soybean (Kawashima et al. 2016). This transfer of *CcRpp1* resulted in soybean plants acquiring complete resistance against *P. pachyrhizi*, showcasing the valuable impact of the pigeonpea reference genome in soybean breeding efforts. Such high-quality genomes also greatly aid in developing single-guided RNA (sgRNA) and the placement of sequencing reads, which are utilized for characterizing edited sections of the crop genome (He et al. 2021). Notably, the advancement of genome assemblies for plants has experienced a significant rise from the release of the rice genome in the year 2005. As of 2020, there were over 200 chromosome-level whole genome assemblies of terrestrial plants available, thanks to the advancements in sequencing capabilities and improved computational technologies (Varshney et al. 2012b, 2021a). With the recent advances and ongoing developments in these fields, we anticipate the sequencing and assembly of good quality genomes for all crops, as well as their wild counterparts, in the upcoming years. Moreover, although the genome assemblies of crops frequently exhibit incompleteness, transcriptomes can serve as vital gateway to genome editing. The dawn of third-generation sequencing, capable of generating long reads, has alleviated previous challenges associated with *de novo* genome assembly, which predominantly relied on short reads (Jiao and Schneeberger 2017). As a result, there is an anticipated enhancement in the accuracy and comprehensiveness of genome assemblies as well as their annotations.

The resequencing of numerous crop germplasm accessions now paves the way to access pangenomes, which signifies the genomic diversity at the species level as opposed to the sequence variation of a single genotype (Varshney et al. 2021b). However, to obtain a comprehensive gene repertoire for a specific crop, it is crucial to incorporate genic diversity from wild species. By assimilating various pangenomes from all species within a particular genus, a 'super-pangenome' can be established, which proves beneficial in harnessing genetic variability from crop wild species (Khan et al. 2020). The obtainability of pangenomes and super-pangenomes is particularly important for developing improved varieties, as crops often exhibit substantial differences within species for presence/absence variations (PAVs) in genes. For instance, in *Brassica oleracea* crops, the PAVs influenced nearly 20% of all the genes (Golicz et al. 2016). Employing a pangenome as a reference provides easy access to genome-wide differences in gene PAVs and copy numbers for editing purposes. Importantly, a comprehensive understanding of the diversity present within a particular species is vital for genome editing, as it facilitates designing cultivar-specific sgRNAs that are essential when the PAM (protospacer adjacent motif) site varies between cultivars. Consequently, the expanding availability of pangenome and super-pangenome will act as a vital reserve for enabling effective genome editing as well as precise breeding in crops.

20.4 Genomic Innovations for Uncovering Novel Targets and Enhancing Precision in Genome Editing

To date, genomic innovations have played a vital role in elucidating gene function, expression, and interactions in various plant species. However, there is still a substantial knowledge gap regarding the functions, expression patterns, and interaction networks of numerous plant genes, which lack sufficient experimental validation (Radivojac et al. 2013). It is crucial to note that genome editing heavily relies on a comprehensive understanding of gene regulation to discover suitable marks for editing, determine sites of gene expression, and enable engineering of regulatory sequences.

Unraveling the aftermath of genome editing via CRISPR/Cas system typically yields a heterogeneous mix of cell populations, wherein only a small subset carries the desired modifications. It becomes crucial to pinpoint the cells that exhibit the intended knockout or targeted mutation. Recent advancements in NGS technologies offer a comprehensive and high-resolution overview of qualitative and quantitative information regarding all modifications, shedding light on potential off-target effects. For instance, in rice, high-throughput sequencing based on NGS has demonstrated cost-effectiveness in confirming CRISPR-induced gene edits by focusing on specific regions designated for modification (Fraiture et al. 2023). However, the successful implementation of CRISPR/Cas techniques necessitates the incorporation of schemes to find and mitigate off-target effects or undesired modifications at non-intended sites. Several computational tools are utilized during genome editing studies to estimate RNA specificity and forecast off-target sites (Li et al. 2022). Nevertheless, a genome-wide analysis utilizing methods like whole-genome sequencing plays a pivotal role in uncovering off-target sites that might elude prediction algorithms (Li et al. 2019; Wang et al. 2021).

Furthermore, methods like targeting induced local lesions in genomes (TILLING) offer a valuable tool for detecting functional gene variants, aiding in the identification of mutations and genes closely linked to desirable agronomic traits. TILLING has been used, for example, to identify 46 gene variants controlling meiosis and recombination in barley (Schreiber et al. 2019), as well as to detect mutations in sunflower genes controlling flowering time, centromere function, auxin transport, and others (Fanelli et al. 2021). Another avenue for understanding functional loci and their interaction networks within the genome involves unraveling the three-dimensional configuration of genomes and utilizing genome-wide chromatin accessibility techniques like Hi-C (Garg et al. 2022), single-cell RNA sequencing (Hwang et al. 2018), and the assay for transposase-accessible chromatin using sequencing (Buenrostro et al. 2013). By employing these approaches, it becomes possible to obtain comprehensive, high-resolution maps of epigenetic elements and binding sites for transcription regulators across the genome. This is critical for comprehending the underlying processes that reign transcriptional regulation and harnessing them for improving crops.

Gene expression atlases (GEAs) serve as invaluable resources for understanding the expression patterns of various genes across different plant tissues and under diverse conditions (Kudapa et al. 2018; Pazhamala et al. 2017). They play a major part in detecting gene function and assessing the potential impacts of genome editing. The dynamic fluctuations in gene expression as crops develop and in response to abiotic or biotic stresses offer valuable insights into gene function, facilitating the modeling of the effects of genome editing across the crop's entire life cycle. Given the limited availability of protein-protein interaction data for most crops, these GEAs can be useful in identifying functionally relevant genes through co-expression analysis (Kudapa et al. 2018). Despite the progress made, the limited comprehensiveness in understanding the functional aspects of genes continues to be a key obstacle in leveraging genome editing for improving future crops. To surmount this challenge, it is crucial to integrate genomic datasets, including GEAs, with accurate computational models to predict gene function accurately. The adoption of such comprehensive and integrated approach will be crucial for tackling the bottleneck in gene function and improving crops through genome editing.

20.5 Innovative Breeding Schemes that Combine Genomic Resources with Genome Editing for Designing Future Crops

As advancements in genome sequencing and assembly, functional omics, as well as CRISPR/Cas efficacy continue at a rapid pace, we propose an innovative breeding scheme that can assume a progressively significant role in shaping future crops (Fig. 20.1). This breeding scheme involves the amalgamation of genomics, transcriptomics, and phenotypic data obtained from crop germplasm resources like wild relatives, landraces, and varieties. By establishing connections between agronomic traits of interest and specific genomic regions, novel genes/alleles can be discovered. These identified genomic targets can then undergo genome editing, wherein undesired sections of the genome, for example CRISPR/Cas genes, are removed from the first generation transformant (T_0) through crosses. The advancement of new allelic variants can be accomplished by selecting only individuals that meet specific thresholds for characters that can be accurately evaluated under speed breeding conditions, followed by evaluation in multi-location field trials. By repeatedly applying editing and selection processes, improved varieties will be developed, which can subsequently be released for commercial cultivation by farmers. This innovative breeding scheme, which relies on genomic resources, iterative editing and selection, and rapid generation advancement offers enhanced predictability in breeding outcomes and can hasten the improvement of varieties, thereby eliminating the need to remove unfavourable alleles from the genome. Finally, while we anticipate that genetically modified (GM) techniques and molecular breeding strategies including marker-assisted selection and GS will continue to offer advantages to breeders in the foreseeable future, these methods will eventually be surpassed by novel breeding schemes that incorporate genome editing. This transition will occur as our understanding of functional genomics in crop genomes expands, opening up new opportunities for efficient crop improvement strategies.

20.6 Exploring the Boundless Applications of Genomics-Powered Genome Editing for Crop Improvement and Breeding

Genome editing has become an important technology in agriculture, thanks to its unmatched ability to accurately modify crop genomes facilitated by ground-breaking advancements in genomics. This technology has revolutionized the field by not only enabling the development of novel crop varieties featuring beneficial attributes like increased yield and disease resistance, but also by significantly improving current breeding technologies and expediting the crop domestication process.

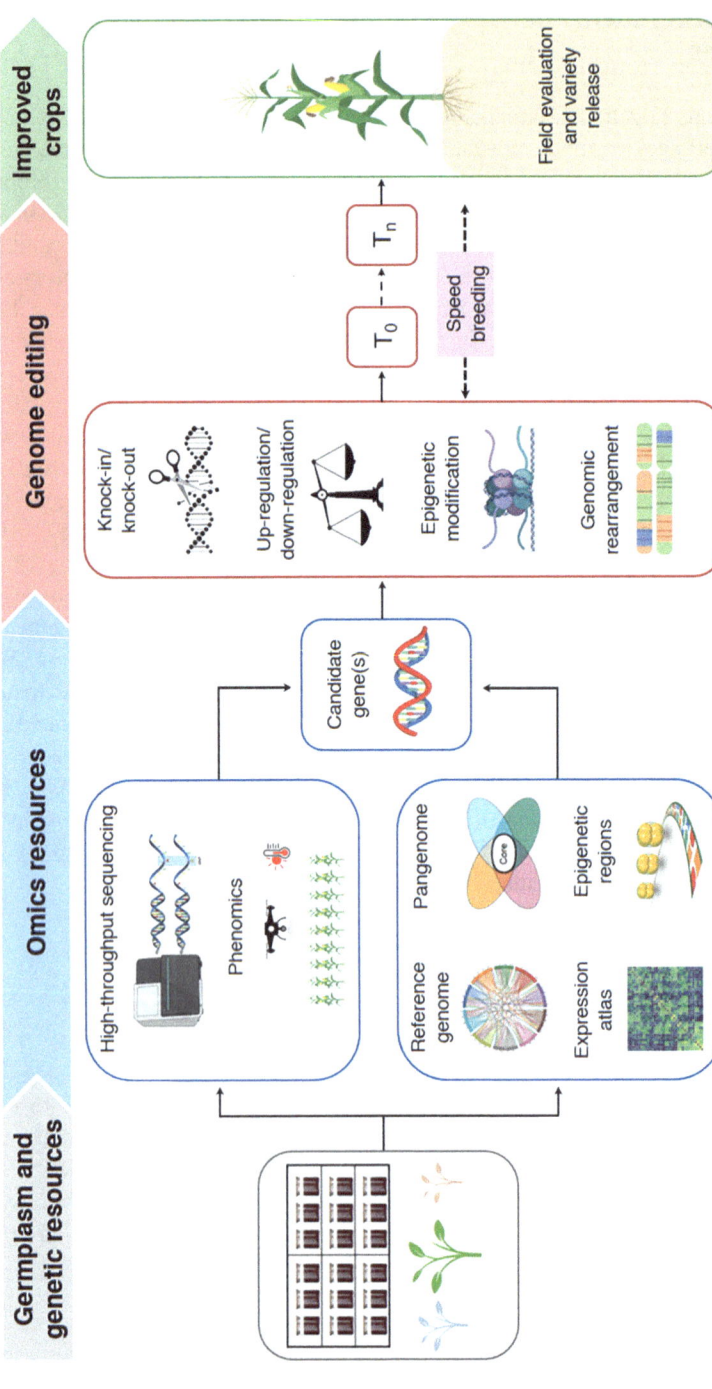

Fig. 20.1 Innovative breeding scheme integrating genomic resources with genome editing for tailoring future crops. The candidate genes identified through diverse omics data on crop germplasm can undergo genome editing within speed breeding conditions. The optimization of new allelic variants involves selecting the most effective alleles within the target population of environments. Repeated cycles of editing and selection will facilitate the development of climate-smart crops to address future food security

20.6.1 Improving Crop Characteristics

CRISPR/Cas has been extensively employed to improve numerous crop characteristics. Among the various factors influencing yield, engineering cytokinin homeostasis has proven to be a practical approach for improving cereal productivity. For instance, the deletion of a gene encoding cytokinin oxidase/dehydrogenase (CKX) was shown to generate high-yielding wheat varieties (Zhang et al. 2019). Similarly, in the case of rice, altering the C-terminus of the *LOGL5* gene, responsible for encoding a cytokinin activation enzyme, resulted in an augmented grain yield (Wang et al. 2020). Additionally, numerous genes, such as *GS3* in rice (controlling grain size), *GW2* in rice and wheat (controlling grain weight), and *PIN5b* in rice (controlling panicle size), have been edited to elevate grain yield (Zhang et al. 2018; Zhou et al. 2019; Zeng et al. 2019; Liu et al. 2017). However, crop production relies not only on yield but also on other important traits. Cereal grains characterized by reduced amylose content display enhanced attributes for both consumption and culinary purposes, and also find widespread utility in the textile sector. In a recent study, waxy maize varieties were developed by precisely disrupting the granulose-bound starch synthase 1 (*GBSS1*) gene, an important factor in the biosynthesis of amylose, using CRISPR/Cas9 (Gao et al. 2020). Precise CRISPR/Cas9-facilitated editing has enabled the development of better-quality crops with enhanced caroten-oid content (Dong et al. 2020), decreased phytic acid levels (Khan et al. 2019), and increased oleic acid concentrations (Do et al. 2019). Further, utilizing CRISPR/Cas to disrupt host susceptibility genes, as opposed to introducing dominant resistance genes, presents a potential strategy for safeguarding crops against various diseases. An illustrative case is bacterial blight, a condition that markedly reduces worldwide rice production as a consequence of the infection initiated by *Xanthomonas oryzae* pv. *oryzae*. By employing CRISPR/Cas to mutate the promoter regions of rice SWEET genes (*SWEET11, SWEET13,* and *SWEET14*), researchers successfully developed rice varieties that exhibit comprehensive resistance against *X. oryzae* pv. *oryzae* (Oliva et al. 2019; Varshney et al. 2019). Similarly, the concurrent alteration of three *mildew-resistance locus O* genes led to the development of a wheat cultivar that displays broad-spectrum resistance against powdery mildew (Wang et al. 2014).

20.6.2 Fast-Tracking Breeding Technologies

While CRISPR/Cas has demonstrated tremendous potential in crop improvement, integrating it with conventional breeding approaches can bring significant advantages to crop production. Recently, several innovative breeding strategies have emerged, focusing on reproduction-related genes through the utilization of CRISPR/Cas. One such strategy involves harnessing hybrid vigor, a widely employed technique in plant breeding to enhance crop yield and quality. However, the industrial production of hybrid seeds requires prevention of self-pollination in the female parent line to exclude homozygous seeds. Among various approaches to address this challenge, the most promising method is to create male sterility in maternal lines. By utilizing CRISPR/Cas to engineer the *male sterile 1 (Ms1)* and *Ms45* genes, scientists successfully incorporated male sterility into hexaploid wheat (Okada et al. 2019; Singh et al. 2018). Similarly, the manipulation of a potential strictosidine synthase gene with the CRISPR/Cas tool led to the creation of a sterile male tomato line (Du et al. 2020). Importantly, although procedures for developing hybrid seeds through male-sterile lines have been established, they still remain expensive and painstaking for certain crops. In this regard, inducing apomixis represents a promising alternative to stabilize elite hybrid backgrounds. Prior research has demonstrated that rice and Arabidopsis, using MiMe lines created through CRISPR/Cas technology with triple mutations in the *PAIR1, REC8*, and *OSD1* genes, produced clonal diploid gametes and tetraploid seeds. Furthermore, ectopically expressing *BABY BLOOM1* in *MiMe* rice triggered parthenogenesis, resulting in the development of progeny that were genetically similar to their maternal line (Khanday et al. 2019). While these synthetic-apomictic lines may not be suitable for large-scale hybrid seed production due to reduced fertility and minimal apomixis rate, these strategies can be readily implemented to vegetable crops where seed production is of lesser importance.

20.6.3 Accelerating Crop Domestication

CRISPR/Cas, with its precise genome engineering capabilities, holds significant potential in expediting the domestication process of crops. To illustrate, cultivated tomatoes are highly susceptible to environmental fluctuations, whereas a putative ancestor of tomatoes, known as *Solanum pimpinellifolium*, exhibits remarkable resistance to salinity and pathogenic bacteria. However, to transform *S. pimpinellifolium* into a commercially viable crop, certain undesirable traits such as small fruit size, low nutritional content, sprawling growth habit, and sensitivity to day length need to be addressed. Expanding upon a thorough comprehension of these traits, scientists utilized a multiplexed CRISPR/Cas system to specifically target essential genes, such as *CLV3* and *WUS* (regulating fruit size), *CycB* (influencing lycopene content), *GGP1* (controlling vitamin C content), *SP* (modulating plant growth pattern), and *SP5G* (governing floral induction), among others. Through these targeted edits, *S. pimpinellifolium* moved closer to becoming an attractive tomato variety (Zsögön et al. 2018; Li et al. 2018). Importantly, these domesticated plants retained the salt and bacterial spot disease resistance properties of *S. pimpinellifolium*. Furthermore, research has also begun to explore the domestication of African rice (*Oryza glaberrima*) using CRISPR/Cas (Ran et al. 2017). These investigations have opened up avenues for accelerating the domestication of various crops, thereby highlighting the transformative potential of CRISPR/Cas.

20.7 Conclusions and Future Perspectives

The CRISPR/Cas technology has arisen as a ground-breaking instrument for both basic and practical research in crops. With its unmatched capability to modify genomes, this technology has played a pivotal role in developing numerous crop varieties that exhibit enhanced agronomic performance, bringing about a revolution in breeding efforts. In the years to come, genome editing is expected to assume a crucial role in expediting crop improvement programs and enhancing their effectiveness. This advancement holds immense potential in addressing food and nutrition security in developed and developing nations alike, as it enables designing crops with increased yields, heightened resistance against diseases and pests, and improved resilience to environmental pressures. Numerous plant biotechnologies linked to CRISPR/Cas have emerged, including delivery systems that facilitate crop genome editing, precise gene regulation techniques at various steps, and the development of multiplexed and high-throughput genome editing methods that have empowered genome alterations at multiple locations, along with plant-directed evolution.

However, despite the versatility of these methods, they have not fully met all the requirements for manipulating crop genomes, and further advancements are crucial for the deployment of CRISPR/Cas for improving crops. Progress in genomics research, including the development of high-quality reference genomes, pangenomes, and super-pangenomes, along with systematic identification of potential editing sites through functional genomics, will be essential in meeting these needs. Moreover, it is vital to boost the efficiency of delivering the CRISPR/Cas system and minimize off-target editing. To ensure the success of novel breeding schemes employing genome editing, it is essential to incorporate various methods, including genomics, transcriptomics, phenomics, and biotechnology. Taken together, the implementation of genomic innovations in genome editing has undoubtedly brought about a transformative impact on agriculture and crop biotechnology and will persistently drive revolutionary changes in these fields.

Acknowledgements RKV acknowledges the support from Food Futures Institute, Murdoch University, Australia.

Conflict of Interest Statement The authors declare that they have no conflict of interest.

References

Bansal KC, Molla KA, Chinnusamy V (2022) Genome editing: a boon for plant biologists, breeders and farmers. Curr Sci 123:15–19

Barmukh R, Roorkiwal M, Dixit GP et al (2022) Characterization of *"QTL-hotspot"* introgression lines reveals physiological mechanisms and candidate genes associated with drought adaptation in chickpea. J Exp Bot 73:7255–7272

Beyene Y, Semagn K, Crossa J et al (2016) Improving maize grain yield under drought stress and non-stress environments in sub-Saharan Africa using marker-assisted recurrent selection. Crop Sci 56:344–353

Buenrostro JD, Giresi PG, Zaba LC et al (2013) Transposition of native chromatin for fast and sensitive epigenomic profiling of open chromatin, DNA-binding proteins and nucleosome position. Nat Methods 10:1213–1218

Crossa J, Perez-Rodriguez P, Cuevas J et al (2017) Genomic selection in plant breeding: methods, models, and perspectives. Trends Plant Sci 22:961–975

van Dijk M, Morley T, Rau ML et al (2021) A meta-analysis of projected global food demand and population at risk of hunger for the period 2010–2050. Nat Food 2:494–501

Do PT, Nguyen CX, Bui HT et al (2019) Demonstration of highly efficient dual gRNA CRISPR/Cas9 editing of the homeologous *GmFAD2-1A* and *GmFAD2-1B* genes to yield a high oleic, low linoleic and alpha-linolenic acid phenotype in soybean. BMC Plant Biol 19:311

Dong OX, Yu S, Jain R et al (2020) Marker-free carotenoid-enriched rice generated through targeted gene insertion using CRISPR-Cas9. Nat Commun 11:1178

Doudna JA, Charpentier E (2014) Genome editing: the new frontier of genome engineering with CRISPR-Cas9. Science 346:1258096

Du M, Zhou K, Liu Y et al (2020) A biotechnology-based male-sterility system for hybrid seed production in tomato. Plant J 102:1090–1100

Epstein R, Sajai N, Zelkowski M et al (2023) Exploring impact of recombination landscapes on breeding outcomes. Proc Natl Acad Sci USA 120:e2205785119

Fanelli V, Ngo KJ, Thompson VL et al (2021) A TILLING by sequencing approach to identify induced mutations in sunflower genes. Sci Rep 11:9885

Fraiture MA, D'aes J, Guiderdoni E et al (2023) Targeted high-throughput sequencing enables the detection of single nucleotide variations in CRISPR/Cas9 gene-edited organisms. Foods 12:455

Gao H, Gadlage MJ, Lafitte HR et al (2020) Superior field performance of waxy corn engineered using CRISPR-Cas9. Nat Biotechnol 38:579–581

Garg V, Dudchenko O, Wang J et al (2022) Chromosome-length genome assemblies of six legume species provide insights into genome organization, evolution, and agronomic traits for crop improvement. J Adv Res 42:315–329

Golicz AA, Bayer PE, Barker GC et al (2016) The pangenome of an agronomically important crop plant *Brassica oleracea*. Nat Commun 7:13390

Hasan MM, Rafii MY, Ismail MR et al (2015) Marker-assisted backcrossing: a useful method for rice improvement. Biotechnol Biotechnol Equip 29:237–254

He C, Liu H, Xie WZ et al (2021) CRISPR-cereal: a guide RNA design tool integrating regulome and genomic variation for wheat, maize and rice. Plant Biotechnol J 19:2141–2143

Hwang B, Lee JH, Bang D (2018) Single-cell RNA sequencing technologies and bioinformatics pipelines. Exp Mol Med 50:1–14

Jiao WB, Schneeberger K (2017) The impact of third generation genomic technologies on plant genome assembly. Curr Opin Plant Biol 36:64–70

Jinek M, Chylinski K, Fonfara I et al (2012) A programmable dual-RNA-guided DNA endonuclease in adaptive bacterial immunity. Science 337:816–821

Kawashima C, Guimarães G, Nogueira S et al (2016) A pigeonpea gene confers resistance to Asian soybean rust in soybean. Nat Biotechnol 34:661–665

Khan A, Garg V, Roorkiwal M et al (2020) Super-pangenome by integrating the wild-side of a species for accelerated crop improvement. Trends Plant Sci 25:148–158

Khan MSS, Basnet R, Islam SA et al (2019) Mutational analysis of *OsPLDα1* reveals its involvement in phytic acid biosynthesis in rice grains. J Agric Food Chem 67:11436–11443

Khanday I, Skinner D, Yang B et al (2019) A male-expressed rice embryogenic trigger redirected for asexual propagation through seeds. Nature 565:91–95

Knudsen S, Wendt T, Dockter C et al (2022) FIND-IT: accelerated trait development for a green evolution. Sci Adv 8:eabq2266

Kudapa H, Garg V, Chitikineni A et al (2018) The RNA-Seq-based high resolution gene expression atlas of chickpea (*Cicer arietinum* L.) reveals dynamic spatio-temporal changes associated with growth and development. Plant Cell Environ 41:2209–2225

Li C, Chu W, Gill RA et al (2022) Computational tools and resources for CRISPR/Cas genome editing. Genomics Proteomics Bioinformatics. https://doi.org/10.1016/j.gpb.2022.02.006

Li J, Manghwar H, Sun L et al (2019) Whole genome sequencing reveals rare off-target mutations and considerable inherent genetic or/and somaclonal variations in CRISPR/Cas9-edited cotton plants. Plant Biotechnol J 17:858–868

Li T, Yang X, Yu Y et al (2018) Domestication of wild tomato is accelerated by genome editing. Nat Biotechnol 36:1160–1163

Liu J, Chen J, Zheng X et al (2017) GW5 acts in the brassinosteroid signalling pathway to regulate grain width and weight in rice. Nat Plants 3:17043

Molla KA, Sretenovic S, Bansal KC et al (2021) Precise plant genome editing using base editors and prime editors. Nat Plants 7:1166–1187

Okada A, Arndell T, Borisjuk N et al (2019) CRISPR/Cas9-mediated knockout of Ms1 enables the rapid generation of male-sterile hexaploid wheat lines for use in hybrid seed production. Plant Biotechnol J 17:1905–1913

Oliva R, Ji C, Atienza-Grande G et al (2019) Broad-spectrum resistance to bacterial blight in rice using genome editing. Nat Biotechnol 37:1344–1350

Pazhamala LT, Purohit S, Saxena RK et al (2017) Gene expression atlas of pigeonpea and its application to gain insights into genes associated with pollen fertility implicated in seed formation. J Exp Bot 68:2037–2054

Pixley KV, Falck-Zepeda JB, Paarlberg RL et al (2022) Genome-edited crops for improved food security of smallholder farmers. Nat Genet 54:364–367

Radivojac P, Clark WT, Oron TR et al (2013) A large-scale evaluation of computational protein function prediction. Nat Methods 10:221–227

Ran Y, Liang Z, Gao C (2017) Current and future editing reagent delivery systems for plant genome editing. Sci China Life Sci 60:490–505

Ray DK, Mueller ND, West PC et al (2013) Yield trends are insufficient to double global crop production by 2050. PLoS One 8:e66428

Schreiber M, Barakate A, Uzrek N et al (2019) A highly mutagenised barley (cv. Golden promise) TILLING population coupled with strategies for screening-by-sequencing. Plant Methods 15:99

Singh M, Kumar M, Albertsen MC et al (2018) Concurrent modifications in the three homeologs of Ms45 gene with CRISPR-Cas9 lead to rapid generation of male sterile bread wheat (*Triticum aestivum* L.). Plant Mol Biol 97:371–383

Singh M, Nara U, Kumar A et al (2022) Enhancing genetic gains through marker-assisted recurrent selection: from phenotyping to genotyping. Cereal Res Commun 50:523–538

Uffelmann E, Huang QQ, Munung NS et al (2021) Genome-wide association studies. Nat Rev Methods Primers 1:59

Varshney RK, Bohra A, Yu J et al (2021a) Designing future crops: genomics-assisted breeding comes of age. Trends Plant Sci 26:631–649

Varshney RK, Chen W, Li Y et al (2012a) Draft genome sequence of pigeonpea (*Cajanus cajan*), an orphan legume crop of resource-poor farmers. Nat Biotechnol 30:83–89

Varshney RK, Godwin ID, Mohapatra T et al (2019) A SWEET solution to rice blight. Nat Biotechnol 37:1280–1282

Varshney RK, Ribaut JM, Buckler E et al (2012b) Can genomics boost productivity of orphan crops? Nat Biotechnol 30:1172–1176

Varshney RK, Roorkiwal M, Sun S et al (2021b) A chickpea genetic variation map based on the sequencing of 3,366 genomes. Nature 599:622–627

Wang C, Wang G, Gao Y et al (2020) A cytokinin-activation enzyme-like gene improves grain yield under various field conditions in rice. Plant Mol Biol 102:373–388

Wang X, Tu M, Wang Y et al (2021) Whole-genome sequencing reveals rare off-target mutations in CRISPR/Cas9-edited grapevine. Hortic Res 8:114

Wang Y, Cheng X, Shan Q et al (2014) Simultaneous editing of three homoeoalleles in hexaploid bread wheat confers heritable resistance to powdery mildew. Nat Biotechnol 32:947–951

Zeng Y, Wen J, Zhao W et al (2019) Rational improvement of rice yield and cold tolerance by editing the three genes OsPIN5b, GS3, and OsMYB30 with the CRISPR-Cas9 system. Front Plant Sci 10:1663

Zhang Y, Li D, Zhang D et al (2018) Analysis of the functions of TaGW2 homoeologs in wheat grain weight and protein content traits. Plant J 94:857–866

Zhang Z, Hua L, Gupta A et al (2019) Development of an agrobacterium-delivered CRISPR/Cas9 system for wheat genome editing. Plant Biotechnol J 17:1623–1635

Zhou J, Xin X, He Y et al (2019) Multiplex QTL editing of grain-related genes improves yield in elite rice varieties. Plant Cell Rep 38:475–485

Zhu H, Li C, Gao C (2020) Applications of CRISPR-Cas in agriculture and plant biotechnology. Nat Rev Mol Cell Biol 21:661–677

Zsögön A, Čermák T, Naves E et al (2018) De novo domestication of wild tomato using genome editing. Nat Biotechnol 36:1211–1216

Gene Editing in Soybean: Promise to Products

21

Robert M. Stupar and Shaun J. Curtin

Abstract

Legume crops are a vital feedstock for food, feed, fuel, and industrial products. Soybean, as the most cultivated legume crop in the world, has been subject to intense selection and breeding over the past century. The recent emergence of gene editing technology has enabled more efficient and focused studies of soybean gene function, along with the potential to develop new traits and/or novel functions. This chapter provides a brief history of gene editing applications in soybean research, including initial and ongoing applications in gene functional characterization. Furthermore, it discusses new gene editing modalities that may increase the range and flexibility of traits development. Related to these advances in gene knowledge and editing capacity, we discuss the traits that have been targeted for product development in recent years and the traits most likely to be targeted in the future.

Keywords

Soybean · Legume · CRISPR · Prime editing · Seed composition · Yield components · Architecture

R. M. Stupar (✉)
Department of Agronomy and Plant Genetics, University of Minnesota, Saint Paul, MN, USA

Center for Plant Precision Genomics, University of Minnesota, St. Paul, MN, USA
e-mail: stup0004@umn.edu

S. J. Curtin
Department of Agronomy and Plant Genetics, University of Minnesota, Saint Paul, MN, USA

Center for Plant Precision Genomics, University of Minnesota, St. Paul, MN, USA

Plant Science Research Unit, United States Department of Agriculture, Saint Paul, MN, USA
e-mail: shaun.curtin@usda.gov

© National Academy of Agricultural Sciences, under exclusive license to Springer Nature Singapore Pte Ltd. 2023
K. C. Bansal et al. (eds.), *Transformation of Agri-Food Systems*,
https://doi.org/10.1007/978-981-99-8014-7_21

21.1   Introduction

Gene editing is a branch of biotechnology that focuses on the introduction of DNA sequence changes in a targeted and/or gene specific manner. The change(s) of interest occur in the host genome and are typically targeted to native genes, as opposed to the introduction of foreign DNA *per se* (i.e., the classical definition of genetic engineering). Gene editing in crop species tends to focus on four distinct areas of activity: (1) Characterization of known genes and gene families through the development of new allelic variants and combinations of variants; (2) Discovery and candidate validation of previously unknown/uncharacterized genes; (3) Technological breakthroughs that focus on new ways to do, think, and develop the tools of biotechnology; and (4) Product development (e.g., commercializable traits).

The first report of gene editing in whole soybean plants occurred approximately 11 years ago (Curtin et al. 2011), in the wake of pioneering reports from model plant systems (e.g., Shukla et al. 2009; Townsend et al. 2009; Zhang et al. 2010). This initial study (Curtin et al. 2011) used an early-generation technology in which an encoded transgene was specified to fuse Zinc-finger DNA binding domains to the catalytic domain of the *Fok*I endonuclease. When introduced to the cell, the translated protein (known as a zinc-finger nuclease (ZFN)) was capable of binding the DNA sequence specified by the ZF domain and generating a double stranded break. The broken DNA would subsequently be repaired, sometimes resulting in desired results, such as a frameshift mutation or a directed repair (e.g., introduction of a desired nucleotide alteration). The ZF domain could be modified among constructs to recognize different genomic DNA sequences, but the design was inefficient and recognition domains were relatively infrequent throughout the genome. Nonetheless, ZFN-based gene editing paved the way for sequence-specific gene editing, and more efficient technologies quickly followed.

An engineered DNA recognition system based on transcription activator-like effector (TALE) proteins from the pathogenic bacteria *Xanthomonas* (Moscou and Bogdanove 2009) was the next widely-used system. Similar to ZFNs, TALE transgenes were engineered to fuse with the *Fok*I endonuclease, thereby generating expressed proteins capable of generating double stranded breaks at specific locations in the genome (Christian et al. 2010; Čermák et al. 2011). This system, known as TALE-nucleases (TALENs), provided a greater abundance of potential recognition domains compared to the ZFN system – it was possible to target almost any gene. Despite this advantage, TALENs were used in a relatively small number of published soybean studies (e.g., Haun et al. 2014; Curtin et al. 2018).

The brevity of TALEN's popularity is mostly attributed to the development of engineering systems based on Clustered Regularly Interspaced Short Palindromic Repeats (CRISPR). Prokaryotic organisms acquired CRISPR sequences in their genomes as a mechanism of immunity against bacteriophage infections (Barrangou et al. 2007). The biological mechanism of this resistance is based on guide-RNAs (gRNA) encoded within the CRISPR sequence that are able to recognize complementary DNA sequences and target their cleavage by a family of CRISPR-associated (Cas) enzymes. The most well-known of these enzymes is Cas9 from *Streptococcus pyogenes*. Genetic engineers paired elements of this system, namely the essential CRISPR sequences and the Cas9 enzyme, that could be introduced into cells and generate targeted DNA double stranded breaks (Jinek et al. 2012; Cong et al. 2013; Mali et al. 2013). This editing system is highly efficient and modifiable compared to previous versions (e.g., ZFN and TALEN), as the CRISPR gRNA sequences can be quickly changed in a transgenic construct to recognize new genes and sequences. The rest of this chapter focuses on reviewing some of the published works of CRISPR/Cas in soybean—how researchers are using this technology to better understand soybean genes, enhance its utility, and develop useful traits.

21.2 CRISPR as a Functional Validation Tool in Soybean

The first studies of CRISPR/Cas technology in soybean focused on targeting mutations in somatic hairy root tissues. Such work was published by at least three groups in 2015 (Jacobs et al. 2015; Michno et al. 2015; Sun et al. 2015). That same year, the first soybean paper demonstrating whole plant CRISPR/Cas9 editing was published (Li et al. 2015). This landmark work demonstrated multiple editing outcomes, including targeted mutagenesis, gene integration, and gene editing derived from homology-directed recombination of donor template DNA.

Numerous studies in the past few years have used soybean CRISPR/Cas mutagenesis to better characterize known genes or validate the function of candidate genes. Several such studies have focused on genes involved in flowering and development, such as the maturity (E) genes and *FLOWER LOCUS T (FT)* homologs (Cai et al. 2018; Han et al. 2019; Wang et al. 2020a; Qin et al. 2023). Seed development and composition traits have also been the focus of numerous studies using CRISPR technologies for purposes of gene characterization or validation. This includes studies focusing on seed size (Nguyen et al. 2021), numbers of seed per pod (Cai et al. 2021), storage proteins (Bai et al. 2022), fatty acid distribution (Ma et al. 2021), carbohydrates (Le et al. 2020; Virdi et al. 2020), and flavor (Wang et al. 2020b). Moreover, CRISPR-modification has been used to study soybean genes involved in plant architecture components and tissue-specific development processes. Characterization of architecture types have included variants for branching (Bao et al. 2019; Liang et al. 2022) and internode length (Cheng et al. 2019; Wang et al. 2021) traits. Genes involved in other developmental processes include trichome (Campbell et al. 2019) and nodule (Ren et al. 2019; Nguyen et al. 2023) development. Likewise, genes involved in disease and pest resistance have also been characterized using CRISPR/Cas methodologies (Fan et al. 2022; Zhang et al. 2022; Liu et al. 2023; Zhang et al. 2023).

Utility of CRISPR/Cas tools to discover, validate, and characterize gene functions in soybean has become so commonplace that a thorough review of all such published works is beyond the scope of this chapter. Four years ago, such papers principally featured the CRISPR/Cas aspects of the work, oftentimes including the term "CRISPR" in the title and/or abstract. Today, numerous soybean papers include a CRISPR/Cas aspect for gene functional characterization, but no longer feature the

"CRISPR" technology by name in the paper title or abstract, indicating the current widespread use of this methodology.

21.3 Beyond the Knockout: Emerging Tools for Gene Editing

CRISPR/Cas editing reagents have been used for targeted mutagenesis of soybean for close to a decade. In that time there has been minimal effort at optimization partly due to the initial success of these reagents in legumes. This is in contrast to gene editing in *Arabidopsis* which found canonical promoters such as the Cauliflower Mosaic Virus (35S) had low activity and generated a higher incidence of somatic mutations (Yan et al. 2015). Optimization studies were essential in identifying reagent combinations that edited the *Arabidopsis* germline efficiently. Numerous factors that improved efficiency including choice of promoter, terminator, NLS signals, modified sgRNA molecule, Cas9 codon-optimization and the presence of an intron as well as T-DNA architecture were identified (Castel et al. 2019). Successful legume reagents typically have a Cas9 with both N- and C- termini nuclear localization signals (NLS) from either the simian virus 40 (SV40 NLS) (Čermák et al. 2017; Liu et al. 2019) or an endogenous soybean NLS (Glyma.06g207800) (Campbell et al. 2019), use of an intron (Curtin et al. 2018), are highly expressed by constitutive soybean promoters such as the *Glycine max* polyubiquitin 3 promoter (Gmubi3) (Glyma.20G141600) (Chiera et al. 2007; Campbell et al. 2019; Curtin et al. 2018) or the soybean elongation factor 1A (eEF1A) promoter (GmScreamM4) (Glyma.05g114900) (Zhang et al. 2015; Li et al. 2015) and terminators that promote high expression (Čermák et al. 2017). Other promoters that have been used to successfully edit soybean include the 35S (Čermák et al. 2017; Liu et al. 2019; Michno et al. 2020), the parsley ubiquitin (PcUbi) (Kanazashi et al. 2018), and the maize ubiquitin (Zmubi) (Du et al. 2016).

To take soybean mutagenesis beyond the knockout, further improvements to genetic transformation of elite, wild and un-adapted soybean genotypes will be required along with continued incremental improvement to reagent efficiencies. Recent efforts to improve soybean transformation include the use of optimized protocols that have respectable increases of transformation efficiencies and more importantly are less labor intensive. These efficiencies could be further bolstered with strategies to engineer *Agrobacterium* strains such as the use of the Pseudomonas Type III secretion system to suppress plant defenses (Raman et al. 2022) or the deployment of developmental regulators (Wang et al. 2023a; Maher et al. 2020). Improvements to transformation will be essential for carrying out more advanced gene editing modalities such as base editing (BE) and prime editing (PE) whose functional efficiencies can already be lower than targeted mutagenesis strategies (Čermák 2021). Base editing can enable precise nucleotide changes without the requirement for a double-strand break (DSB) or a DNA repair template (Čermák 2021; Gao 2021). It has been successfully applied to plants and will likely be an important tool for soybean crop improvement, such as enhancing disease resistance or improving agronomic traits. It combines the CRISPR/Cas9 system with a cytosine or adenine deaminase enzyme and allows direct conversion of one base into another at a specific target site, usually limited to C:G to T:A and A:T to G:C substitutions (Gao 2021). Base editing reagents can be efficient at generating edits, but their target range can be restricted by its PAM sequence. Fortunately, there are Cas9 orthologs with different PAM sequence requirements that can expand the base editors target range (Xu et al. 2020). Some examples including *Streptococcus pyogenes* (SpCas9) NGG, *S. canis* (ScCas9) NNG, *Staphylococcus aureus* (SaCas9) NNNRRT and *Lachnospiraceae bacterium* (LbCpf1) TTTV (Zhang et al. 2019). A more exciting technological advance is the prime editor. Although PE efficiency in crops is low, it can potentially make all types of base pair modifications as well as repair or insert small base-pair deletions (Čermák 2021).

CRISPR/Cas gene editing modalities like base editing are reshaping the field of crop genetics, offering precise and efficient genetic modification of traits of interest. These techniques will continue to improve, become more accurate and efficient, supporting efforts to address challenges in agriculture and accelerating the development of new traits and products.

21.4   From Promise to Products

While most work published to date on soybean gene editing has focused on basic researcher questions (both in terms of gene function and technology development), there have been a small number of studies reporting new traits that may be of commercial interest. The first such study was published by the company Calyxt and focused on increasing the health quality of soybean oil (Haun et al. 2014). The authors leveraged knowledge from previously published work on the fatty acid desaturase genes FAD2-1A and FAD2-1B in soybean (Pham et al. 2010), and targeted those genes for knockout mutations using TALENs in the 'Bert' cultivar background. Seeds from the resulting double-mutant soybean plants exhibited a tremendous increase in monounsaturated oleic acid, accompanied by a decrease in linoleic and linolenic acid (Haun et al. 2014). The Calyxt team used TALENs again to stack the double-mutant with a knockout of FAD3A, further reducing the linolenic acid content (Demorest et al. 2016). The resulting product, known as Calyno Oil, was the first gene-edited food product commercialized in the United States, following approval by the national regulatory bodies.

Another study that leveraged prior knowledge about soybean seed composition was recently published by Wang et al. (2023b). Kunitz Trypsin Inhibitors (KTi) are anti-nutritional proteins that affect digestibility and are encoded by endogenous genes in soybean. Genes encoding KTi3 and KTi1 are expressed in seeds, necessitating that soybean meal undergo a heat pre-treatment process to inactivate these enzymes before being used for animal feed purposes (Chang et al. 1987). Gillman et al. (2015) previously characterized naturally-occurring frameshift alleles for KTi3 and KTi1 which exhibited reductions in seed trypsin activity. Wang et al. (2023b) recently used a CRISPR/Cas9 system to develop new knockout alleles for these KTi3 and KTi1 genes in the 'Williams 82' cultivar background. The specific line developed by this study is not a released product. However, the authors argued that marker assisted selection can be used to efficiently introgress the reduced KTi trait into modern soybean cultivars.

At the recent World Soybean Research Conference 11, the company GDM Seeds presented two new traits developed from gene knockouts using CRISPR/Cas9. This includes reduced raffinose and stachyose for increased metabolizable energy, and increased drought tolerance (Martins et al. 2023).

The examples above illustrate the desire to use gene editing as a tool for soybean genetic improvement, particularly for seed composition traits. Perhaps these have been targeted by early adopters because the genes controlling their respective phenotypes were previously well-characterized. However, it is not difficult to imagine a wide range of other possible traits that may be targeted for trait and product development in the future. This may include editing of genes that control yield component traits, abiotic stress response traits, disease resistance traits, and quality (e.g., flavor) traits. While most of the first-generation gene edits have targeted knockout alleles of relatively few genes, it is possible that different types of edits (e.g., editing events that repress, activate, or overexpress genes) and/or larger genetic pathways may be targeted in the future for various desired phenotypic outcomes.

21.5 Acknowledgments

Funding for soybean biotechnology research in the Stupar lab is currently provided by grants from the Minnesota Soybean Research and Promotion Council and the United States Department of Agriculture (USDA). The Curtin Lab is supported by the USDA - Agricultural Research Service. Mention of any trade names or commercial products in this article is solely for the purpose of providing specific information and does not imply recommendation or endorsement by the U. S. Department of Agriculture. USDA is an equal opportunity provider and employer, and all agency services are available without discrimination. The authors are grateful to the numerous researchers who have published and shared their research findings on the topics discussed in this work. We apologize to any authors who we failed to cite due to word limits.

References

Bai M, Yuan C, Kuang H, Sun Q, Hu X, Cui L, Lin W, Peng C, Yue P, Song S, Guo Z, Guan Y (2022) Combination of two multiplex genome-edited soybean varieties enables customization of protein functional properties. Mol Plant 15:1081–1083. https://doi.org/10.1016/j.molp.2022.05.011

Bao A, Chen H, Chen L, Chen S, Hao Q, Guo W, Qiu D, Shan Z, Yang Z, Yuan S, Zhang C, Zhang X, Liu B, Kong F, Li X, Zhou X, Tran LP, Cao D (2019) CRISPR/Cas9-mediated targeted mutagenesis of GmSPL9 genes alters plant architecture in soybean. BMC Plant Biol 19:131. https://doi.org/10.1186/s12870-019-1746-6

Barrangou R, Fremaux C, Deveau H, Richards M, Boyaval P, Moineau S, Romero DA, Horvath P (2007) CRISPR provides acquired resistance against viruses in prokaryotes. Science 315:1709–1712. https://doi.org/10.1126/science.1138140

Cai Y, Chen L, Liu X, Guo C, Sun S, Wu C, Jiang B, Han T, Hou W (2018) CRISPR/Cas9-mediated targeted mutagenesis of GmFT2a delays flowering time in soya bean. Plant Biotechnol J 16:176–185. https://doi.org/10.1111/pbi.12758

Cai Z, Xian P, Cheng Y, Ma Q, Lian T, Nian H, Ge L (2021) CRISPR/Cas9-mediated gene editing of GmJAGGED1 increased yield in the low-latitude soybean variety Huachun 6. Plant Biotechnol J 19:1898–1900. https://doi.org/10.1111/pbi.13673

Campbell BW, Hoyle JW, Bucciarelli B, Stec AO, Samac DA, Parrott WA, Stupar RM (2019) Functional analysis and development of a CRISPR/Cas9 allelic series for a CPR5 ortholog necessary for proper growth of soybean trichomes. Sci Rep 9:14757. https://doi.org/10.1038/s41598-019-51240-7

Castel B, Tomlinson L, Locci F, Yang Y, Jones JDG (2019) Optimization of T-DNA architecture for Cas9-mediated mutagenesis in Arabidopsis. PLoS One 14:e0204778. https://doi.org/10.1371/journal.pone.0204778

Čermák T (2021) Sequence modification on demand: search and replace tools for precise gene editing in plants. Transgenic Res 30:353–379. https://doi.org/10.1007/s11248-021-00253-y

Čermák T, Curtin SJ, Gil-Humanes J, Čegan R, Kono TJY, Konečná E, Belanto JJ, Starker CG, Mathre JW, Greenstein RL, Voytas DF (2017) A multipurpose toolkit to enable advanced genome engineering in plants. Plant Cell 29:1196–1217. https://doi.org/10.1105/tpc.16.00922

Čermák T, Doyle EL, Christian M, Wang L, Zhang Y, Schmidt C, Baller JA, Somia NV, Bogdanove AJ, Voytas DF (2011) Efficient design and assembly of custom TALEN and other TAL effector-based constructs for DNA targeting. Nucleic Acids Res 39:e82. https://doi.org/10.1093/nar/gkr218

Chang CJ, Tanksley TD, Knabe DA, Zebrowska T (1987) Effects of different heat treatments during processing on nutrient digestibility of soybean meal in growing swine. J Anim Sci 65:1273–1282

Cheng Q, Dong L, Su T, Li T, Gan Z, Nan H, Lu S, Fang C, Kong L, Li H, Hou Z, Kou K, Tang Y, Lin X, Zhao X, Chen L, Liu B, Kong F (2019) CRISPR/Cas9-mediated targeted mutagenesis of GmLHY genes alters plant height and internode length in soybean. BMC Plant Biol 19:562. https://doi.org/10.1186/s12870-019-2145-8

Chiera JM, Bouchard RA, Dorsey SL, Park E, Buenrostro-Nava MT, Ling PP, Finer JJ (2007) Isolation of two highly active soybean (Glycine max (L.) Merr.) promoters and their characterization using a new automated image collection and analysis system. Plant Cell Rep 26:1501–1509. https://doi.org/10.1007/s00299-007-0359-y

Christian M, Čermák T, Doyle EL, Schmidt C, Zhang F, Hummel A, Bogdanove AJ et al (2010) Targeting DNA double-strand breaks with TAL effector nucleases. Genetics 186:757–761. https://doi.org/10.1534/genetics.110.120717

Cong L, Ran FA, Cox D, Lin S, Barretto R, Habib N, Hsu PD, Wu X, Jiang W, Marraffini LA, Zhang F (2013) Multiplex genome engineering using CRISPR/Cas systems. Science 339:819–823. https://doi.org/10.1126/science.1231143

Curtin SJ, Xiong Y, Michno JM, Campbell BW, Stec AO, Čermák T, Starker C, Voytas DF, Eamens AL, Stupar RM (2018) CRISPR/Cas9 and TALENs generate heritable mutations for genes involved in small RNA processing of Glycine max and Medicago truncatula. Plant Biotechnol J 16:1125–1137. https://doi.org/10.1111/pbi.12857

Curtin SJ, Zhang F, Sander JD, Haun WJ, Starker C, Baltes NJ, Reyon D, Dahlborg EJ, Goodwin MJ, Coffman AP, Dobbs D, Joung JK, Voytas DF, Stupar RM (2011) Targeted mutagenesis of duplicated genes in soybean with zinc-finger nucleases. Plant Physiol 156:466–473. https://doi.org/10.1104/pp.111.172981

Demorest ZL, Coffman A, Baltes NJ, Stoddard TJ, Clasen BM, Luo S, Retterath A, Yabandith A, Gamo ME, Bissen J, Mathis L, Voytas DF, Zhang F (2016) Direct stacking of sequence-specific nuclease-induced mutations to produce high oleic and low linolenic soybean oil. BMC Plant Biol 16:225. https://doi.org/10.1186/s12870-016-0906-1

Du H, Zeng X, Zhao M, Cui X, Wang Q, Yang H, Cheng H, Yu D (2016) Efficient targeted mutagenesis in soybean by TALENs and CRISPR/Cas9. J Biotechnol 217:90–97. https://doi.org/10.1016/j.jbiotec.2015.11.005

Fan S, Zhang Z, Song Y, Zhang J, Wang P (2022) CRISPR/Cas9-mediated targeted mutagenesis of GmTCP19L increasing susceptibility to phytophthora sojae in soybean. PLoS One 17: e0267502. https://doi.org/10.1371/journal.pone.0267502

Gao C (2021) Genome engineering for crop improvement and future agriculture. Cell 184:1621–1635. https://doi.org/10.1016/j.cell.2021.01.005

Gillman JD, Kim WS, Krishnan HB (2015) Identification of a new soybean kunitz trypsin inhibitor mutation and its effect on bowman-birk protease inhibitor content in soybean seed. J Agric Food Chem 63:1352–1359. https://doi.org/10.1021/jf505220p

Han J, Guo B, Guo Y, Zhang B, Wang X, Qiu LJ (2019) Creation of early flowering germplasm of soybean by CRISPR/Cas9 technology. Front Plant Sci 10:1446. https://doi.org/10.3389/fpls.2019.01446

Haun W, Coffman A, Clasen BM, Demorest ZL, Lowy A, Ray E, Retterath A, Stoddard T, Juillerat A, Cedrone F, Mathis L, Voytas DF, Zhang F (2014) Improved soybean oil quality by targeted mutagenesis of the fatty acid desaturase 2 gene family. Plant Biotechnol J 12:934–940. https://doi.org/10.1111/pbi.12201

Jacobs TB, LaFayette PR, Schmitz RJ, Parrott WA (2015) Targeted genome modifications in soybean with CRISPR/Cas9. BMC Biotechnol 15:16. https://doi.org/10.1186/s12896-015-0131-2

Jinek M, Chylinski K, Fonfara I, Hauer M, Doudna JA, Charpentier E (2012) A programmable dual-RNA-guided DNA endonuclease in adaptive bacterial immunity. Science 337:816–821. https://doi.org/10.1126/science.1225829

Kanazashi Y, Hirose A, Takahashi I, Mikami M, Endo M, Hirose S, Toki S, Kaga A, Naito K, Ishimoto M, Abe J, Yamada T (2018) Simultaneous site-directed mutagenesis of duplicated loci in soybean using a single guide RNA. Plant Cell Rep 37:553–563. https://doi.org/10.1007/s00299-018-2251-3

Le H, Nguyen NH, Ta DT, Le TNT, Bui TP, Le NT, Nguyen CX, Rolletschek H, Stacey G, Stacey MG, Pham NB, Do PT, Chu HH (2020) CRISPR/Cas9-mediated knockout of galactinol synthase-encoding genes reduces raffinose family oligosaccharide levels in soybean seeds. Front Plant Sci 11:612942. https://doi.org/10.3389/fpls.2020.612942

Li Z, Liu ZB, Xing A, Moon BP, Koellhoffer JP, Huang L, Ward RT, Clifton E, Falco SC, Cigan AM (2015) Cas9-guide RNA directed genome editing in soybean. Plant Physiol 169:960–970. https://doi.org/10.1104/pp.15.00783

Liang Q, Chen L, Yang X, Yang H, Liu S, Kou K, Fan L, Zhang Z, Duan Z, Yuan Y, Liang S, Liu Y, Lu X, Zhou G, Zhang M, Kong F, Tian Z (2022) Natural variation of Dt2 determines branching in soybean. Nat Commun 13:6429. https://doi.org/10.1038/s41467-022-34153-4

Liu J, Gunapati S, Mihelich NT, Stec AO, Michno JM, Stupar RM (2019) Genome editing in soybean with CRISPR/Cas9. Methods Mol Biol 1917:217–234. https://doi.org/10.1007/978-1-4939-8991-1_16

Liu T, Ji J, Cheng Y, Zhang S, Wang Z, Duan K, Wang Y (2023) CRISPR/Cas9-mediated editing of GmTAP1 confers enhanced resistance to phytophthora sojae in soybean. J Integr Plant Biol 65:1609–1612. https://doi.org/10.1111/jipb.13476

Ma J, Sun S, Whelan J, Shou H (2021) CRISPR/Cas9-mediated knockout of GmFATB1 significantly reduced the amount of saturated fatty acids in soybean seeds. Int J Mol Sci 22:3877. https://doi.org/10.3390/ijms22083877

Maher MF, Nasti RA, Vollbrecht M et al (2020) Plant gene editing through de novo induction of meristems. Nat Biotechnol 38:84–89. https://doi.org/10.1038/s41587-019-0337-2

Mali P, Yang L, Esvelt KM, Aach J, Guell M, DiCarlo JE, Norville JE, Church GM (2013) RNA-guided human genome engineering via Cas9. Science 339:823–826. https://doi.org/10.1126/science.1232033

Martins P, Beló A, Malone G, Quiroga M (2023) The first two elite soybean varieties genetically edited in South America. In Abstracts of the World Soybean Research Conference 11, Vienna, Austria, 18–23 June 2023

Michno JM, Virdi K, Stec AO et al (2020) Integration, abundance, and transmission of mutations and transgenes in a series of CRISPR/Cas9 soybean lines. BMC Biotechnol 20:10. https://doi.org/10.1186/s12896-020-00604-3

Michno JM, Wang X, Liu J, Curtin SJ, Kono TJ, Stupar RM (2015) CRISPR/Cas mutagenesis of soybean and Medicago truncatula using a new web-tool and a modified Cas9 enzyme. GM Crops Food 6:243–252. https://doi.org/10.1080/21645698.2015.1106063

Moscou MJ, Bogdanove AJ (2009) A simple cipher governs DNA recognition by TAL effectors. Science 326:1501. https://doi.org/10.1126/science.1178817

Nguyen CX, Dohnalkova A, Hancock CN, Kirk KR, Stacey G, Stacey MG (2023) Critical role for uricase and xanthine dehydrogenase in soybean nitrogen fixation and nodule development. Plant Genome 16:e20171. https://doi.org/10.1002/tpg2.20172

Nguyen CX, Paddock KJ, Zhang Z, Stacey MG (2021) GmKIX8-1 regulates organ size in soybean and is the causative gene for the major seed weight QTL qSw17-1. New Phytol 229:920–934. https://doi.org/10.1111/nph.16928

Pham AT, Lee JD, Shannon JG, Bilyeu KD (2010) Mutant alleles of FAD2-1A and FAD2-1B combine to produce soybeans with the high oleic acid seed oil trait. BMC Plant Biol 10:195. https://doi.org/10.1186/1471-2229-10-195

Qin C, Li H, Zhang S, Lin X, Jia Z, Zhao F, Wei X, Jiao Y, Li Z, Niu Z, Zhou Y, Li X, Li H, Zhao T, Liu J, Li H, Lu Y, Kong F, Liu B (2023) GmEID1 modulates light signaling through the evening complex to control flowering time and yield in soybean. Proc Natl Acad Sci U S A 120:e2212468120. https://doi.org/10.1073/pnas.2212468120

Raman V, Rojas CM, Vasudevan B et al (2022) Agrobacterium expressing a type III secretion system delivers Pseudomonas effectors into plant cells to enhance transformation. Nat Commun 13:2581. https://doi.org/10.1038/s41467-022-30180-3

Ren B, Wang X, Duan J, Ma J (2019) Rhizobial tRNA-derived small RNAs are signal molecules regulating plant nodulation. Science 365:919–922. https://doi.org/10.1126/science.aav9907

Shukla VK, Doyon Y, Miller JC, DeKelver RC, Moehle EA, Worden SE, Mitchell JC, Arnold NL, Gopalan S, Meng X et al (2009) Precise genome modification in the crop species Zea mays using zinc-finger nucleases. Nature 459:437–441. https://doi.org/10.1038/nature07992

Sun X, Hu Z, Chen R, Jiang Q, Song G, Zhang H, Xi Y (2015) Targeted mutagenesis in soybean using the CRISPR-Cas9 system. Sci Rep 5:10342. https://doi.org/10.1038/srep10342

Townsend JA, Wright DA, Winfrey RJ, Fu F, Maeder ML, Joung JK, Voytas DF (2009) High-frequency modification of plant genes using engineered zinc-finger nucleases. Nature 459:442–445. https://doi.org/10.1038/nature07845

Virdi KS, Spencer M, Stec AO, Xiong Y, Merry R, Muehlbauer GJ, Stupar RM (2020) Similar seed composition phenotypes are observed from CRISPR-generated in-frame and knockout alleles of a soybean KASI ortholog. Front Plant Sci 11:1005. https://doi.org/10.3389/fpls.2020.01005

Wang J, Kuang H, Zhang Z, Yang Y, Yan L, Zhang M, Song S, Guan Y (2020b) Generation of seed lipoxygenase-free soybean using CRISPR-Cas9. Crop J 8:432–439. https://doi.org/10.1016/j.cj.2019.08.008

Wang L, Sun S, Wu T, Liu L, Sun X, Cai Y, Li J, Jia H, Yuan S, Chen L, Jiang B, Wu C, Hou W, Han T (2020a) Natural variation and CRISPR/Cas9-mediated mutation in GmPRR37 affect photoperiodic flowering and contribute to regional adaptation of soybean. Plant Biotechnol J 18:1869–1881. https://doi.org/10.1111/pbi.13346

Wang N, Ryan L, Sardesai N, Wu E, Lenderts B, Lowe K, Che P, Anand A, Worden A, van Dyk D, Barone P, Svitashev S, Jones T, Gordon-Kamm W (2023a) Leaf transformation for efficient random integration and targeted genome modification in maize and sorghum. Nat Plants 9:255–270. https://doi.org/10.1038/s41477-022-01338-0

Wang X, Li MW, Wong FL, Luk CY, Chung CY, Yung WS, Wang Z, Xie M, Song S, Chung G, Chan TF, Lam HM (2021) Increased copy number of gibberellin 2-oxidase 8 genes reduced trailing growth and shoot length during soybean domestication. Plant J 107:1739–1755. https://doi.org/10.1111/tpj.15414

Wang Z, Shea Z, Rosso L, Shang C, Li J, Bewick P, Li Q, Zhao B, Zhang B (2023b) Development of new mutant alleles and markers for KTI1 and KTI3 via CRISPR/Cas9-mediated mutagenesis to reduce trypsin inhibitor content and activity in soybean seeds. Front Plant Sci 14:1111680. https://doi.org/10.3389/fpls.2023.1111680

Xu H, Zhang L, Zhang K, Ran Y (2020) Progresses, challenges, and prospects of genome editing in soybean (Glycine max). Front Plant Sci 11:571138. https://doi.org/10.3389/fpls.2020.571138

Yan L, Wei S, Wu Y, Hu R, Li H, Yang W, Xie Q (2015) High-efficiency genome editing in Arabidopsis using YAO promoter-driven CRISPR/Cas9 system. Mol Plant 8:1820–1823. https://doi.org/10.1016/j.molp.2015.10.004

Zhang F, Maeder ML, Unger-Wallace E, Hoshaw JP, Reyon D, Christian M, Li X, Pierick CJ, Dobbs D, Peterson T et al (2010) High frequency targeted mutagenesis in Arabidopsis thaliana using zinc finger nucleases. Proc Natl Acad Sci U S A 107:12028–12033. https://doi.org/10.1073/pnas.0914991107

Zhang N, McHale LK, Finer JJ (2015) Isolation and characterization of "GmScream" promoters that regulate highly expressing soybean (Glycine max Merr.) genes. Plant Sci 241:189–198. https://doi.org/10.1016/j.plantsci.2015.10.010

Zhang Y, Blahut-Beatty L, Zheng S, Clough SJ, Simmonds DH (2023) The role of a soybean 14-3-3 gene (Glyma05g29080) on white mold resistance and nodulation investigations using CRISPR-Cas9 editing and RNA silencing. Mol Plant-Microbe Interact 36:159–164. https://doi.org/10.1094/MPMI-07-22-0157-R

Zhang Y, Guo W, Chen L, Shen X, Yang H, Fang Y, Ouyang W, Mai S, Chen H, Chen S, Hao Q, Yuan S, Zhang C, Huang Y, Shan Z, Yang Z, Qiu D, Zhou X, Cao D, Li X, Jiao Y (2022) CRISPR/Cas9-mediated targeted mutagenesis of GmUGT enhanced soybean resistance against leaf-chewing insects through flavonoids biosynthesis. Front Plant Sci 13:802716. https://doi.org/10.3389/fpls.2022.802716

Zhang Y, Malzahn AA, Sretenovic S, Qi Y (2019) The emerging and uncultivated potential of CRISPR technology in plant science. Nat Plants 5:778–794. https://doi.org/10.1038/s41477-019-0461-5

Genomics and Genome Editing for Crop Improvement

22

Satendra K. Mangrauthia, Kutubuddin A. Molla,
Raman M. Sundaram, Viswanathan Chinnusamy, and K. C. Bansal

Abstract

The sustainable growth of agriculture requires continuous efforts towards developing new plant varieties that can yield more with lesser inputs, specifically, water, fertilizers, and pesticides. Of late the objectives of plant breeding have shifted from yield-centered breeding to environmentally sustainable breeding including the new plant concept that can help decarbonization and reduce greenhouse gas emissions. These widened goals of plant breeding are achievable through smart exploitation of genetic information using powerful breeding technologies such as genomic assisted breeding (GAB), transgenic, and genome editing etc. The past two decades have been crucial in terms of simultaneous advancement in crop genomics and the discovery of precise and efficient breeding tools. The sequencing and re-sequencing of crops and their diverse accessions have helped in marking the gene sequences and their potential association with breeding traits. Similarly, the shift of breeding from random introgression by cross-breeding to targeted introgression of genes by marker-assisted breeding and transgenic can be considered as a landmark progress of breeding tools. Most importantly, the finest and cleanest breeding through genome editing has the

Satendra K. Mangrauthia and Kutubuddin A. Molla contributed equally.

S. K. Mangrauthia · R. M. Sundaram
ICAR-Indian Institute of Rice Research, Hyderabad, India

K. A. Molla
ICAR-National Rice Research Institute, Cuttack, India

V. Chinnusamy
ICAR-Indian Agricultural Research Institute, New Delhi, India

K. C. Bansal (✉)
Formerly at ICAR-National Bureau of Plant Genetic Resources, New Delhi, India

© National Academy of Agricultural Sciences, under exclusive license to Springer
Nature Singapore Pte Ltd. 2023
K. C. Bansal et al. (eds.), *Transformation of Agri-Food Systems*,
https://doi.org/10.1007/978-981-99-8014-7_22

potential to revolutionize agriculture, and the next decade should see remarkable progress in the delivery of environmentally sustainable plant varieties developed through new breeding technologies. The degree of success of these breeding tools would heavily rely on the discovery of new genes/alleles/haplotypes and their functional analysis through classical wet lab experiments. This chapter provides a comprehensive overview of the structural and functional genomics of crops and their applications for cultivar improvement through cross-breeding, transgenic, and genome editing approaches. The priority traits and crops for improvement through genome editing and new technological advancements such as based editing and prime editing have also been discussed.

Keywords

MAB · GWAS · *Oryza sativa* · CRISPR/Cas · Genome editing · Prime editing · Base editing · MAGIC · Transgenic

22.1   Introduction

India is the heartland of agricultural diversity. With a projected population of 1.67 billion by 2050, India faces the daunting challenge of sustainably feeding its people with nutritious food, while grappling with dwindling arable land, diminishing ground water and erratic climatic patterns. In this changing landscape, genomics and genome engineering emerge as powerful tools, offering disruptive solutions to many recalcitrant challenges of crop improvement.

From the Green Revolution that ushered in an era of increased cereal production to the contemporary era of precision agriculture, India is always rapid in adopting new technologies to meet the needs of its growing population. Today, in the era of climate change, resource scarcity, and changing consumer preferences, the tools and techniques provided by genomics and genome engineering are vital.

Developing crops to suffice human needs is one of the remarkable things done by *homo sapiens* to ensure their survival and propagation. However, methods used to develop desired crops have been modified over the years. In the distant past, desired varieties were bred by raising the frequency of desired plants in a heterogeneous population by practicing simple phenotypic selection. Later on, continued research and developments in agriculture resulted in the concept of uniform varieties, such as pure line varieties. This was then followed by development of hybrids, synthetic varieties, composite varieties, and so on. However, developing varieties and hybrids was constrained by the extent of available variations that could be utilized. Unfortunately, these efforts could only tap into the existing natural variations, and that too, is limited to the sexually crossable gene pool. Through the advancement of various biotechnological methods, these constraints were successfully surmounted.

In this chapter, we discuss the remarkable advancements of structural and functional genomics of crops, how genomic tools evolved over time, and how those tools assisted in trait discovery and varietal development. Next, we delve deeper into the available genome engineering technologies, including transgenesis, cisgenesis, and genome editing. We also deliberate on the contribution of biotech crops for ensuring food security in a changing climate.

22.2 Crop Genomics

The completion of genome sequencing of different crops, availability of high-quality reference genome sequences and lowering costs of high-throughput sequencing technologies, re-sequencing of crop varieties, landraces, and wild relatives have improved our understanding of genetic diversity, evolution, and domestication of various crop plants, leading towards their targeted and precise improvement. The advanced bioinformatics tools and computation approaches help mining the massive genomics data and develop the high-density SNP genotyping platforms used for linkage mapping, genome-wide association studies and genomic assisted breeding. Genomics of important crops like maize (Jackson et al. 2022), wheat (Walkowiak et al. 2020), legumes (Ojiewo et al. 2019; Jha et al. 2022), and brassica crops (He et al. 2021) has been well described in previous studies. In the following sections, we discuss the progress on structural and functional genomics of rice, a model crop plant.

22.2.1 Progress on Rice Structural Genomics

Among the cultivated crop plants, the genomics of *Oryza sativa* and its wild relatives have been studied most extensively. Since the sequencing of the first full genome of the japonica rice cultivar Nipponbare, massive genome data of rice accessions have been generated through different studies. The gold standard genome of Nipponbare cultivar of *Oryza sativa* (japonica group) serves as reference genome for majority of genomics studies (International Rice Genome Sequencing Project 2005). In addition to reference japonica cultivar, the indica rice cultivar 93–11 was also sequenced to generate the reference genome of indica genome (Yu et al. 2002). Tremendous progress on genomics has been witnessed due to advancement of sequencing technologies and bioinformatics pipelines. Several genomics databases have been developed to store and manage the genome sequence data (Wang and Han 2022). For example, RAP-DB database (https://rapdb.dna.affrc.go.jp/index.html) encompasses the information of rice genes, their expression patterns, chromosomal position, gene structure, and annotated function, etc. Similarly, RiceXPro databases include detailed expression behavior of rice genes throughout its growth stages and tissues (Sato et al. 2011). After deciphering of the genome size of ~389 Mb and 37,544 annotated protein coding genes by International Rice Genome Sequencing Project (2005), information about the alleles of these genes and their haplotypes has been generated due to extensive structural genomic studies. Recently, 3000 rice accessions representing 89 countries were sequenced to understand the genetic diversity within *Oryza sativa* gene pool, which led to discovery of more than 18 million single nucleotide polymorphisms (SNPs) (The 3,000 rice genomes project 2014). Among the 3000 rice accessions, more than 90,000 structural variations were identified along with 2.4 million InDels and 29 million SNPs helping rice genomics and breeding programs (Wang et al. 2018). Based on these genomic sequences, Rice SNP-Seek Database (https://snpseek.irri.org/_download.zul) has been developed and made available to public and this tool is being extensively utilized in studying the structural genomics, allele and haplotype variations, and breeding programs.

In another such effort, 1143 Indica rice parental lines of superior hybrids of China were re-sequenced to identify the genetic loci associated with heterosis. They identified 3.86 million high-quality nuclear SNPs and 0.717 million InDels with 98 loci that span 3218 SNPs and 539 indels, which are possibly associated with heterosis (Lv et al. 2020). The heterotic loci included genes such as *mads3, Rf4, hd3a, Ehd2, Ehd4, Ghd7, Gn1a, LAX1* and *Sd1*. In addition to significant progress observed in structural genomics of cultivated rice, the genome sequencing and characterization of wild rice species including *Oryza rufipogon, Oryza granulata, O. nivara, O. barthii, O. glumaepatula, O. meridionalis, O. punctata,* and *L. perrieri* (Stein et al. 2018; Shi et al. 2020; Xie et al. 2021) have led to discovery of structural variations of genomes associated with domestication, diseases and insect resistance and climate resilience. The whole genome sequencing of an elite rice restorer line KMR3 and *Oryza rufipogon*-derived IL50–13 (Chinsurah Nona 2/Gosaba 6) helped identification of new alleles and genes associated with yield and salinity tolerance in rice (Thummala et al. 2022). The continuous progress on sequencing technologies has led to identification of several unknown genomic regions. Recently, the long read sequencing of 75 rice genomes was performed with Oxford Nanopore Technologies long-read and Illumina short-read platforms. A total of 111 rice genomes with long read sequences were analyzed and compared with 3 K rice genomes, leading to the identification of 879 Mb novel sequences and 19,000 novel genes in the rice pan-genome (Zhang et al. 2022). Another such effort by Shang et al. (2022) on long read sequencing to 251 rice accessions including the *Oryza sativa* and wild derivatives led to construction of super pan genome of rice. The study revealed several unknown structural variations in rice genome and genes associated with agronomic traits including grain size and weight, and adaptation and domestication.

22.2.2 Progress on Rice Functional Genomics

Functional genomics with an aim to understand the function of protein coding genes, non-coding RNAs, regulatory nucleotide sequences, or any intergenic regions in various biological processes is the key to utilize the genes for crop improvement programs. While structural genomics has witnessed tremendous progress in recent years, the slow pace of functional genomics has been the major bottleneck in harnessing the full potential of gene deployment in rice improvement programs. Appreciable efforts have been made to map the QTLs and genes for various traits using bi-parental or multi-parental mapping populations. Association of a specific genetic locus with the particular trait through molecular mapping has been the major driver of functional genomics that helped introgression of these mapped QTLs/genes into cultivated rice varieties. QTLs for low phosphorus deficiency (*Pup1*) (Wissuwa et al. 1998; Kale et al. 2021), submergence tolerance-SUBMERGENCE 1 (*SUB1*) (Xu and Mackill 1996), salinity tolerance (*Saltol*) (Thomson et al. 2010), drought tolerance (*qDTY* 1.1) (Vikram et al. 2011), grain quality traits specific for Basmati rice (Amarawathi et al. 2008), blast disease resistance (Sharma et al. 2005), bacterial blight resistance (Kumar et al. 2012; Sinha et al. 2023), sheath blight resistance (Channamallikarjuna et al. 2010), etc.. Few of these QTLs were dissected and the causal genes were functionally characterized. For example, phosphorus-starvation tolerance 1 (*PSTOL1*) gene, a serine/threonine receptor-like kinases of the LRK10L-2 subfamily, was identified and characterized in detail from Pup1 QTL that confers the low P tolerance (Gamuyao et al. 2012). Similarly, the *Gn1a* QTL for high grain number/panicle mapped on chromosome 1 of rice was characterized to identify the gene contributing to the trait, viz., cytokinin oxidase 2. The fine-mapping exercise was followed by development of near isogenic lines and transgenic overexpression studies revealed that recessive allele of cytokinin oxidase 2 gene underlying *Gn1a* is responsible for high grain number trait in rice (Ashikari et al. 2005). Many such QTLs identified for grain size, grain weight, panicle density, and other yield traits have been characterized, the causal genes have been identified and their mechanism of action have been deciphered (Li et al. 2021a, b).

In addition to molecular mapping, the genome-wide expression analysis of genes and non-coding RNAs through microarray and RNAseq experiments along with targeted expression analysis through northern hybridization and quantitative PCR heled identification of probable candidate genes associated with various useful traits (Lenka et al. 2011; Agarwal et al. 2014; Mangrauthia et al. 2016; Bhogireddy et al. 2021). Efforts have also been made to integrate large-scale RNA-seq data for improved annotation of protein coding genes and non-coding RNAs (Sang et al. 2020). The whole-genome analysis of microRNAs in 16 rice samples helped identification of species and tissue specific microRNAs (miRNAs) induced during high temperature stress, and few of the candidate miRNAs for heat tolerance were suggested (Sailaja et al. 2014; Mangrauthia et al. 2017). The detailed expression analysis of miRNAs and their target genes in different growth stages and heat treatments suggested highly dynamic gene regulation through miRNAs during heat stress (Bhogireddy et al. 2022). Agarwal et al. (2015) analyzed targeted expression of miRNAs and genes to decipher the candidate genes associated with grain iron and zinc traits. RNAseq based identification of miRNAs in cultivated and wild rice species and combination of proteomics and transcriptomics approaches helped identification of promising genes and miRNAs for sheath blight disease resistance (Prathi et al. 2018; Chopperla et al. 2020). The heterosis related genes from wild rice species *Oryza rufipogon* were analyzed using whole genome transcriptome approach (Guttikonda et al. 2020). Although the studies based on transcriptome or proteome have shown great promise for identification of genes and pathways associated with almost all the agronomically important traits of rice, the downstream experiments with conclusive evidence for trait-gene association have been very limited. Moreover, the discovery of hundreds and thousands of promising genes in such studies makes it difficult for authors to choose a specific gene among large numbers for downstream experiments. Among the most successful functionally characterized genes identified through RNAseq approach is *DREB1C* which showed 41.3–68.3% enhanced yield in rice by overexpression through transgenic approach, while the knockout lines showed significant decrease in rice yield. In addition to high grain yield, the DREB1C over-expression lines showed better nitrogen use efficiency and 13–19 days early maturity (Wei et al. 2022). Similarly, a gene encoding methyl transferase involved in conversion of carlactonoic acid to methyl carlactonoate was identified in rice using the RNAseq approach. The functional analysis of the gene showed its role in phosphate starvation response and strigolactone biosynthesis (Haider et al. 2023).

The genome annotations of 16 platinum standard reference genomes of rice accessions revealed 41,346 average number genes in rice (Yu et al. 2023), among which only ~4500 gene have been functionally characterized (Huang et al. 2022). The phenotypic effect of a large number of uncharacterized genes is still unknown. Moreover, the inclusion of different alleles of a gene, their haplotypes, alternate transcripts, non-coding RNAs, and regulatory elements for functional characterization makes this task gigantic. In addition to functional analysis of these elements to study their individual effects on phenotype, it would be necessary to upscale the study by understanding the protein and metabolites interactions to fully appreciate their role in plant growth and metabolism. In order to develop rice genotypes suitable for changing climate and sustainable agriculture growth, it is very important to focus attention of functional genomics research to unravel new genetic players that can help researchers to bring another revolution for the benefit of humanity. Genome-wide approaches to develop T-DNA insertion mutant populations, EMS mutants, or CRISPR/Cas9 library of rice genes (Yi and An 2013; Mohapatra et al. 2014; Meng et al. 2017) will hasten the progress of functional genomics, however, there is need of continued efforts and perseverance to succeed the mammoth goal of rice functional genomics. Reddy et al. (2020) utilized activation-tagging' mutagenesis approach to unravel the function of *OsPAP90* that affects D1 protein stability and Photosystem II efficiency, while Poli et al. (2021) utilized one of the EMS mutants of rice cultivar Nagina22 for understanding the basis phosphorus use efficiency trait induced by mutation. In yet another path-breaking research, the EMS mutant population of indica rice cultivar 9311 was screened to identify the NGR5 (nitrogen-mediated tiller growth response 5) that encodes APETALA2 (AP2)–domain transcription factor. The NGR5 regulates histone H3 lysine 27 trimethylation and downstream expression of genes/transcription factors associated with nitrogen use efficiency, tillering, and grain yield (Wu et al. 2020). Many such studies utilizing targeted or random mutants are required to better understand the physiology of rice plant and gene-trait associations.

22.3 Genomics Applications for Plant Breeding

Many plant varieties have been developed and released using conventional plant breeding which relied on selection of individuals from a segregating population through phenotypic selection. Later, selection based on genomics information has not only facilitated phenotypic selection, it helped pooling of favorable alleles and haplotypes for significant increase of breeding value. Thus, genomics played a vital role in marker assisted selection, breeding by design, and genomic selection to provide a great leap forward in plant breeding.

22.3.1 Technological Advancement

In the past 30 years, various molecular marker techniques like RAPDs, RFLPs, AFLP, and SSRs have been employed in marker-assisted breeding. However, these markers often involve substantial labor and time investments making their implementation impractical in large scale breeding programs. The advancement in NGS, coupled with whole genome re-sequencing, RNA sequencing and haplotyping helped identification, validation and evaluation of molecular markers in plants (Lai et al. 2010). The application of re-sequencing witnessed a growing embrace in the creation of diagnostic markers that align with the target genes sequences and help in marker assisted selection (MAS). In the past, the formation of core collections primarily relied on factors such as geographical distribution and pedigrees. However, the application of genomics helped in efficient utilization of crop core collections (Jia et al. 2017). Genomic tools like single nucleotide polymorphism (SNP) markers derived from genotyping arrays or whole genome sequence data furnish insights at both DNA and RNA offering a detailed understanding which in turn is less affected by environmental variables.

Owing to the advancements in the genomics and the evolution of high-throughput genotyping technologies, the orientation of plant breeding has gradually transitioned from conventional phenotypic driven to genotype centered selection. The utilization of MAS could introduce a certain degree of enhancement in breeding efficiency and stood in a dominant position in breeding programs for a sustainable period (Xu and Crouch 2008). Later in the past two decades, the swift progress of whole genome sequencing and marker development technologies has facilitated the utilization of high-density SNP markers for comprehensive multi-omic analysis at remarkably low expense that bridges the gap between genotype and phenotype.

In contemporary times, Genomic-Assisted Breeding (GAB) has emerged as a potent strategy within the realm of plant breeding. GAB offers the integration of genomic tools alongside the high-throughput phenotyping thereby multiplying the breeding methodologies through molecular markers to predict phenotype from the genotypic information. By employing GAB, breeders are developing populations and performing analysis solely using genotypic characterized progenies, and subsequently narrowing down their focus on the phenotypic assessment (Cooper et al. 2014). Moreover, the evaluation of genotypic trait data can be conveniently conducted during off season periods that accelerates the plant breeding process. Notably, GAB proves to be advantageous in advancing many qualitative and quantitative traits due to its high-quality precision, shortened breeding cycles and increased selection efficiency. Thus, GAB holds a significant promise in expediting the creation of novel plant varieties.

The successful construction of crop genetic maps combined with high throughput marker detection methods paved a way for widespread adoption of genomic selection (GS) for plant breeders. GS is operated by harnessing the collective information encapsulated within an individual's genetic makeup across the entire genome, rather than focusing solely on a limited set of molecular markers associated with known genes. It provides a strategic advantage in overcoming the hurdles posed by multiple QTL and offers a promising path towards advanced crop improvement. In the near future, the widespread integration of MAS and GS will help in getting notable enhancements in plant breeding on a genomic scale. This will encompass a comprehensive set of practices like genomic surveys to get the allelic variations, strategies towards exploring genomic loci with GWAS, molecular documentation of germplasm repositories, utilization of haplotype-based techniques and computational based genomic decision making (Peng-fei et al. 2017). Together these components converge to form GAB, a paradigm poised for significant crop improvement strategies.

22.3.2 Varietal Development through MAS

The genomics driven MAS has helped development and release of several rice cultivars. Improved Samba Mahsuri was developed by introgression of three bacterial blight resistant genes (xa5 + xa13 + Xa21) in the fine grain popular rice cultivar Samba Mahsuri (BPT5204). Similarly, DRR Dhan 42 was developed by introgression of drought tolerant QTLs (*qDTY 2.2, QDTY 4.1 and qDTY 10.1*) and DRR Dhan 50 was developed by introgression of drought and submergence tolerant QTLs (*qDTY2.1, qDTY3.1, and Sub1*) in the background of IR64 and Samba Mahsuri, respectively. Wheat cultivar PBW 761 (Unnat PBW 550) was developed through markers assisted introgression of gene Yr15 for Stripe rust resistance. Vivek QPM9 maize hybrid was developed through markers assisted introgression of gene opaque2 for increased lysine & tryptophan while Pusa Vivek QPM9 Improved maize hybrid with increased Provitamin-A was developed through introgression of *crtRB1* gene. The cultivars through MAS have been developed in various other crops like pearl millet, soybean, chickpea, and groundnut etc. (Yadava et al. 2022). With more than 5 mha area affected by bacterial blight disease, the MAS derived resistant cultivars like Improved Samba Mahsuri (ISM) have played a significant role in minimizing the farmers loss due to disease as well as reducing the pesticide consumption. The ISM is being grown in ~1,50,000 ha area in India. Similarly other crop varieties developed through have played significant roles in enhancing farmer's income.

22.3.3 Trait Discovery

Genome-wide association study (GWAS) plays a pivotal role in exploration of genome-wide SNPs associated with complex traits thereby unraveling the genetic architecture of traits. SNP based heritability through GWAS involves the interpretation of phenotypic variation by elucidating the additive and dominance effects. The exploration of these SNPs can be made possible by including multi-omic based research which helps in understanding their functional implications (Loos 2020). Genome wide association mapping was conducted to identify QTLs for sheath blight resistance, plant height and days to heading among 417 accessions from rice diversity Panel 1 (RDP1). *Rhizoctonia solani* was inoculated on 417 accessions at two field locations Arkansas, USA (AR) and Nanning, China (NC). In GWAS with 3.4 million SNPs, 18 SNPs showed association with sheath blight resistance. These QTLs can be used along with tiller and panicle count for improved yield potential and disease resistance (Li et al. 2022a, b).

A genome wide association study is an unbiased research approach which helps to identify the common genomic variants that are associated with quantitative traits. Its primary objective is to identify the associations of phenotypes by testing the allelic frequencies of genetic variants between the individuals that are ancestrally similar but differ in their phenotypic expression (Uffelman et al. 2021). GWAS is made easy and popular now-a-days as there is a decreasing cost in genome-wide genotyping and the continuous updating of association studies which increased the number of variants associated with research. This evolution of association studies led to an expansion in the repertoire of considered variants under specialized studies. The field of genomics has undergone substantial progression, marked by improvements in data resources, genotyping efficiency, and the reservoir of large datasets. These advancements coupled with the availability of big data have efficiently facilitated the study of even the rare variants through GWAS. The computational efficiency by the usage of statistical packages and models has remained instrumental in sustaining the forward momentum of GWAS based research spanning areas like multi-variant GWAS, enhancing genetic potential off horticultural crops, delving into cluster analysis of many traits for heterogeneity testing and also scrutinize the interactions between genes and the environment across the genome. In brief, GWAS includes a potent avenue for comprehensive exploration in the field of genomics, revealing an interplay between genetic makeup and phenotypic data across a diverse spectrum of traits.

GWAS emerges as a potent approach that effectively handles extensive population scale data. Leveraging the advanced computational methodologies like FarmCPU, EMMAX within the GAPIT and rMVP has proven instrumental in mitigating spurious effects of population structure and relatedness thereby curbing false positives. The integration of GWAS with functional genomics stands as a strategic maneuver, allowing for identification of candidate genes for specific traits. The ongoing progress in molecular biology enhances the contributions of tissue and cell type studies of multiple traits data and paves a valuable ally in GWAS research as in many studies, availability of data is a major constraint. Moreover, co-localization of genomic regions (SNPs) pinpointed by GWAS along with QTL data confers robust means for identifying potential candidate gene(s) of interest. In essence GWAS, showcases its importance as a multidimensional tool that not only enhances our understanding of complex traits with common variants but also serves as a springboard for discovering novel genetic rare variants which are useful in crop improvement.

The trait discovery has also been accelerated by new advancements in development of mapping populations. The utilization of experimental and germplasm populations has demonstrated significant importance in the identification of QTLs associated with targeted agronomic traits. An innovative and distinctive form of multi-parent population termed as Multi-Parent Advanced Generation Intercross (MAGIC) consists a collection of stable/fixed lines (RILs) each harboring a genome that is an admixture of genetic compositions of various ancestral parents (Mackay and Powell 2007). MAGIC populations display their potential to significantly enhance the detection of QTLs, facilitate the creation of superior breeding materials, and notable genetic advancements for crop improvement. It helps accumulation of recombination events attained through multiple cycles of inter-crossing and selfing significantly elevates the precision of QTL mapping, and uncover correlations between genes and traits with high resolution (Scott et al. 2020). The elevated recombination rate contributes to additional reduction in linkage disequilibrium while simultaneously yielding minimal population structure which results in false positives.

MAGIC populations present a potential alternative to the inherent limitations of bi-parental and germplasm populations. These MAGIC populations arise from crossing of more than two distinct inbred founder lines, offering a bridge that connects traditional linkage mapping and association mapping. This strategy brings about a noteworthy enhancement in precision mapping by integrating multiple genetic resources, thereby enriching genotypic and phenotypic diversity. MAGIC populations occupy an intermediate position, balancing the genetic diversity found in natural accessions of a diallelic RIL system. The unique recombination of RILs and diversity enables them to study the complex trait structures and help in advancing the crop breeding programs. MAGIC populations have been successfully established in numerous species and helps in pinpointing polymorphisms linked to QTLs or genes governing desirable traits in plants (Arrones et al. 2020). Currently MAGIC and other multi-parent populations (MPPs) have found their place across a diverse array of species including cereals, legumes, vegetables and fruit crops.

22.4   Genomics Applications for Genome Engineering

22.4.1 Technological Advancement

With the availability of enormous amount of genomics information, including both structural and functional genomics, it has become easier to manipulate genome of a crop for incorporating a desired trait and removing an undesired one. Discovery of

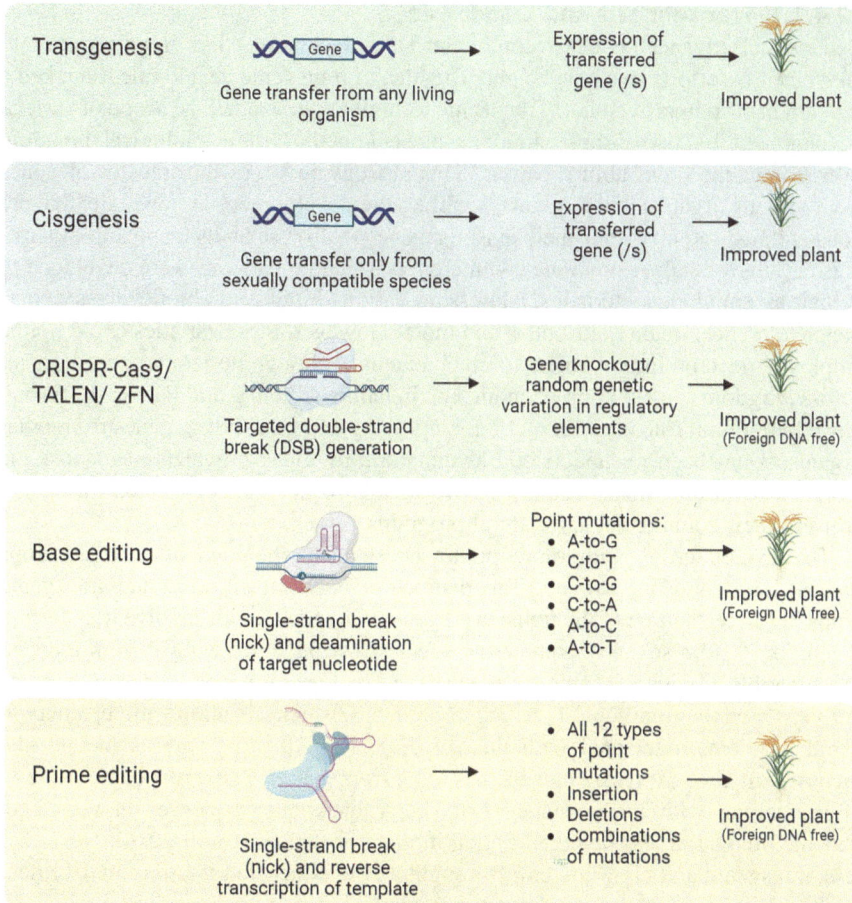

Fig. 22.1 Schematic of different genome engineering technologies available for basic research and crop improvement. For conventional genome editing through CRISPR-Cas9/TALEN/ZFN, the option of gene knock-in and other complex editing through homology-directed repair (HDR) is not shown here. For base editing, dCas version is not shown

DNA/RNA and exploring different facets of nucleic acids in the cell have enabled us to invent multiple powerful tools to artificially manipulate them. Significant progress has been made in accurately transferring advantageous DNA fragments, genes or regulatory elements, into the genome of a target crop, thereby creating novel traits. Recently, precise editing of existing DNA letters in target regions of the genome has enabled creating desired traits to meet the current needs (Karmakar et al. 2022). Below, we briefly delve into the technologies that gave us unparalleled capabilities to enhance crop genomes, rendering them more valuable in today's context for crop improvement (Fig. 22.1).

22.4.1.1 Transgenesis and Cisgenesis

Exponential advancement in recombinant DNA technology has empowered us to take breeding efforts to a new height. The idea of transgenic development marked a new era in crop improvement. Transgenic technology expanded the scope of varietal development by harnessing variations present in the whole biological kingdom overcoming the cross-ability barrier. This strategy involves the transfer of genes from various organisms into plants, enabling the development of new traits that are not naturally found within their existing gene pool or sexually compatible germplasm. Subsequently, numerous commercially valuable varieties were developed to exhibit essential characteristics, addressing various biotic and abiotic stresses, herbicide tolerance, quality attributes, and more. However, these varieties could not be employed for crop improvement to much extent. Although horizontal gene transfer across kingdom is a common phenomenon in nature (Keeling and Palmer 2008) and played important role in the evolution, employing and transferring genes from other organisms has been misunderstood as unnatural by many organizations across the world. Due to this, many countries were unable to deregulate selected transgenic crops in their commercial agricultural programs.

To develop crops with more public acceptance, the idea of cisgenic crops provided a promising avenue. The distinction between cisgenic and transgenic crops lies in the source of the transferred genes. Cis-genic crops involve transferring a gene only from sexually compatible species instead of deriving it from a cross-incompatible organism. Moreover, the variety developed is considered similar to one derived from the traditional breeding approach. Cis-genesis allows the breeders to avoid problems associated with linkage drag. Nevertheless, various factors like inconsistent gene expression levels stemming from genetic background and position effects, along with the persistence of vector backbone sequences in the target genome, make it a less favoured option in many instances (Cardi 2016). Nevertheless, transgenic and cis-genic crops brought a new dimension in crop improvement programs, employing variation that is not available in existing germplasm.

22.4.1.2 Conventional Genome Editing (GE)

The advent of genome editing has been proven successful in circumventing several limitations of transgenesis to further aid crop improvement programs. Instead of inserting a whole gene, GE involves the addition or removal of one or multiple bases at specific target sites. Although there are several genome editing technologies like ZFN, TALEN, and CRISPR-Cas available with us, CRISPR-Cas has become the method of choice in recent times. The conventional CRISPR/Cas-based GE relies on inducing double-strand breaks (DSBs) at target site. The targeted DSB can be generated in the genome by designing a small guide RNA. Guide RNA directs and specifies Cas nuclease to bind and cleave the target DNA (Jinek et al. 2012). Subsequently, the DSB is mended by the cell's internal repair machinery. During the repairing process, these DSBs can undergo either non-homologous end joining (NHEJ) or homology-directed repair (HDR) depending on the availability of donor fragments. NHEJ is the predominant pathway that higher eukaryotes employ to repair DSB. NHEJ often generates random insertion or deletion (Indel) while re-ligating the broken ends. Random Indels often disrupt the open reading frame of a gene. This technique has primarily been employed to disrupt gene function by generating nonsense codons and truncated proteins. Since this method is completely random in terms of the nature of induced mutations, it cannot be used to induce desired base substitutions or to install predefined indels. To address this limitation, modified versions of CRISPR/Cas-based GE were developed, including base editing and prime editing techniques.

22.4.1.3 Base Editing (BE)

BE is especially known for inducing substitution mutations at targeted sites without any DSBs and donor templates (Komor et al. 2016; Gaudelli et al. 2017). It is known to induce both transition and transversion base substitutions depending on the type of base editors. Base editors are made up of a catalytically impaired Cas protein (nCas9), a deaminase domain, and sometimes a repair protein. Cytosine base editors (CBEs) and adenine base editors (ABEs) enable us to modify C-to-T and A-to-G, respectively, in the genome. These editors are distinguished by their ability to deaminate cytosine and adenine, respectively. CBEs and ABEs have been quickly adopted for plant systems to install functional SNPs (Molla and Yang 2019; Molla et al. 2020a). In recent times, several new classes of BEs were developed which can induce transversion mutations. For example, C-to-G base editor (CGBE), C-to-A base editor (CABE), A-to-Y base editor (AYBE) (where Y = C/T) (Molla et al. 2020b; Chen et al. 2023; Tong et al. 2023). Moreover, dual BEs with the ability to carry out simultaneous editing of A and C have also been developed by fusing both adenine and cytosine deaminase with nCas9 (Li et al. 2020a).

Along with the rapid development of BEs for nuclear DNA, BEs for organellar DNA have recently been developed. Unlike nuclear BEs, organellar BEs are based on TALE arrays. Mitochondrial and plastid DNA of different organisms were edited with single base resolution with those BEs (Mok et al. 2020; Kang et al. 2021; Lee et al. 2022; Li et al. 2021a, b). The rich toolbox of base editors for nuclear and organellar DNA offers unprecedented opportunities for crop improvement.

22.4.1.4 Prime Editing (PE)

Though BEs have empowered us to introduce point mutations, they cannot produce precise indels at target sites. For introducing precise indels and complex editing, researchers rely on homology-directed repair (HDR) by creating DSB and supplying a donor template. However, HDR efficiency is extremely low in higher plants (Molla et al. 2022). Prime editing (PE) tool can introduce the desired stretch of nucleotides at any desired location (Anzalone et al. 2019) and this renders it with the ability to induce all eight transversions and four transition mutations. Structurally, PE consists of nCas9 fused with reverse transcriptase (RT). A modified guide RNA, known as prime editing guide RNA (pegRNA), is used to instruct PE where to make edits and specify what to edit. Collectively, it can be employed to introduce sense, missense, and nonsense mutations, small indels, and combinations of mutations at virtually any location in the genome, depending on the pegRNA's specificity. Although the benefit served by PE is unparallel, further research to increase its editing efficiency is required. In plant systems, PE suffers from low efficiency, which hinders its regular use in crop improvement (Molla et al. 2021). However, recent reports showed significant technical advancements in PE for increasing efficiency and versatility (Lin et al. 2021; Jiang et al. 2022; Li et al. 2022a, b).

22.4.2 Varietal Development and Social Impact

Numerous crop varieties have been developed in the past by using transgenic technologies. Within this group of varieties, the integration and application of genes from *Bacillus thuringiensis* (Bt) can be regarded as a significant milestone in the history of humanity. The *Cry* genes derived from this bacterium have been widely used to safeguard several crops against lepidopteran and coleopteran insects (Peterson et al. 2020). During the mid-1990s, Bt-corn, Bt-cotton, and Bt-maize were introduced for commercial use among American farmers as a means to combat harmful insects (Mendelsohn et al. 2003). Moreover, China has also adopted Bt-cotton to protect the crop (Li et al. 2020a, b). Interestingly, Bt-cotton is the only transgenic crop allowed in India. Since its approval in 2002, Bt-cotton has gradually replaced a substantial portion of non-Bt cotton and is now grown on more than 90% of the total cotton acreage in India (Research Outreach 2021). Notably, the adoption of Bt cotton increased cotton production by 24% and farmer's profits by 50% (Qaim 2020). Cultivation of Bt cotton has also resulted in reduced pesticide consumption, helping to decrease environmental hazards (Krishna and Qaim 2012). Insecticide usage, particularly to counteract *Helicoverpa armigera*, was recorded to decline from 71% in 2001 to 3% in 2011 (Choudhary and Gaur 2015). Bt-cotton adoption has also positively impacted household living standards; consumption expenditure of Bt farmers increased by 18% increase over non-Bt farmers (Peshin et al. 2021).

Transgenic crops have been cultivated in various parts of the world, including the United States, India, China, Canada, Brazil, Argentina, and many other countries. In 2019, a total of 190.4 million hectares were reported to be under transgenic crop cultivation across 29 countries (ISAAA 2019). Around 72 countries are involved in either producing or consuming transgenic crops. In 2019, it was estimated that over 70% of the worldwide cotton and soybean cultivation area was occupied by transgenic varieties. Additionally, more than 25% of the global canola and maize cultivation area is sown under transgenic crops. Moreover, alfalfa, sugar beets, sugarcane, papaya, safflower, potatoes, and brinjal are other important transgenic crops cultivated in different countries. According to a recent report from USDA, ~55% of U.S. cropland is planted to GM varieties, including corn, soybeans, cotton, canola, sugar beet, potato, apple, alfalfa, papaya, squash (Success Farming 2023). Despite the adoption of transgenics across the world, the European Union has shown reluctance to adopt GM crops consistently over the years. However, only Spain and Portugal are producing Bt-maize to prevent its agriculture from European corn borer. Apart from herbicide tolerance and insect tolerance, transgenics have been also used to develop commercially available drought-tolerant crops. Recently, HB4 wheat developed by incorporating a gene from sunflower, has been approved in several countries to combat yield losses induced by drought (González et al. 2019). Recently, the Indian government has also approved the environmental release of transgenic mustard hybrid 11 (DMH11) and its parental lines based on the Barnase/Barstar system (Ray 2022). This represents a significant step towards achieving self-sufficiency in India's edible oil production. It is estimated that it can yield 25–30% more than zonal check varieties. Transgenic crops help in overall socio-economic development by raising farmers' income and in turn their living standards. Besides this, it also contributes to breeding climate-resilient varieties and the conservation of biodiversity.

Although genome editing technology is a recent invention, the rapid adoption of this technology for crop improvement resulted in several commercially available crop varieties. In recent years several plants and animals have been rolled out in the market for commercial use. In the US, TALEN-edited calyno soybean with high oleic acid was commercialized in 2019. Pungency-free mustard greens have been developed using CRISPR-Cas9 and released in the US in 2023 for use as a salad. Japan has started selling first ever CRISPR-crop, high-GABA tomato, since 2021. The genome-edited tomato possessing high GABA (gamma-aminobutyric acid) content is expected to boost relaxation and reduce blood pressure. Along with this, genome-edited waxy corn with high amylopectin has been approved recently in Japan. Besides plants, Japan allowed two genome-edited fishes for commercial sale. Genome-edited "Madai" red sea bream and "22-seiki fugu" tiger puffer are known to achieve larger sizes compared to their wild-type counterparts in a relatively shorter period (Anon 2022).

22.4.3 Indian Initiatives on Genome Editing for Crop Improvement

To boost the use of genome editing for crop improvement in the nation, the Government of India (GoI) has been providing adequate funds to its researchers through different national funding agencies. Moreover, several national organizations/funding agencies such as the Indian Council of Agricultural Research (ICAR), the Department of Biotechnology (DBT), the Department of Science and Technology (DST), and State Agriculture University (SAUs), and other public and private institutions are also devoted to promoting genome editing research in plant sciences. Interestingly, in a revolutionary step, GoI in 2022 has exempted the genome-edited plants falling under the category of transgene-free SDN1 (Site-directed nuclease 1) and SDN2 (Site-directed nuclease 2) from biosafety regulations. This has opened the avenues towards climate-resilient, nutritionally-rich, and self-reliant food security. In the past decade, a number of crops have been developed by genome editing at different institutes across India. Recently, the *OsDST* gene was edited in the mega *indica* rice variety MTU1010 from the Indian Agricultural Research Institute, New Delhi. CRISPR/Cas9 generated knockout alleles were able to produce plants with tolerance to osmotic and salt stress (Kumar et al. 2020). Applying genome editing to oilseeds, a low-seed and high-leaf glucosinolate content mustard was created recently, by targeting glucosinolate transporters (*GTR*) (Mann et al. 2023). This led to improved oil quality besides maintaining defense against different fungal and insect pathogens. The genome-edited mustard has been developed by scientists from NIPGR, New Delhi. Interestingly, a team at the ICAR-Indian Institute of Rice Research (ICAR-IIRR), Hyderabad increased the grain number in the Samba Mahsuri rice variety, while maintaining its characteristic grain and cooking quality. Scientists at ICAR-National Rice Research Institute (ICAR-NRRI), Cuttack have developed rice lines with higher yield by editing the *IPA1* gene and lodging-resistant rice lines by editing the *SD1* gene.

Interestingly, β-carotene-enriched banana has been developed by editing the *lycopene epsilon-cyclase* (LCYε) gene at the National Agri-Food Biotechnology Institute, Mohali, India. In the genome-edited banana, β-carotene content was increased by six-fold over the unedited lines (Kaur et al. 2020). Scientists from ICGEB, New Delhi have also developed several genome-edited lines for different traits including rice with high grain numbers and rice lines with herbicide tolerance.

22.4.4 Priority Crops and Traits for Genome Editing

Climate change has started affecting the food security of India. Climate change can severely impact agriculture, disrupting food production and affecting food quality. India has become the most populous country in the world. As per the IPCC sixth assessment report, India is one of the most climate-vulnerable countries in the world. Staple food crops such as rice, wheat, and maize need to be bred for climate resilience. The changing climate is acting as a driving force for the emergence of new biotic and abiotic stresses to which food crops need to be acclimatized. Therefore, India needs to emphasize developing climate-resilient crops combining resistance power to abiotic and biotic stresses with higher yields. Traits like enhanced photosynthesis and improved use efficiency of water and nutrients are crucial to produce more with a lesser environmental footprint (Bansal et al. 2022). Since CRISPR-Cas9 tools allow us to edit multiple genes simultaneously, they will facilitate the development of crops with pyramided traits required in a fast-changing climate.

Besides the improvement of traits, it is time to increase diversity in our food systems. A diverse diet is a healthy diet. In India, the contribution of whole grains is highest in total calorie consumption, whereas fruits and vegetables are under-consumed (Sharma et al. 2020). Genome editing can be used to make fruits and vegetables more accessible to the Indian diet by manipulating their production constraint traits. Moreover, it can also be used to expand the cultivation of several nutritionally rich orphan crops such as millets outside their natural habitat by removing undesired traits. Genome editing can be used to develop new crops by domesticating crop wild relatives (Pattnaik et al. 2023). Increasing the adaptability and popularity of orphan crops, and de novo domestication of semi-domesticated crops and crop wild relatives through genome editing would be an attractive way for sustainable agriculture in the era of climate change. Table 22.1 summarizes the crops and traits that need to be focused on for genome editing in the current scenario.

22.4.5 Genome-Editing as a High Throughput Functional Genomics Tool

Conventional mutagenesis using physical and chemical mutagenesis creates genome-wide mutations at random. It is used for creating mutant population, and large number of mutant population is screened to identify mutants with desirable phenotype, and then the mutant locus is cloned using map based cloning approaches. This is highly labour, resource and time intensive approach, and function of only a small fraction of genes have been deciphered by this approach. Genome editing offer a new tool to create genome-wide mutants and characterize these mutants for deciphering the gene function. In rice this approach was attempted. A total of 39,045 gRNAs were used for targeted editing of all the genes of rice, 84,384 transgenic plants were generated and a total of 91,004 targeted loss-of-function mutants were generated (Lu et al. 2017). In canola, Brassica napus, a total of 18,414 sgRNAs were designed to target 10,480 genes and an editing efficiency of 52.2% was obtained (He et al. 2023). Similar approach need to be employed in India to identify genes for various traits and product development using genome editing.

22.5 Challenges and Future Prospects

The past two decades have witnessed tremendous progress in DNA and RNA sequencing technologies vis-a-vis high-throughput computational tools and advanced bioinformatics pipelines. The cheaper cost of sequencing has enabled the re-sequencing of thousands of cultivars and landraces, hence, helping the researchers to understand the structure of naturally available alleles and haplotypes, and associating those variations with the phenotype. The biggest challenge for researchers is to mine the massive sequencing data available in the public domain to decipher the useful information that can be utilized by plant breeders and molecular biologists. Genomic-assisted breeding (GAB) is a highly promising breeding methodology that needs to be practiced by common breeders. The cheaper cost of SNP genotyping platforms will certainly accelerate GAB adoption and discovery of genes through GWAS or gene mapping using biparental or MAGIC populations. Functional analysis of genes and alleles through overexpression, knockout, protein-protein-DNA interactions, and localization studies will enable breeders and biotechnologists to make the right choice for gene introgression through cross-breeding or allele improvement through genome editing. While structural genomics has largely been driven by machines, functional genomics requires wet-lab experimentation and a thorough understanding of plant biology. The funding agencies and policy-makers will have to put more emphasis on functional genomics

Table 22.1 Key agricultural crops and desirable genetic traits for enhancement through genome editing in India (adopted from Bansal et al. 2022)

Category	Crop	Prioritized traits	Potential candidate gene(s)
Cereal	Rice	High grain yield	*Gn1a, GS3, GW2, DEP1, TB1, IPA1* (promoter edits), *PYL1/4/6*
		Resistance to diseases: Bacterial leaf blight (BLB), Blast, Sheath blight	BLB: *SWEET11/13/14* Blast: *ERF922, BSRK1*
		Resistance to insect-pests: Brown plant hopper	*CYP71A1*
		Herbicide tolerance for direct-seeded rice	*ALS, ACCase, EPSPS*
		Enhanced water (WUE) and nitrogen use efficiency (NUE)	WUE: *EPF-LIKE9, or STOMAGEN, EPFL10* NUE: *NRT1.1B*
		Tolerance to drought and salinity	*DST, RR22*
		High Fe and Zn content in grains	*GW2, HRZ* (Hemerythrin RING Zinc finger)
Oilseeds	Mustard	Plant type: Increased secondary branching	*MAX1, TT2, TT8*
		Sclerotinia (stem rot) resistance	*Ferulate-5-hydroxylase*
		Resistance to Orobanche	*ALS*
		Resistance to *Alternaria* leaf spot	Not known yet[a]
		Oil quality: Low glucosinolate and low erucic acid	Low Glucosinolate: *MYB28; GTR* Low erucic acid: *Fatty acid elongase1 (FAE1)*
	Ground-nut	Bushy plant type	Not known yet[a]
		Resistance to diseases: Leaf spot, rust and stem rot	Not known yet[a]
		Drought tolerance	Not known yet[a]
		High oleic acid in seeds	*FAD2*
	Soybean	Photo-insensitivity (Early flowering/ early maturity)	*E1, E2,* and *E3* genes
		Herbicide tolerance	*ALS, EPSPS*
		High oleic acid in seeds	*FAD21A* and *FAD21B* genes
Pulses	Chickpea	Resistance to diseases: Dry rot, Botrytis Grey Mould, Ascochyta Blight	Not known yet[a]
		Pod borer resistance	Not known yet[a]
		Tolerance to drought and salt stress	Not known yet[a]
		Herbicide tolerance	*ALS, EPSPS*
	Pigeon pea	Synchronized maturity/early maturity	Not known yet[a]
		Photo- and thermo-insensitivity	Not known yet[a]
		Drought tolerance	Not known yet[a]
		Herbicide tolerance	*ALS, EPSPS*

(continued)

Table 22.1 (continued)

Category	Crop	Prioritized traits	Potential candidate gene(s)
		Resistance to diseases: Alternaria blight, Anthracnose	Not known yet[a]
		Pod borer resistance	Not known yet[a]
	Green gram	Resistance to yellow mosaic virus	Not known yet[a]
		Resistance to powdery mildew	Not known yet[a]
		Drought and heat tolerance	Not known yet[a]
Cash crops	Sugarcane	High Yield	Not known yet[a]
		Resistance to Red rot disease	Not known yet[a]
		Juice quality (>18% sucrose in a 10-month-old crop)	Not known yet[a]
	Cotton	High yield: Increased boll weight and number	Not known yet[a]
		Increased fibre length and strength	Not known yet[a]
		High ginning out-turn (~40%)	Not known yet[a]
		Resistance to pink Bollworm	Not known yet[a]
Fruit and Vegetable crops	Banana	Resistance to Fusarium wilt Race 1 and TR4 (Panama wilt)	Strong Promoter knock-in upstream to *RGA2*
		Resistance to Sigatoka leaf spot	Not known yet[a]
		Nutritional fruit quality: High beta-carotene and iron	Beta Carotene: *Lycopene epsilon-cyclase*
		BBTV (Banana bunchy Top virus) and BSV (Banana Streak virus)	*eIF4* ORF1/2/3 of integrated endogenous BSV
	Tomato	Improved processing quality: Total soluble solids (>5%) and acidity (>0.5%)	*Invertase inhibitor (INVINH)*
		Male sterility for hybrid seed production	*Ms10(35), GSTAA, LeRBOH,* and *LeRBOHE*
		Resistance to Tomato leaf curl virus (TLCV) disease	*SlPelo*
	Potato	Reduce cold-induced sweetening and acrylamide content	Sweetening: *Vacuolar Invertase* Acrylamide: *Asparagine synthetase 1*
		Resistance to diseases: Potato virus Y (PVY) and Potato New Delhi apical leaf curl virus, Bacterial wilt, Late blight	PVY: *P3, CI, Nib, CP* viral genes; *eIF4E* and *coilin host genes* Late Blight: *StDMR6–1* and *StCHL1*
		Tuber starch quality	*GBSS, SBE1* and *SBE2*

[a]Target genes need to be identified

to unravel the hidden genetic treasure for its effective utilization to make agriculture more productive and sustainable for the benefit of humanity.

The development of transgenic and genome-edited crops relies on the availability of efficient transformation protocols. Although a few genotypes of model crops have established protocols, many locally grown genotypes and landraces still suffer from recalcitrancy. The situation is poorer for crops such as pulses, millets, oilseeds, and cash crops. As a result, they are still creeping in harnessing the benefits of transgenesis and genome editing. Establishing tissue culture and transformation systems for more and more genotypes is the need of the hour for using these genome engineering tools in diverse locally grown genotypes.

The availability of advanced biotechnological tools has greatly complemented conventional crop breeding efforts and added new dimensions by increasing pace and accuracy. Biotech crops have contributed to increasing productivity, conserving biodiversity, mitigating challenges related to climate change, and improving economic, health, and social benefits (ISAAA 2019). Although a total of 72 countries adopted biotech crops by means of planting and importing, many countries still have stringent policies that hinder the wide adoption of such crops. The alarming impact of climate change and the growing world population demands a rapid shift in how we produce our food. Biotech crops, including transgenic and genome-edited crops, are being developed to provide adequate nutritious food to consumers. Improved crop varieties have been developed with agronomic traits to mitigate climate change-associated agricultural problems. The realization of the extensive advantages offered by biotech crops to address hunger and poverty critically depends on both public acceptance and government policies that facilitate their adoption.

References

3,000 Rice Genomes Project (2014) The 3,000 rice genomes project. Gigascience 3(1):2047-217X

Agarwal S, Mangrauthia SK, Sarla N (2015) Expression profiling of iron deficiency responsive microRNAs and gene targets in rice seedlings of Madhukar x Swarna recombinant inbred lines with contrasting levels of iron in seeds. Plant Soil 396:137–150. https://doi.org/10.1007/s11104-015-2561-y

Agarwal S, Tripura Venkata VGN, Kotla A, Mangrauthia SK, Neelamraju S (2014) Expression patterns of QTL based and other candidate genes in Madhukar× Swarna RILs with contrasting levels of iron and zinc in unpolished rice grains. Gene 546:430–436

Amarawathi Y, Singh R, Singh AK, Singh VP, Mohapatra T, Sharma TR, Singh NK (2008) Mapping of quantitative trait loci for basmati quality traits in rice (Oryza sativa L.). Mol Breed 21:49–65

Anon (2022) Japan embraces CRISPR-edited fish. Nat Biotechnol 40:10. https://doi.org/10.1038/s41587-021-01197-8

Anzalone AV, Randolph PB, Davis JR, Sousa AA, Koblan LW, Levy JM, Chen PJ, Wilson C, Newby GA, Raguram A, Liu DR (2019) Search-and-replace genome editing without double-strand breaks or donor DNA. Nature 576(7785):149–157

Arrones A, Vilanova S, Plazas M, Mangino G, Pascual L, Diez M, Probens J, Gramazio P (2020) The dawn of the age of multi-parent magic populations in plant breeding: novel powerful next-generation resources for genetic analysis and selection of recombinant elite material. Biology (Basel) 9(8):229

Ashikari M, Sakakibara H, Lin S, Yamamoto T, Takashi T, Nishimura A, Angeles ER, Qian Q, Kitano H, Matsuoka M (2005) Cytokinin oxidase regulates rice grain production. Science 309(5735):741–745

Bansal KC, Molla KA, Chinnusamy V (2022) Genome editing: a boon for plant biologists, breeders and farmers. Curr Sci 123:15–19

Bhogireddy S, Babu MS, Swamy KN (2022) Expression dynamics of genes and microRNAs at different growth stages and heat treatments in contrasting high temperature responsive rice genotypes. J Plant Growth Regul 41:74–91. https://doi.org/10.1007/s00344-020-10282-2

Bhogireddy S, Mangrauthia SK, Kumar R (2021) Regulatory non-coding RNAs: a new frontier in regulation of plant biology. Funct Integr Genomics 21:313–330. https://doi.org/10.1007/s10142-021-00787-8

Cardi T (2016) Cisgenesis and genome editing: combining concepts and efforts for a smarter use of genetic resources in crop breeding. Plant Breed 135(2):139–147

Channamallikarjuna V, Sonah H, Prasad M, Rao GJ, Chand S, Upreti HC, Singh NK, Sharma TR (2010) Identification of major quantitative trait loci qSBR11-1 for sheath blight resistance in rice. Mol Breed 25:155–166. https://doi.org/10.1007/s11032-009-9316-5

Chen L, Hong M, Luan C, Gao H, Ru G, Guo X, Zhang D, Zhang S, Li C, Wu J, Randolph PB (2023) Adenine transversion editors enable precise, efficient A• T-to-C• G base editing in mammalian cells and embryos. Nat Biotechnol:1–13

Chopperla R, Mangrauthia SK, Bhaskar Rao T, Balakrishnan M, Balachandran SM, Prakasam V, Channappa G (2020) A comprehensive analysis of MicroRNAs expressed in susceptible and resistant rice cultivars during rhizoctonia solani AG1-IA infection causing sheath blight disease. Int J Mol Sci 21:7974. https://doi.org/10.3390/ijms21217974

Choudhary B, Gaur K (2015) Biotech cotton in India, 2002 to 2014. ISAAA Series of Biotech Crop Profiles. ISAAA, Ithaca, NY, pp 1–34

Cooper M, Messina CD, Podlich D, Totir LR, Baumgarten A, Hausmann NJ, Wright D, Graham G (2014) Predicting the future of plant breeding: complementing empirical evaluation with genetic prediction. Crop Pasture Sci 65:311–336

Gamuyao R, Chin JH, Pariasca-Tanaka J, Pesaresi P, Catausan S, Dalid C, Slamet-Loedin I, Tecson-Mendoza EM, Wissuwa M, Heuer S (2012) The protein kinase Pstol1 from traditional rice confers tolerance of phosphorus deficiency. Nature 488(7412):535–539. https://doi.org/10.1038/nature11346

Gaudelli NM, Komor AC, Rees HA, Packer MS, Badran AH, Bryson DI, Liu DR (2017) Programmable base editing of A• T to G• C in genomic DNA without DNA cleavage. Nature 551(7681):464–471

González F, Capella M, Ribichich K, Curin F, Giacomelli J, Ayala F, Watson G, Otegui ME, Chan RL (2019) Wheat transgenic plants expressing the sunflower gene HaHB4 significantly outyielded their controls in field trials. J Exp Bot 70:1669–1681

Guttikonda H, Thummala SR, Agarwal S (2020) Genome-wide transcriptome profile of rice hybrids with and without Oryza rufipogon introgression reveals candidate genes for yield. Sci Rep 10: 4873. https://doi.org/10.1038/s41598-020-60922-6

Haider I, Yunmeng Z, White F, Li C, Incitti R, Alam I, Gojobori T, Ruyter-Spira C, Al-Babili S, Bouwmeester HJ (2023) Transcriptome analysis of the phosphate starvation response sheds light on strigolactone biosynthesis in rice. Plant J 114(2):355–370

He J, Zhang K, Yan S, Tang M, Zhou W, Yin Y, Chen K, Zhang C, Li M (2023) Genome-scale targeted mutagenesis in Brassica napus using a pooled CRISPR library. Genome Res 33(5): 798–809

He Z, Ji R, Havlickova L et al (2021) Genome structural evolution in Brassica crops. Nat Plan Theory 7:757–765. https://doi.org/10.1038/s41477-021-00928-8

Huang F, Jiang Y, Chen T, Li H, Fu M, Wang Y, Xu Y, Li Y, Zhou Z, Jia L, Ouyang Y, Yao W (2022) New data and new features of the FunRiceGenes (functionally characterized rice genes). Rice (N Y) 15(1):23. https://doi.org/10.1186/s12284-022-00569-1

International Rice Genome Sequencing Project (2005) The map-based sequence of the rice genome. Nature 11;436(7052):793–800. https://doi.org/10.1038/nature03895

ISAAA (2019) ISAAA Brief 55-2019: Executive summary. Biotech crops drive socio-economic development and sustainable environment in the new frontier. https://www.isaaa.org/resources/publications/briefs/55/executivesummary/default.asp

Jackson D, Tian F, Zhang Z (2022) Maize genetics, genomics, and sustainable improvement. Mol Breed 42:2. https://doi.org/10.1007/s11032-021-01266-5

Jha UC, Nayyar H, Parida SK, Bakır M, von Wettberg EJB, Siddique KHM (2022) Progress of genomics-driven approaches for sustaining underutilized legume crops in the post-genomic era. Front Genet 13:831656. https://doi.org/10.3389/fgene.2022.831656

Jia J, Li H, Zhang X, Li Z, Qiu L (2017) Genomics-based plant germplasm research (GPGR). Crop J 5:166–174

Jiang Y, Chai Y, Qiao D, Wang J, Xin C, Sun W, Cao Z, Zhang Y, Zhou Y, Wang XC, Chen QJ (2022) Optimized prime editing efficiently generates glyphosate-resistant rice plants carrying homozygous TAP-IVS mutation in EPSPS. Mol Plant 15(11):1646–1649

Jinek M, Chylinski K, Fonfara I, Hauer M, Doudna JA, Charpentier E (2012) A programmable dual-RNA–guided DNA endonuclease in adaptive bacterial immunity. Science 337(6096): 816–821

Kale RR, Durga Rani CV, Anila M, Mahadeva Swamy HK, Bhadana VP, Senguttuvel P (2021) Novel major QTLs associated with low soil phosphorus tolerance identified from the Indian rice landrace, Wazuhophek. PLoS ONE 16(7):e0254526. https://doi.org/10.1371/journal.pone.0254526

Kang BC, Bae SJ, Lee S, Lee JS, Kim A, Lee H, Baek G, Seo H, Kim J, Kim JS (2021) Chloroplast and mitochondrial DNA editing in plants. Nat Plants 7(7):899–905

Karmakar S, Das P, Panda D, Xie K, Baig MJ, Molla KA (2022) A detailed landscape of CRISPR-Cas-mediated plant disease and pest management. Plant Sci 323:111376

Kaur N, Alok A, Kumar P, Kaur N, Awasthi P, Chaturvedi S, Pandey P, Pandey A, Pandey AK, Tiwari S (2020) CRISPR/Cas9 directed editing of lycopene epsilon-cyclase modulates metabolic flux for β-carotene biosynthesis in banana fruit. Metab Eng 59:76–86

Keeling PJ, Palmer JD (2008) Horizontal gene transfer in eukaryotic evolution. Nat Rev Genet 9(8): 605–618

Komor AC, Kim YB, Packer MS, Zuris JA, Liu DR (2016) Programmable editing of a target base in genomic DNA without double-stranded DNA cleavage. Nature 533(7603):420–424

Krishna VV, Qaim M (2012) Bt cotton and sustainability of pesticide reductions in India. Agric Syst 107:47–55

Kumar PN, Sujatha K, Laha GS, Rao KS, Mishra B, Viraktamath BC, Hari Y, Reddy CS, Balachandran SM, Ram T, Madhav MS (2012) Identification and fine-mapping of Xa33, a novel gene for resistance to Xanthomonas oryzae pv. oryzae. Phytopathology 102(2):222–228

Kumar VV, Verma RK, Yadav SK, Yadav P, Watts A, Rao MV, Chinnusamy V (2020) CRISPR-Cas9 mediated genome editing of drought and salt tolerance (OsDST) gene in indica mega rice cultivar MTU1010. Physiol Mol Biol Plants 26:1099–1110

Lai J, Li R, Xu X, Jin W, Xu M, Zhao H, Xiang Z, Song W, Ying K, Zhang M, Jiao Y, Ni P, Zhang J, Li D, Guo X, Ye K, Jian M, Wang B, Zheng H, Liang H (2010) Genome-wide patterns of genetic variation among elite maize inbred lines. Nat Genet 42:1027–1030

Lee S, Lee H, Baek G, Kim JS (2022) Precision mitochondrial DNA editing with high-fidelity DddA-derived base editors. Nat Biotechnol 41(3):378–386

Lenka SK, Katiyar A, Chinnusamy V, Bansal KC (2011) Comparative analysis of drought-responsive transcriptome in Indica rice genotypes with contrasting drought tolerance. Plant Biotechnol J 3:315–327. https://doi.org/10.1111/j.1467-7652.2010.00560

Li C, Zhang R, Meng X, Chen S, Zong Y, Lu C, Qiu JL, Chen YH, Li J, Gao C (2020a) Targeted, random mutagenesis of plant genes with dual cytosine and adenine base editors. Nat Biotechnol 38(7):875–882

Li D, Zhang F, Pinson RM, Edwards D, Jackson K, Xia X, Eizenga GC (2022a) Assessment of rice sheath blight resistance including associations with plant architecture, as revealed by genome-wide association studies. Rice 15:31

Li G, Tang J, Zheng J, Chu C (2021a) Exploration of rice yield potential: decoding agronomic and physiological traits. Crop J 9:577–589

Li J, Chen L, Liang J, Xu R, Jiang Y, Li Y, Ding J, Li M, Qin R, Wei P (2022b) Development of a highly efficient prime editor 2 system in plants. Genome Biol 23(1):1–9

Li R, Char SN, Liu B, Liu H, Li X, Yang B (2021b) High-efficiency plastome base editing in rice with TAL cytosine deaminase. Mol Plant 14(9):1412–1414

Li Y, Hallerman EM, Wu K, Peng Y (2020b) Insect-resistant genetically engineered crops in China: development, application, and prospects for use. Annu Rev Entomol 65:273–292

Lin Q, Jin S, Zong Y, Yu H, Zhu Z, Liu G, Kou L, Wang Y, Qiu JL, Li J, Gao C (2021) High-efficiency prime editing with optimized, paired pegRNAs in plants. Nat Biotechnol 39(8): 923–927

Loos RJF (2020) 15 years of genome-wide association studies and no signs of slowing down. Nat Commun 11:5900. https://doi.org/10.1038/s41467-020-19653-5

Lu Y, Ye X, Guo R, Huang J, Wang W, Tang J, Tan L, Zhu JK, Chu C, Qian Y (2017) Genome-wide targeted mutagenesis in rice using the CRISPR/Cas9 system. Mol Plant 10(9):1242–1245

Lv Q, Li W, Sun Z, Ouyang N, Jing X, He Q, Wu J, Zheng J, Zheng J, Tang S, Zhu R (2020) Resequencing of 1,143 indica rice accessions reveals important genetic variations and different heterosis patterns. Nat Commun 11(1):4778. https://doi.org/10.1038/s41467-020-18608-0

Mackay I, Powell W (2007) Methods for linkage disequilibrium mapping in crops. Trends Plant Sci 12(2):57–63

Mangrauthia SK, Agarwal S, Sailaja B (2016) Transcriptome analysis of Oryza sativa (rice) seed germination at high temperature shows dynamics of genome expression associated with hormones signaling and abiotic stress pathways. Trop Plant Biol 9:215–228. https://doi.org/10.1007/s12042-016-9170-7

Mangrauthia SK, Bhogireddy S, Agarwal S, Prasanth VV, Voleti SR, Neelamraju S, Subrahmanyam D (2017) Genome-wide changes in microRNA expression during short and prolonged heat stress and recovery in contrasting rice cultivars. J Exp Bot 68(9):2399–2412

Mann A, Kumari J, Kumar R, Kumar P, Pradhan AK, Pental D, Bisht NC (2023) Targeted editing of multiple homologues of GTR1 and GTR2 genes provides the ideal low-seed, high-leaf glucosinolate oilseed mustard with uncompromised defence and yield. Plant Biotechnol J

Mendelsohn M, Kough J, Vaituzis Z, Matthews K (2003) Are Bt crops safe? Nat Biotechnol 21(9): 1003–1009

Meng X, Yu H, Zhang Y, Zhuang F, Song X, Gao S, Gao C, Li J (2017) Construction of a genome-wide mutant library in rice using CRISPR/Cas9. Mol Plant 10(9):1238–1241. https://doi.org/10.1016/j.molp.2017.06.006

Mohapatra T, Robin S, Sarla N, Sheshashayee M, Singh AK, Singh K (2014) EMS induced mutants of upland rice variety Nagina22: generation and characterization. Proc Indian Natl Sci Acad 80: 163–172. https://doi.org/10.16943/ptinsa/2014/v80i1/55094

Mok BY, de Moraes MH, Zeng J, Bosch DE, Kotrys AV, Raguram A, Hsu F, Radey MC, Peterson SB, Mootha VK, Mougous JD (2020) A bacterial cytidine deaminase toxin enables CRISPR-free mitochondrial base editing. Nature 583(7817):631–637

Molla KA, Qi Y, Karmakar S, Baig MJ (2020b) Base editing landscape extends to perform transversion mutation. Trends Genet 36(12):899–901

Molla KA, Shih J, Wheatley MS, Yang Y (2022) Predictable NHEJ insertion and assessment of HDR editing strategies in plants. Front Genome Ed 4:825236. https://doi.org/10.3389/fgeed.2022.825236

Molla KA, Shih J, Yang Y (2020a) Single-nucleotide editing for zebra3 and wsl5 phenotypes in rice using CRISPR/Cas9-mediated adenine base editors. Abiotech 1:106–118

Molla KA, Sretenovic S, Bansal KC, Qi Y (2021) Precise plant genome editing using base editors and prime editors. Nat Plants 7(9):1166–1187

Molla KA, Yang Y (2019) CRISPR/Cas-mediated base editing: technical considerations and practical applications. Trends Biotechnol 37(10):1121–1142

Ojiewo C, Monyo E, Desmae H, Boukar O, Mukankusi-Mugisha C, Thudi M, Pandey MK, Saxena RK, Gaur PM, Chaturvedi SK, Fikre A, Ganga Rao N, SameerKumar CV, Okori P, Janila P, Rubyogo JC, Godfree C, Akpo E, Omoigui L, Nkalubo S, Fenta B, Binagwa P, Kilango M, Williams M, Mponda O, Okello D, Chichaybelu M, Miningou A, Bationo J, Sako D, Diallo S, Echekwu C, Umar ML, Oteng-Frimpong R, Mohammed H, Varshney RK (2019) Genomics, genetics and breeding of tropical legumes for better livelihoods of smallholder farmers. Plant Breed 138(4):487–499. https://doi.org/10.1111/pbr.12554

Pattnaik D, Avinash SP, Panda S, Bansal KC, Chakraborti M, Kar MK et al (2023) Accelerating crop domestication through genome editing for sustainable agriculture. J Plant Biochem Biotechnol 1-17. https://doi.org/10.1007/s13562-023-00837-1

Peng-fei L, Lubberstedt T, Ming-liang X (2017) Genomics-assisted breeding—a revolutionary strategy for crop improvement. J Integr Agric 16(12):2674–2685

Peshin R, Hansra BS, Singh K, Nanda R, Sharma R, Yangsdon S, Kumar R (2021) Long-term impact of Bt cotton: an empirical evidence from North India. J Clean Prod 312:127575

Peterson JA, Obrycki JJ, Harwood JD (2020) Bacillus thuringiensis: transgenic crops. In Managing biological and ecological systems (pp. 7–24). CRC Press

Poli Y, Nallamothu V, Hao A (2021) NH787 EMS mutant of rice variety Nagina22 exhibits higher phosphate use efficiency. Sci Rep 11:9156. https://doi.org/10.1038/s41598-021-88419-w

Prathi NB, Palit P, Madhu P, Ramesh M, Laha GS (2018) Balachandran SM, Madhav MS, Sundaram RM, Mangrauthia SK. Proteomic and transcriptomic approaches to identify resistance and susceptibility related proteins in contrasting rice genotypes infected with fungal pathogen Rhizoctonia solani. Plant Physiol Biochem 130:258–266

Qaim M (2020) Bt cotton, yields and farmers' benefits. Nat Plants 6(11):1318–1319

Ray K (2022) https://www.deccanherald.com/india/india-gives-environmental-approval-for-genetically-modified-mustard-1157037.html

Reddy MR, Mangrauthia SK, Reddy SV, Manimaran P, Yugandhar P, Babu PN, Vishnukiran T, Subrahmanyam D, Sundaram RM, Balachandran SM (2020) PAP90, a novel rice protein plays a critical role in regulation of D1 protein stability of PSII. J Adv Res 30:197–211. https://doi.org/10.1016/j.jare.2020.11.008

Research Outreach (2021) Genetically modified cotton: How has it changed India? https://researchoutreach.org/articles/genetically-modified-cotton-how-changed-india/

Sailaja B, Voleti S, Subrahmanyam D, Sarla N, Prasanth VV, Bhadana V (2014) Prediction and expression analysis of miRNAs associated with heat stress in oryza sativa. Rice Sci 21:3–12. https://doi.org/10.1016/S1672-6308(13)60164-X

Sang J, Zou D, Wang Z, Wang F, Zhang Y, Xia L, Li Z, Ma L, Li M, Xu B, Liu X, Wu S, Liu L, Niu G, Li M, Luo Y, Hu S, Hao L, Zhang Z (2020) IC4R-2.0: Rice genome reannotation using massive RNA-seq data. Genomics Proteomics Bioinformatics 2:161–172. https://doi.org/10.1016/j.gpb.2018.12.011

Sato Y, Antonio BA, Namiki N, Takehisa H, Minami H, Kamatsuki K, Sugimoto K, Shimizu Y, Hirochika H, Nagamura Y (2011) RiceXPro: a platform for monitoring gene expression in japonica rice grown under natural field conditions. Nucleic Acids Res 39:D1141-8

Scott MF, Ladejobi O, Amer S, Bentley AR, Biernaskie J, Boden SA, Clark M, Dell'Acqua M, Dixon LE, Filippi CV, Fradgley N, Gardner KA, Mackay IJ, O'Sullivan D, Percival-Alwyn L, Roorkiwal M, Singh RK, Thudi M, Varshney RK, Venturini L, Whan A, Cockram J, Mott R (2020) Multi-parent populations in crops: a toolbox integrating genomics and genetic mapping with breeding. Heredity (Edinb) 125(6):396–416. https://doi.org/10.1038/s41437-020-0336-6

Shang L, Li X, He H, Yuan Q, Song Y, Wei Z, Lin H, Hu M, Zhao F, Zhang C, Li Y, Gao H, Wang T, Liu X, Zhang H, Zhang Y, Cao S, Yu X, Zhang B, Zhang Y, Tan Y, Qin M, Ai C, Yang Y, Zhang B, Hu Z, Wang H, Lv Y, Wang Y, Ma J, Wang Q, Lu H, Wu Z, Liu S, Sun Z, Zhang H, Guo L, Li Z, Zhou Y, Li J, Zhu Z, Xiong G, Ruan J, Qian Q (2022) A super pan-genomic landscape of rice. Cell Res 32(10):878–896. https://doi.org/10.1038/s41422-022-00685-z

Sharma M, Kishore A, Roy D, Joshi K (2020) A comparison of the Indian diet with the EAT-lancet reference diet. BMC Public Health 20(1):1–13

Sharma TR, Madhav MS, Singh BK (2005) High-resolution mapping, cloning and molecular characterization of the Pi-k h gene of rice, which confers resistance to Magnaporthe grisea. Mol Gen Genomics 274:569–578. https://doi.org/10.1007/s00438-005-0035-2

Shi C, Li W, Zhang QJ, Zhang Y, Tong Y, Li K, Liu YL, Gao LZ (2020) The draft genome sequence of an upland wild rice species, Oryza granulata. Sci Data 7(1):131. https://doi.org/10. 1038/s41597-020-0470-2

Sinha P, Kumar D, Shaik H, Manish S, Gokulan CG, Das A, Anila M (2023) Fine mapping and sequence analysis reveals a promising candidate gene encoding a novel NB-ARC domain derived from wild rice (Oryza officinalis) confers bacterial blight resistance. Front Plant Sci 14:1173063

Stein JC, Yu Y, Copetti D (2018) Genomes of 13 domesticated and wild rice relatives highlight genetic conservation, turnover and innovation across the genus Oryza. Nat Genet 50:285–296. https://doi.org/10.1038/s41588-018-0040-0

Success Farming (2023). https://www.agriculture.com/gm-crops-grown-on-55-of-u-s-cropland-says-usda-7644573?utm_source=social2&utm_medium=social&utm_campaign= shareurlbuttons

Thomson MJ, de Ocampo M, Egdane J (2010) Characterizing the saltol quantitative trait locus for salinity tolerance in rice. Rice 3:148–160. https://doi.org/10.1007/s12284-010-9053-8

Thummala SR, Guttikonda H, Tiwari S, Ramanan R, Baisakh N, Neelamraju S, Mangrauthia SK (2022) Whole-genome sequencing of KMR3 and Oryza rufipogon-derived introgression line IL50-13 (Chinsurah Nona 2/Gosaba 6) identifies candidate genes for high yield and salinity tolerance in rice. Front Plant Sci 30(13):810373

Tong H, Wang X, Liu Y, Liu N, Li Y, Luo J, Ma Q, Wu D, Li J, Xu C, Yang H (2023) Programmable A-to-Y base editing by fusing an adenine base editor with an N-methylpurine DNA glycosylase. Nat Biotechnol:1–5

Uffelman E, Huang QQ, Munung NS, de Vries J, Okada Y, Martin AR, Martin HC, Lappalainen T, Posthuma D (2021) Genome-wide association studies. Nat Rev Methods Primers 1:59

Vikram P, Swamy BM, Dixit S (2011) qDTY 1.1, a major QTL for rice grain yield under reproductive-stage drought stress with a consistent effect in multiple elite genetic backgrounds. BMC Genet 12:89. https://doi.org/10.1186/1471-2156-12-89

Walkowiak S, Gao L, Monat C et al (2020) Multiple wheat genomes reveal global variation in modern breeding. Nature 588:277–283. https://doi.org/10.1038/s41586-020-2961-x

Wang C, Han B (2022) Twenty years of rice genomics research: from sequencing and functional genomics to quantitative genomics. Mol Plant 15(4):593–619. https://doi.org/10.1016/j.molp. 2022.03.009

Wang W, Mauleon R, Hu Z, Chebotarov D, Tai S, Wu Z, Li M, Zheng T, Fuentes RR, Zhang F, Mansueto L (2018) Genomic variation in 3,010 diverse accessions of Asian cultivated rice. Nature 557(7703):43–49. https://doi.org/10.1038/s41586-018-0063-9

Wei S, Li X, Lu Z, Zhang H, Ye X, Zhou Y, Li J, Yan Y, Pei H, Duan F, Wang D (2022) A transcriptional regulator that boosts grain yields and shortens the growth duration of rice. Science 377:6604

Wissuwa M, Yano M, Ae N (1998) Mapping of QTLs for phosphorus deficiency tolerance in rice (Oryza sativa L.). Theor Appl Genet 97:777–783

Wu K, Wang S, Song W, Zhang J, Wang Y, Liu Q, Yu J, Ye Y, Li S, Chen J, Zhao Y, Wang J, Wu X, Wang M, Zhang Y, Liu B, Wu Y, Harberd NP, Fu X (2020) Enhanced sustainable green revolution yield via nitrogen-responsive chromatin modulation in rice. Science 367:6478. https://doi.org/10.1126/science.aaz2046

Xie X, Du H, Tang H, Tang J, Tan X, Liu W, Li T, Lin Z, Liang C, Liu YG (2021) A chromosome-level genome assembly of the wild rice Oryza rufipogon facilitates tracing the origins of Asian cultivated rice. Sci China Life Sci 64:282–293

Xu K, Mackill DJ (1996) A major locus for submergence tolerance mapped on rice chromosome 9. Mol Breed 2:219–224

Xu Y, Crouch JH (2008) Marker-assisted selection in plant breeding: from publications to practice. Crop Sci 48:391–407

Yadava DK, Hossain F, Choudhury PR, Kumar D, Singh AK, Sharma TR, Mohapatra T (2022) Crop cultivars developed through molecular breeding, 2nd edn. Indian Council of Agricultural Research, New Delhi

Yi J, An G (2013) Utilization of T-DNA tagging lines in rice. J Plant Biol 56:85–90. https://doi.org/10.1007/s12374-013-0905-9

Yu J, Hu S, Wang J, Wong GK, Li S, Liu B, Deng Y, Dai L, Zhou Y, Zhang X, Cao M (2002) A draft sequence of the rice genome (Oryza sativa L. ssp. indica). Science 296(5565):79–92. https://doi.org/10.1126/science.1068037

Yu Z, Chen Y, Zhou Y, Zhang Y, Li M, Ouyang Y, Chebotarov D, Mauleon R, Zhao H, Xie W, McNally KL, Wing RA, Guo W, Zhang J (2023) Rice Gene Index: a comprehensive pan-genome database for comparative and functional genomics of Asian rice. Mol Plant 16(5):798–801. https://doi.org/10.1016/j.molp.2023.03.012

Zhang F, Xue H, Dong X, Li M, Zheng X, Li Z, Xu J, Wang W, Wei C (2022) Long-read sequencing of 111 rice genomes reveals significantly larger pan-genomes. Gen Res 32 (5):853–863. https://doi.org/10.1101/gr.276015.121. Epub 2022 Apr 8. PMID: 35396275; PMCID: PMC9104699

Trade in Gene-Edited Crops; International Perspectives

23

Michael G. K. Jones

Abstract

More food will need to be produced from less land to feed the growing world population in a sustainable manner, whilst preserving biodiversity for future generations. This will require the application of the best science and technology, including in the areas of AgriBio, Agritech and Food Tech. The genetic potential of crop plants in the field is what underlies all these technologies, and the major new suite of technologies described as Genome- or Gene-Editing (abbreviated here as 'GEd') are providing exciting new opportunities for genetic crop improvement.

Significantly, the potential now exists to break the nexus between Genetically Modified (GM) plants and GEd plants, and to avoid the issues that have prevented wider use of GM crops. National and international policies and regulations on GEd crop produce are advancing rapidly. Their alignment or harmonisation is a pre-requisite for achieving the full benefits and enabling international trade in GEd produce. Many countries in North and South America and the Asia-Pacific region, more recently in Africa and potentially in Europe, have already or are now reassessing their regulatory regimes. The general principle being followed is that if the GEd undertaken could have been achieved by conventional breeding practices (e.g. mutagenesis, wide crosses within a species' gene pool), then there is no reason why GEd produce which achieves the same ends more rapidly and precisely should be regulated any differently from conventionally bred produce.

M. G. K. Jones (✉)
Western Australian State Agricultural Biotechnology Centre, Food Futures Institute, School of Agricultural Sciences, Murdoch University, Perth, WA, Australia
e-mail: m.jones@murdoch.edu.au

© National Academy of Agricultural Sciences, under exclusive license to Springer Nature Singapore Pte Ltd. 2023
K. C. Bansal et al. (eds.), *Transformation of Agri-Food Systems*,
https://doi.org/10.1007/978-981-99-8014-7_23

323

In this Chapter, the current regulatory status of GEd crops is described, with a focus on trade and the Asia-Pacific region, since this is where two-thirds of the world's population will reside.

Keywords

Crops · Gene editing · Genome editing · GEd · Cas9 · Regulations · Asia-Pacific · Path-to-market · Trade · Harmonisation · Science diplomacy · Trade

23.1 Introduction

Plant breeding relies on the availability of genetic variation, and breeders try to combine the best combinations of alleles to develop improved varieties. This can involve crossing varieties with more distant species within their gene pool to introduce a new trait or gene conferring resistance to pests or diseases, or inducing random mutations using chemical or physical methods. The latter include applying chemical mutagens such as ethyl methanesulfonate (EMS), sodium azide, gamma irradiation (Hassan and Abd-El-Haleem 2014), fast neutron bombardment (Li et al. 2001), T-DNA insertions, or from somaclonal variation if plants are regenerated via a callus phase. Mutagenic treatments induce many double-stranded breaks (DSBs) at random sites in the genome, and breeders then try to find mutants with useful properties.

The process of generating targeted mutations or 'gene edits' has now been made much more precise with the discovery and development of programmable nucleases which can induce DSBs at precise sites in the genome. This can be described as 'targeted mutagenesis'. The development of programmable nucleases has progressed through various technologies – Meganucleases, Zinc Finger Nucleases (ZFNs), TALENS and CRISPR-based systems. The latter is much the most widely used system for crop improvement, and associated new technologies are emerging rapidly (Jones et al. 2022). The position of the DSB, made by the enzyme Cas9, is directed by the binding of a single guide RNA (gRNA). New CRISPR-based tools, Cas enzymes and approaches enable the regulation of endogenous genes, up-regulation or down-regulation of gene expression, multiple edits using more than one gRNA, base editing, prime editing and epigenetic modifications. These tools enable GEd in plants to include precise insertions, sequence replacements and base substitutions.

The CRISPR-based systems have been termed NBTs—'New Breeding Technologies' or 'Techniques', or in the UK they are referred to as 'Precision Breeding'. These terms and the underlying principles now differentiate GM technologies from GEd technologies.

23.2 Terminology—What's in a Name?

There is now an opportunity to differentiate the targeted edits obtained from gene-editing technology from products of transgenic technologies. This difference is important, since there are 20 years of 'baggage' and misinformation associated with transgenic (Genetically Modified—GM/GMO or Genetically Engineered—GE, or Biotech) crops. This is despite the fact that GM crops constitute 10% of the world's food, and all the evidence to date supports their safety as food and feed. The definition of transgenic crops is that they contain genetic material from beyond the crop's gene pool—for example, the endotoxins from *Bacillus thuringienses* (Bt) which confer specific resistance to limited lepidopteran insect pests. Bt crops such as cotton and maize have been extremely successful in reducing the need for many insecticide sprays to control cotton bollworm, reducing exposure of farm workers to insecticides, and benefitting the environment. Similarly, herbicide-tolerant plants have been critical in weed control, enabling very large areas of land to be cultivated with zero tillage, so reducing loss of topsoil and helping retain groundwater.

Despite these major benefits, current global regulatory environments ensure that, in most cases, it is now prohibitively expensive to develop and release new GM crops.

In contrast, Ged produce either does not contain external genetic material, or if so, then that genetic material is from within the plant's gene pool, and so could have been introduced by conventional breeding methods, although with a much longer time frame.

Hence it is important not to use the abbreviation 'GE' for both Genetic Engineering (i.e. GM) and for Genome/Gene Editing, since this creates confusion. Using the abbreviation Ged for Gene Editing is preferred since this enables differentiation of Ged from GM/GE technologies.

23.3 Definitions of Gene-Editing—Site Directed Nucleases (SDN)

Since Ged is based on making DSBs at a specific site (or sites) in a DNA sequence, the term site-directed nuclease (SDN) is the most widely used acronym for describing Ged events (although it is not used in China). The definitions of SDN events and the spectrum of gene-editing applications are provided below (Fig. 23.1), including how 'natural' or detectable the gene-edits would be. (NHEJ: non-homologous end joining, HR: homology-dependent repair, modified from Jones 2016).

The SDN terminology used is defined in the following Table (Table 23.1). SDN-0 involves the use of deactivated or 'dead' Cas9 (dCas9), which cannot generate DSBs, but with gRNA still binds DNA with the same specificity. Targeted epigenetic changes ('marks') can be made with dCas9 fusion proteins to modify methylation or acetylation marks which may enable modulation of gene expression without altering the DNA sequence.

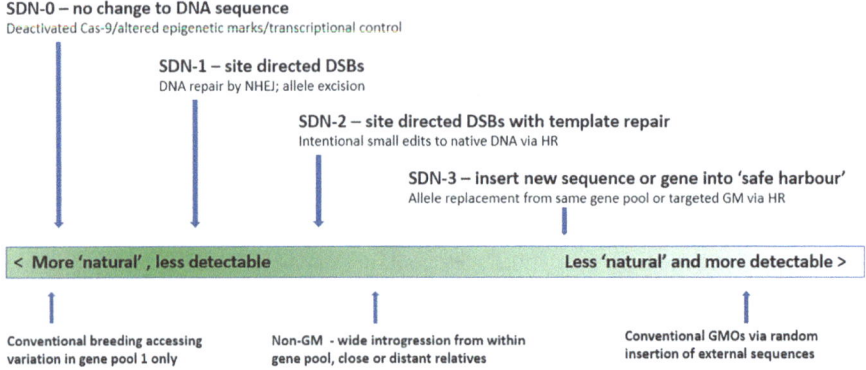

Fig. 23.1 The spectrum of GEd changes obtained from Site Directed Nuclease (SDN) applications. SDN-0 applies when there is no change in base sequence but epigenetic changes occur. SDN-1 refers to DNA repair after repair of DSBs by NHEJ; SDN-2 indicates a repair template has been used to make intentional edits at a DSB sites by HR; SDN-3 indicates insertion of a new sequence at a DSB site by HR which can be from outside the plant's gene pool, but at a specific site, referred to here as a 'safe haven'

Table 23.1 Definitions of SDN categories

SDN Application	Targeted to a specific genome location?	Use of repair template? Origin of the repair template	Type of targeted sequence change(s)
SDN-0	YES Epigenetic change defined by nuclease specificity	NO	**Epigenetic changes only** (e.g. transcriptional control)
SDN-1	YES Defined by nuclease specificity	NO	**Edit(s)** (Spontaneous mutations: deletions, substitutions, addition of bases)
SDN-2	YES Defined by nuclease specificity	YES Species own gene pool only	**Edit(s)** (Predefined mutations, sequence optimisation, allele replacement)
SDN-3	YES Defined by nuclease specificity	YES Any source, including species own gene pool only	**Insertions** (Addition of sequence at the targeted location in the genome)

23.4 A Summary of the Applications of GEd Technology

A comprehensive database of the published applications of GEd technology for crop plants is provided in the 'EU-SAGE' database (eu-sage.eu, see below), compiled by the organisation 'European Sustainable Agriculture Through Genome Editing'. EU-SAGE is a network representing plant scientists at 134 European plant science institutes and societies that have combined to provide information about GEd and to

Fig. 23.2 The number of
GEd publications based on
trait categories

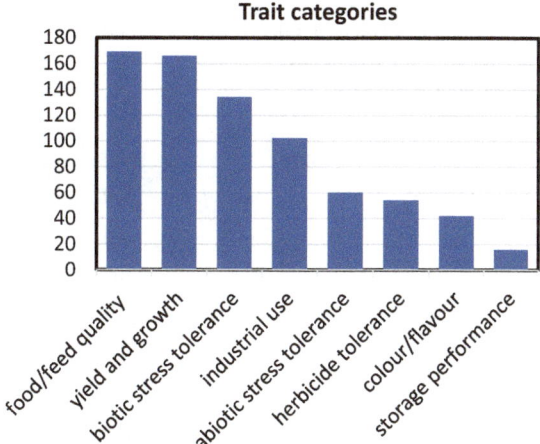

Fig. 23.3 The relative
number of Ged publications
from different countries

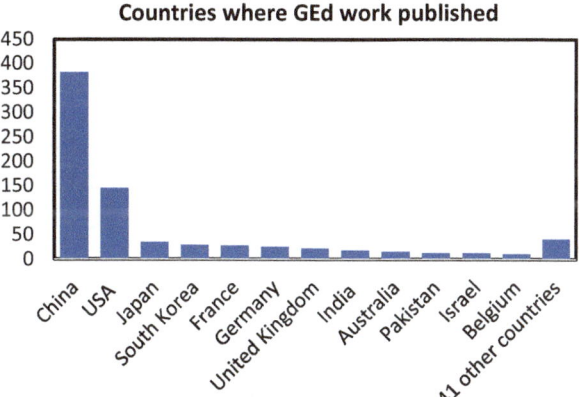

promote the development of policies in European and EU member states that enable the use of GEd for sustainable agriculture and food production.

The EU-SAGE database provides information on GEd applications by plant species, traits, the GEd techniques used, the country of origin, and the nature of each study/application. Each publication is referenced, and the database is searchable. The trait categories covered include food and feed quality, yield and growth, tolerance to biotic and abiotic stresses, herbicide tolerance, colour/flavour and storage performance. A summary of the current data on GEd crops in the EU-SAGE database (accessed 13 July 2023) is provided in the following figures.

The current database includes 743 references for trait categories. It is evident that the traits studied include all standard plant breeding targets: from quality, yield and growth, tolerance to biotic and abiotic stresses, industrial applications, herbicide tolerance, colour/flavour to storage performance (Fig. 23.2).

Looking at the countries undertaking GEd work on crop plants (Fig. 23.3), it is clear that by far the most R&D on GEd for crops is being done in China (382 reports),

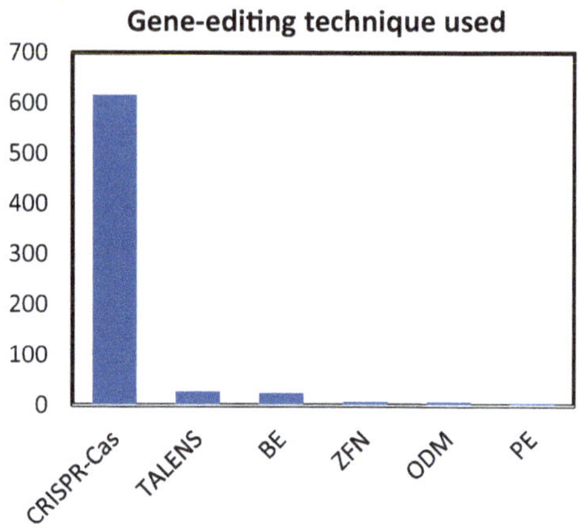

Fig. 23.4 The number of publications based on the GEd method used (see text for the method abbreviations)

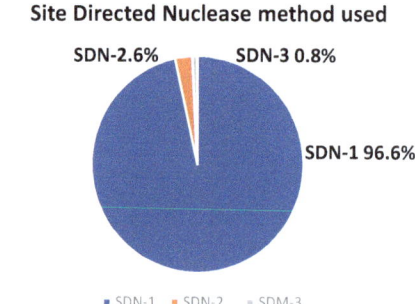

Fig. 23.5 The relative proportion of publications using SDN-1, 2 or 3 technology

followed by the USA (145), Japan (35), South Korea (29), France (28), Germany (25), UK (22), India (18), Australia (15), Pakistan (13), Israel (13), Belgium (11) with research also reported from 41 other countries.

Of the Ged technologies being used, CRISPR/Cas systems dominate the methodology (Fig. 23.4), with 90.05% of the reports using CRISPR/Cas technology, 3.94% using TALENS, 3.5% Base Editing, 0.87% ODM and 0.58% Prime Editing.

For those using the CRISPR/Cas systems, an overwhelming 96.6% of the reported products used SDN-1 technology, 2.6% SDN-2 and only 0.8% SDN-3 (Fig. 23.5).

GEd has been applied to 69 different plant species, with most research done on rice, followed by tomato, maize, soybean, wheat, canola, potato, tobacco and barley (Fig. 23.6).

23.5 What Factors Limit Commercialisation of GEd Produce?

It is now clear that GEd systems can generate intended, targeted changes in DNA sequences to deliver desired phenotypes in most crop plants. The major factor limiting commercialisation and trade in GEd produce is the need to align national

Fig. 23.6 The relative number of GEd publications in different crop plants

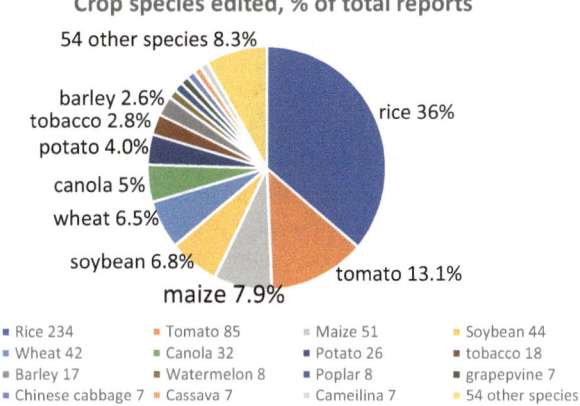

Crop species edited, % of total reports

- 54 other species 8.3%
- barley 2.6%
- tobacco 2.8%
- potato 4.0%
- canola 5%
- wheat 6.5%
- soybean 6.8%
- maize 7.9%
- tomato 13.1%
- rice 36%

■ Rice 234	■ Tomato 85	▪ Maize 51	■ Soybean 44
■ Wheat 42	■ Canola 32	■ Potato 26	■ tobacco 18
■ Barley 17	■ Watermelon 8	■ Poplar 8	■ grapevine 7
■ Chinese cabbage 7	■ Cassava 7	▪ Cameilina 7	▪ 54 other species

and international regulations relating to GEd produce. This aspect is discussed in more detail in Sect. 23.6.

Other factors to consider that relate to commercial application include:

- Confirming that no 'external' nucleic acid sequences are present
- The possibility of off-target edits
- Licensing GEd technology
- Unintentional low-level presence of GEd seeds/produce in bulk trade

23.5.1 Absence of External Nucleic Acid Sequences

When the editing machinery (Cas protein/gRNAs) is introduced as a T-DNA and a successful SDN-1 edit has been identified, the plant can then be selfed and selected for null segregants (i.e., plants with the edit but which do not contain T-DNA). How much effort should be made to determine whether there is any residual introduced DNA? Reasonable tests include:

- Sequencing the site of gene-editing
- Quantitative PCR to show the absence of backbone T-DNA
- Checking the sequence at the edited site to determine that no new ORFs are present, or if so, check any resultant peptide against allergen databases to ensure that there is no new allergen

Whole genome sequencing of an SDN-1 edited plant is unnecessary. The cost for plants like wheat, with a 17GB genome, would be prohibitive and prevent translation of products to end-users. Suggestions of this type from genome sequencing laboratories interested in academic publications should be resisted.

For SDN-2 edits, the above tests are also appropriate, given that wide crosses in conventional breeding are not subject to any such rigorous evaluation.

23.5.2 Off-Target Edits—Are they of any Biological Significance?

Possible off-target edits should also be viewed in context. Spontaneous mutations occur naturally at rates of $\sim 10^{-8}$ to 10^{-9} per site per generation, and pan-genome sequencing has revealed that many sequence differences exist within a species. There are also more than 3200 commercially sold varieties generated by induced mutagenesis, and also varieties obtained from somaclonal variation (EFSA Panel on Genetically Modified Organisms: Mullins et al. 2021).

Off-target edits can be minimised by improving the choice of gRNAs using available new bioinformatic tools (also now using Artificial Intelligence, AI), or using more specific CRISPR editing systems. Good design of gRNAs leads to undetectable levels of off-target edits (Young et al. 2019), with evidence from rice (Tang et al. 2018a, b), cotton (Li et al. 2019), and maize (Young et al. 2019; Lee et al. 2019) attributing nearly all of the variation in re-sequenced GEd plants to tissue culture-induced somaclonal variation.

There is also a history of the safe consumption of plant foods which contain many mutations, regardless of origin, and most have no phenotypic or biological effect. Most mutations are neutral in effect and are not identified or removed in conventional or mutation breeding. Any off-target edits that result in an undesirable phenotype would be eliminated in conventional breeding. Possible off-target edits in plants present much-reduced safety concerns compared to those that might arise in medical applications of GEd (Graham et al. 2020).

Indeed, the long history of safe use of foods in which mutations/edits are present, with many different mutations, shows that the vast majority off-target mutations/edits are of no biological significance to consumers.

23.5.3 Patents and Licensing of GEd Technologies

The area of patents for GEd technologies is complex and detailed discussion is beyond the scope of this article. The major Licensee of GEd technology or crop plants in Corteva Agriscience, in association with Dow Dupont/Pioneer and the Broad Institute. Any group considering commercialising a GEd crop or its products should seek professional and legal advice concerning the potential gene targets, the need for a license, licensing terms and geographical restrictions.

23.5.4 Unintentional Low-Level Presence

Because of the current international misalignment of GEd regulations, it is possible that GEd produce which has been deregulated and can be grown as a standard crop in one country, may still be regulated as a GMO in another. During trans-national transport, usually of bulk grains, there could be a small amount of SDN-1 or SDN-2 seeds deregulated in the producing country but present unintentionally in a non-GEd shipment. A group of 15 countries combined to form the Global Low-Level Presence

Initiative (https://llp-gli.org) to address this issue for GM produce, and the same issue could occur for GEd produce. A sensible agreement on accepting low level presence of GEd produce (e.g. 1%) in bulk shipments would solve this possible trade issue.

23.6   The Current Global Status of GEd Regulations

Most jurisdictions worldwide recognise the need for food security. The United Nations Sustainable Development Goals, especially Goal 2, Zero Hunger, reflects this need. There is general recognition that new GEd technologies promise to make a major contribution to increasing the world's food supply for farmers and consumers, and will benefit human health and the environment. Developing regulations are now treating GEd crops more as plants which have been mutagenised in a targeted and precise manner, rather than at random (Heffron and Herring 2017).

Global regulations relating to GEd plants in agriculture are evolving rapidly. The following image (Fig. 23.7) reflects the global regulatory status in 2020 (Schmidt et al. 2020). Most countries in North and South America do not regulate Ged crops as GMOs. In contrast, the EU and New Zealand regulate them as GMOs, and in many other countries there are ongoing discussions to address and progress regulations.

Notably, in 2020, the regulatory status of Ged plants in many Asia-Pacific countries was not clear, as shown in the map below (Schmidt et al. 2020).

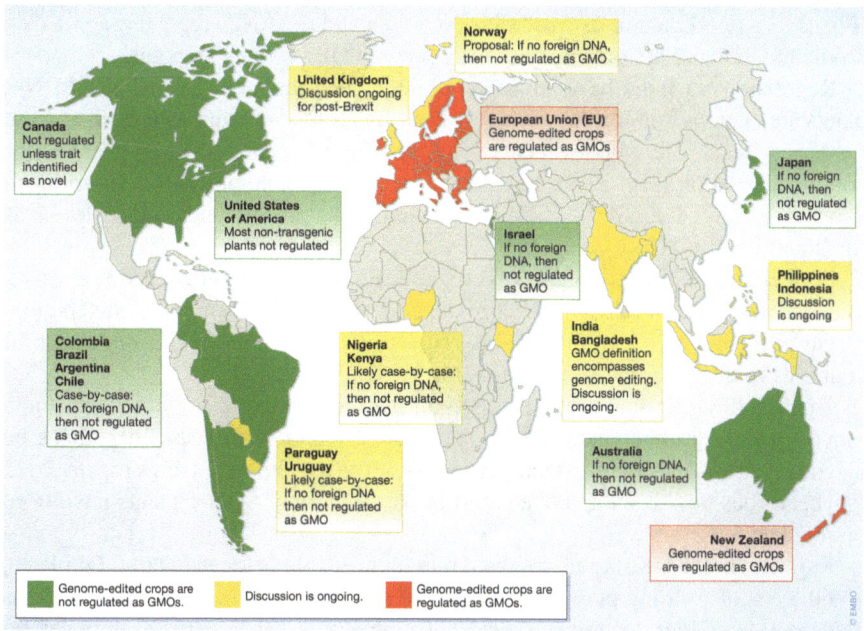

Fig. 23.7   The global status of regulation of GEd crops in 2020

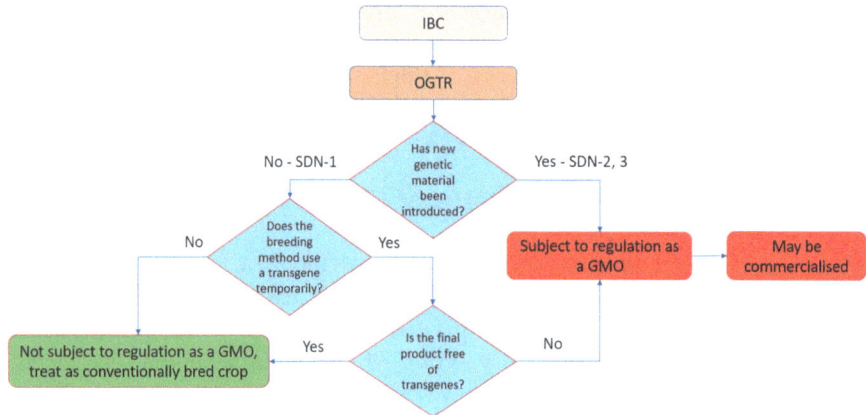

Fig. 23.8 The pathway to deregulation of GEd crop produce in Australia. Note that only SDN-1 produce has been deregulated

In an Australian government-funded project, we set out to interact with scientists and regulators in the Asia-Pacific region, to map existing GEd regulations, and to encourage alignment of national GEd regulations. Two examples of current regulatory pathways in this region are provided, for Australia and India.

In Australia, only SDN-1 plants have been de-regulated and can be treated as conventionally bred plants. However, SDN-2 plants are still regulated as GMOs (Fig. 23.8). Food Standards Australian New Zealand (FSANZ) regulates non-living food and feed, and the outcomes of its review on GEd food is now due.

Regulations in India have advanced rapidly, such that both SDN-1 and SDN-2 plants are exempt from further biosafety trials under field conditions as well as from food and feed safety assessments, and require no approval from GEAC for commercialisation (Fig. 23.9). In contrast, editing technologies with a footprint of exogenous DNA (i.e., SDN-3 is still regulated at the process and product levels as for 'traditional' GMOs).

Further details are provided in the publication by Jones et al. (2022), which includes the path-to-market of GEd produce in many of the countries in this region.

On overview of the results of this project and the current regulatory status of countries in the Asia-Pacific region is provided below (Fig. 23.10).

Japan and the Philippines are leaders in the region, and it is notable that India moved rapidly to deregulate SDN-1 and SDN-2 plants. China should perhaps be hatched yellow and green, since permission to grow the first GEd crop in 2023 (soybean) has now been approved, and in Australia only SDN-1 plants have been de-regulated.

England, in the UK, has also moved rapidly to enable field testing of GEd plants, and there is also strong pressure in the EU to relax its regulations on GEd crops, as evidenced by a draft document of the EU Commission which would exempt certain 'GM' plants from its strict ruling on growing GM crops (EU Commission 2023).

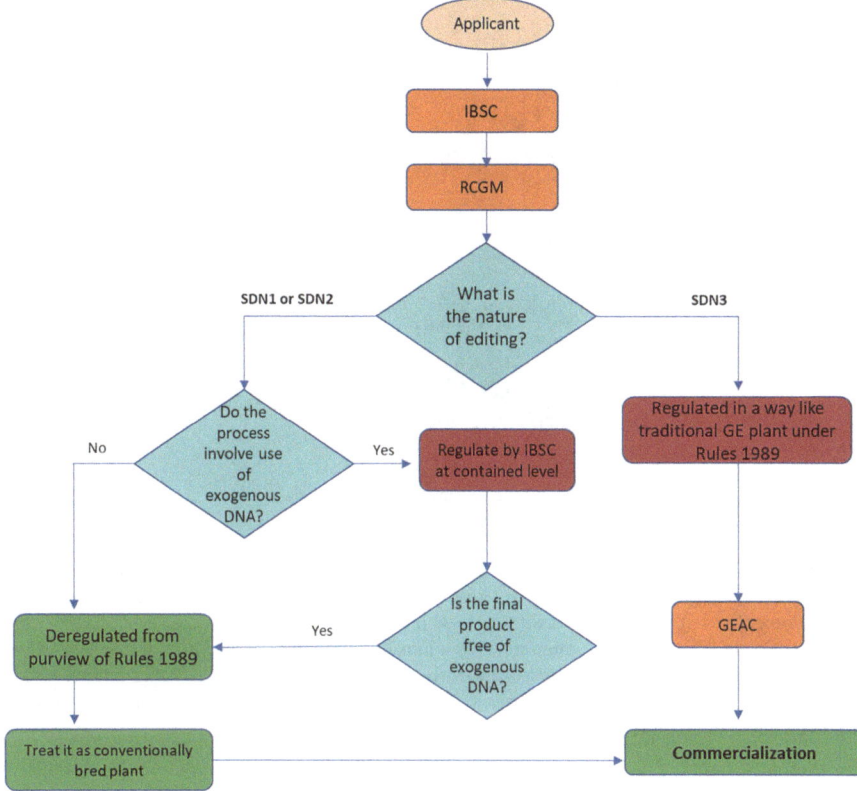

Fig. 23.9 The pathway to deregulation of GEd produce in India

23.7   Conclusions

As explained by Heffron and Herring (2017), nature does not code plants as GM or not GM. These are purely political conventions. Transgenic and mutant plants occur naturally, and contribute to the variation used in conventional plant breeding. In fact, genomic studies reveal that horizontal gene transfer between species is a relatively common event. Randomly-induced mutants have also been generated as a standard breeding tool for more than 50 years: products of mutagenic treatments can be found on supermarket shelves and in 'organic' food shops.

The opportunity now exists to break the regulatory and perception nexus between Genetically Modified (GM) plants and GEd plants, and to avoid the issues that have prevented wider use of GM crops. Simply based on the presence (GM) or absence (GEd) of genetic sequences from outside a plant's gene pool, momentum is now building to treat GEd produce in the same way as products—varieties—developed by conventional breeding. This is a very positive development, since GEd

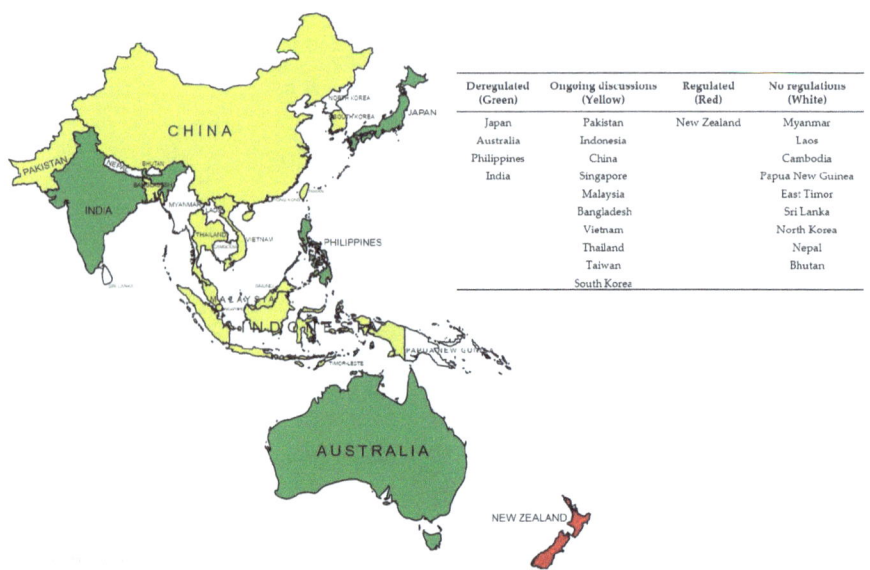

Deregulated (Green)	Ongoing discussions (Yellow)	Regulated (Red)	No regulations (White)
Japan	Pakistan	New Zealand	Myanmar
Australia	Indonesia		Laos
Philippines	China		Cambodia
India	Singapore		Papua New Guinea
	Malaysia		East Timor
	Bangladesh		Sri Lanka
	Vietnam		North Korea
	Thailand		Nepal
	Taiwan		Bhutan
	South Korea		

Fig. 23.10 The regulatory status for GEd crops in countries in the Asia-Pacific region. It is based on the deregulation of SDN-1 crops (green), with some countries also deregulating SDN-1 and SDN-2 products. Countries with ongoing discussions (yellow) and regulated as GMOs (red). Note that regulation of GEd crops in China is under discussion, but does not use SDN terminology: at present GEd is still under GMO product safety management measures, but with less onerous requirements in the pathway to commercial approval

technologies have broad applicability, there is less potential for monopolistic control, and GEd technologies provide a more democratic and level playing field.

To progress sensible international regulations for GEd crops, it is important to avoid complex regulatory assessments (like a requirement for whole genome sequencing), unless the same standards are applied to conventional and mutation breeding. There is an increasing imperative to achieve the potential of increasing food production to meet world food demands, with regulations based on scientific evidence rather than political expediency.

Acknowledgements Support is acknowledged from the Department of Agriculture, Forestry, and Fisheries (formerly DAWE), Government of Australia, Project Assisting Small Exporters (PASE), Building Capacity for Small Exporters to Exploit 'new breeding technologies', Grant number 4-FA4N7WL.

References

EFSA Panel on Genetically Modified Organisms, Mullins E, Bresson JL, Dalmay T, Dewhurst IC, Epstein MM, Firbank LG, Guerche P, Hejatko J, Moreno FJ (2021) In vivo and in vitro random mutagenesis techniques in plants. EFSA J 19:e06611

EU Commission (2023) European Green Deal: more sustainable use of plant and soil natural resources; July 5, 2023, https://ec.europa.eu/commission/presscorner/detail/en/IP_23_3565

Graham N, Patil GB, Bubeck DM, Dobert RC, Glenn KC, Gutsche AT, Kumar S, Lindbo JA, Maas L, May GD et al (2020) Plant genome editing and the relevance of off-target changes. Plant Physiol 183:1453–1471

Hassan MS, Abd-El-Haleem S (2014) Effectiveness of gamma rays to induce genetic variability to improve some agronomic traits of canola (Brassica napus L). Asian J Crop Sci 6:123–132

Heffron KL, Herring RJ (2017) The end of the GMO? Genome editing, gene drives and new Frontiers of plant technology. Rev Agrarian Stud 7:1–32. https://doi.org/10.22004/ag.econ.308366

Jones HD (2016) Are plants engineered with CRISPR technology genetically modified organisms? Biochemist 38:14–17

Jones MGK, Fosu-Nyarko J, Iqbal S, Adeel M, Romero-Aldemita R, Arujanan M, Kasai M, Wei X, Prasetya B, Nugroho S et al (2022) Enabling trade inGene-edited produce in Asia and Australasia: the developing regulatory landscape and future perspectives. Plan Theory 2022(11):2538. (36pp). https://doi.org/10.3390/plants11192538

Lee K, Zhang Y, Kleinstiver BP, Guo JA, Aryee MJ, Miller J, Malzahn A, Zarecor S, Lawrence-Dill CJ, Joung JK et al (2019) Plant Biotechnol J 17:362–372

Li J, Manghwar H, Sun L, Wang P, Wang G, Sheng H, Zhang J, Liu H, Qin L, Rui H, Li B (2019) Whole genome sequencing reveals rare off-target mutations and considerable inherent genetic or/and somaclonal variations in CRISPR/Cas9-edited cotton plants. Plant Biotechnol J 17:858–868

Li X, Song Y, Century K, Straight S, Ronald P, Dong X, Lassner M, Zhang Y (2001) A fast neutron deletion mutagenesis-based reverse genetics system for plants. Plant J 27:235–242. https://doi.org/10.1046/j.1365-313x.2001.01084.x.27

Schmidt SM, Bellisle M, Frommer WB (2020) The evolving landscape around genome editing in agriculture. EMBO Rep 21:e50680

Tang X, Liu G, Zhou J, Ren Q, You Q, Tian L, Xin X, Zhong Z, Liu B, Zheng X, Zhang D (2018b) A large-scale whole-genome sequencing analysis reveals highly specific genome editing by both Cas9 and Cpf1 (Cas12a) nucleases in rice. Genome Biol 19:1–3

Tang X, Liu G, Zhou J, Ren Q, You Q, Tian L, Xin X, Zhong Z, Liu B, Zheng X et al (2018a) A large-scale whole-genome sequencing analysis reveals highly specific genome editing by both Cas9 and Cpf1 (Cas12a) nucleases in rice. Genome Biol 19:84

Young J, Zastrow-Hayes G, Deschamps S, Svitashev S, Zaremba M, Acharya A, Paulraj S, Peterson-Burch B, Schwartz C, Djukanovic V, Lenderts B (2019) CRISPR-Cas9 editing in maize: systematic evaluation of off-target activity and its relevance in crop improvement. Sci Rep 30:6729

New Seed Technologies: Socioeconomic Considerations

24

R. Ramakumar

Abstract

This paper is a discussion on how social scientists can constructively engage with new technological developments in agriculture and use multi-dimensional and multi-disciplinary frameworks in their research. It focuses on the GM seed technology, and specifically addresses questions around the introduction of Bt-Cotton seeds in India. The issues raised vis-à-vis GM crops would equally be applicable to the prospective editions of gene-edited crops. Firstly, the paper argues for a better appreciation of the socioeconomic contexts in the study of new agricultural technologies. Secondly, it outlines a multidimensional framework within which the technological and socioeconomic variables could be analysed together to understand the net social value of a technology. Thirdly, in doing so, it also outlines a set of economic, institutional, and sociocultural factors that need to be factored in and accounted for in any analysis of new seed technologies. Here, the paper tries to distinguish between policy failures and technological failures in the case of Bt-Cotton in India and argues for a nuanced analysis that does not conflate the two.

Keywords

Science policy · GM crops · Indian agriculture · Bt-cotton · Genetic regulation

The author thanks Ashish Kamra for research assistance.

R. Ramakumar (✉)
School of Development Studies, Tata Institute of Social Sciences, Mumbai, India
e-mail: rr@tiss.edu

© National Academy of Agricultural Sciences, under exclusive license to Springer
Nature Singapore Pte Ltd. 2023
K. C. Bansal et al. (eds.), *Transformation of Agri-Food Systems*,
https://doi.org/10.1007/978-981-99-8014-7_24

337

24.1   Introduction

The role of science and technology in improving the conditions of life and work of humankind is extensively documented (see UNESCO 2000). In agriculture, science and technology played a central role in shifting the nature of agricultural growth from being *extensive* to being *intensive*. For instance, if India produced about 50 million tons of food grains per year after close to 10,000 years of crop cultivation till 1950, it could raise it to more than double i.e., 121 million tons per year, in a short period of 25 years between 1950 and 1975 (Swaminathan 1968). Across the world, such rise of crop productivity that accompanied scientific advances – particularly through plant breeding and the external application of chemicals – helped humankind avoid hunger, famines and the devastating effects of indiscriminate deforestation. Cumulative global carbon emissions from 1850 would have been one-third higher in the absence of green revolution (Burney et al. 2010). Further, as crop improvement by *farmers* gave way to crop improvement by *scientists with farmers*, plant breeding became less *uncertain* and more *predictable*—thus hastening the development of seeds with new traits.

If the green revolution from the decade of 1960s was driven by the cross-pollinated semi-dwarf wheat and rice varieties, developments in the late-20th and early-21st centuries were marked by a "gene revolution" (Parayil 2003). The gene revolution had two stages of evolution: the emergence of *transgenic technologies* by the 1990s and of *gene-editing* by the 2000s. If the transgenic technologies based themselves on the recombinant-DNA technique, the gene-editing technologies are based on the CRISPR technique, which makes it possible to add, delete, or change genetic material at specified locations in the plant genome (Fukuda-Parr 2007; Hille and Charpentier 2016).

How must agrarian scholars, with skills in social sciences, understand and analyse these new techniques? This is an important question that the editors of this volume put to me as an economist by training. In this article, I shall try to outline a broad framework within which agrarian scholars can begin such analyses. Firstly, I shall examine how the broader socioeconomic context behind the emergence of new techniques can be better appreciated in history. Secondly, I shall try to outline the major variables that social scientists could study while researching the impacts of the new technologies. I shall also point out that many of these variables are also interdependent in their causes and effects. Thirdly, I shall try to draw attention to a few political economy considerations in the analysis of new technologies in agriculture. Finally, in the concluding section, I shall try to argue for a multi-dimensional and multi-disciplinary perspective in the study of new technologies in agriculture—rather than one-sided and narrowly defined terms of reference.

Unlike pure sciences, social sciences are a heavily contested domain. For this very reason, I must underline at the outset that my views in this paper would not represent any consensus in the social science community. In fact, my views must be seen as emanating from a certain worldview regarding the role of science and technology—a point that I shall try to highlight wherever possible or necessary.

24.2 Studying the Emergence of Technologies

In this section, I attempt to highlight the need to foreground the centrality of the *socioeconomic context* in studying the history and emergence of new technologies. The discussion here is more generic than one that focusses specifically on GM crops or gene-edited crops.

24.2.1 The Socioeconomic Context

The role of science and technology in the history of humankind is inextricably linked to the relationship between man and nature (Engels 1954). Human beings have, for thousands of years, strived to change and transform nature through their *conscious* and *purposive* actions. They did not undertake this effort in the abstract or purely randomly; instead, they actively imagined an outcome prior to embarking upon a project to realise them. Every such effort was a derivative of the accumulated past knowledge about nature; was in correspondence with the nature of the prevailing social formation; and was shaped by the consciousness of people. In other words, there was a historical and material basis to the leaps in knowledge that humankind made at every point of time in history – a topic that has interested social scientists for long.

To cite an example from England in the seventeenth and eighteenth century i.e., prior to the industrial revolution, the household production system and the putting-out system had given way to the manufactories. With the growth of manufactories, which were relatively larger in size, successive stages of production were more closely organised as a unity (Dobb 1947). As a result, the division of labour was extended between different stages of production culminating in the final product. This shortened the time taken to pass raw materials from one stage to another and avoided major coordination problems. It was this division of labour that prepared the ground for new mechanical inventions to emerge during the industrial revolution. In the absence of such a material basis, the motivations for mechanical inventions would have been lacking.

A social scientist would typically be looking for such linkages while studying the role of technology in history. Given that material—and cultural too, to add—forces play a major role in influencing the actions of human beings in specific historical situations, social scientists cannot study scientific actions or events in a vacuum. They study the forward movement of science as a product of accumulated history; see its motivation rooted in the prevailing social formation; study its momentum as driven by the distribution of incentives across various actors; and consequently, infer that the growth of scientific enterprise has always had an element of autonomy embedded within.

Social scientists also do not study science as removed from people. As humankind transformed nature through the instruments of production, their own consciousness was also altered and moulded by their labouring in productive activities. If technology made it possible for human beings to improve lifestyles and reduce drudgery of labour, it also laid the ground for new forms of *control of one group of human beings by another group of human beings* (see Hobsbawm 1975). Thus, with new machinery in the 18th and 19th centuries came the era of Taylorism and Fordism, and their time and motion studies, which were essentially new ways of intensifying the labour process to raise productivity and expand private profits under a capitalist system. The size of production units could now be increasingly enlarged, a process that unfolded alongside the destruction of small/petty production and the consequent creation of a large assetless labour force. The largeness of units led to newer forms of division of labour—and new possibilities of labour control.

This dialectic of technological development has interested social scientists right from the time of Adam Smith and Karl Marx. In short, if new technologies provided the foundation for raising productivity, they also laid the basis for new forms of labour control in a capitalist society. This contradictory aspect of the reality is one reason why discussions on technology must be foregrounded on their social and economic contexts. Focussing singularly on the former would be an apology for the latter outcome; and focussing singularly on the latter outcome would be to be blind to the enormous improvements wrought by the former. A dialectical framework allows social scientists to impart an appropriate element of comprehensiveness to their analysis.

24.2.2 The Case of Agriculture

The importance of studying the socioeconomic context in the emergence of science and technology can be specifically illustrated in the case of agriculture also. Substantive improvements in the use of technology in agriculture followed the industrial revolution in Europe from the eighteenth century (Gard 1931). But even prior to the industrial revolution, the agricultural revolution from the 14th and 15th centuries had led to the large-scale displacement of peasantry from their land, the enlargement in the size of farms and the creation of a large assetless labour force (Byres 1988). During these early years when surplus labour was freely available, the nature of agricultural improvements was nominal: crop rotation, selective breeding, and the adoption of other improved agricultural practices.

But by the early-eighteenth century, the continuing increase in the size of the farms necessitated mechanisation, and this became the motivation for the development of the plough, the seed drill, and the threshing machine. Afterwards, as surplus labour began to dry up and wages for hired labourers rose, cultivation in large farms could not proceed without extensive mechanisation (Hayami and Ruttan 1985). Thus, steam power of the industrial revolution came to be applied to the machines used in agriculture, such as the threshing machines, reaping machines, and pumps.

The responses of agricultural workers to mechanisation were militant. Faced with unemployment, agricultural workers attacked the threshing machines. The Swing Riots of 1830–32, in Kent under the leadership of Captain Swing, was an important episode in English rural history (Hobsbawm and Rude 2001). The English state's response was severe; 19 people were executed, 505 people were transported to Australia and 644 people were imprisoned.

In the brief history narrated above, a social scientist would not just be studying the incentives that led to the adoption of the new technologies. They would begin with the changes in the rural social formation of the time and how it created a material basis in agriculture for the emergence of large farms as well as the necessity for new technologies like the threshing machine to emerge. They would also study the role played by protest movements like the Swing Riots in shaping the evolution of new technologies and the public perceptions therein. In short, how did the new technologies respond to a material need of the time? How did they create a basis for new growth impulses and a further transformation of the social formation of the time? What role did power relations play in the evolution of the technology? How were the contradictions between the growth impulses of the technology and its effects on different sections of the society managed? Did the conditions of agricultural workers become impoverished after the introduction of the threshing machine?

24.3 Studying New Agricultural Technologies

In this section, I come to the discussion on how social scientists could understand the motivations for new agricultural technologies as well as the need for them to take account of the multi-dimensional concerns related to the new technologies. In Fig. 24.1, I have tried to provide a short list of such factors.

24.3.1 The Framework

A new technology, to begin with, must be studied in the context of the prevailing state of agriculture and its specific challenges. In some cases, new technologies may be required to address a problem of food shortage or food insecurity, as was the case of the green revolution in India. In the early-1960s, undernourishment and hunger were pervasive in India. India was largely dependent on food imports and food aid from the West, which weakened the country's political sovereignty. Economists and political scientists have provided important analyses of how India's foreign policy and economic policy in the early-1960s were held hostage under its "ship-to-mouth" existence vis-à-vis food grains. These studies also provide important insights into the conduct and evolution of national science policies in a food-deficit nation.

Social scientists must also contextualise—within a multi-disciplinary framework—the need for new technologies in relation to various challenges in agriculture. Higher yields are necessary if a country must become able to raise agricultural incomes. Higher yields are also an indicator of a growing rural economy, which in turn influences the extent of rural employment generation and rural poverty reduction. A low productivity-low income system of production is not just economically inefficient but also vulnerable to external shocks. A production system with

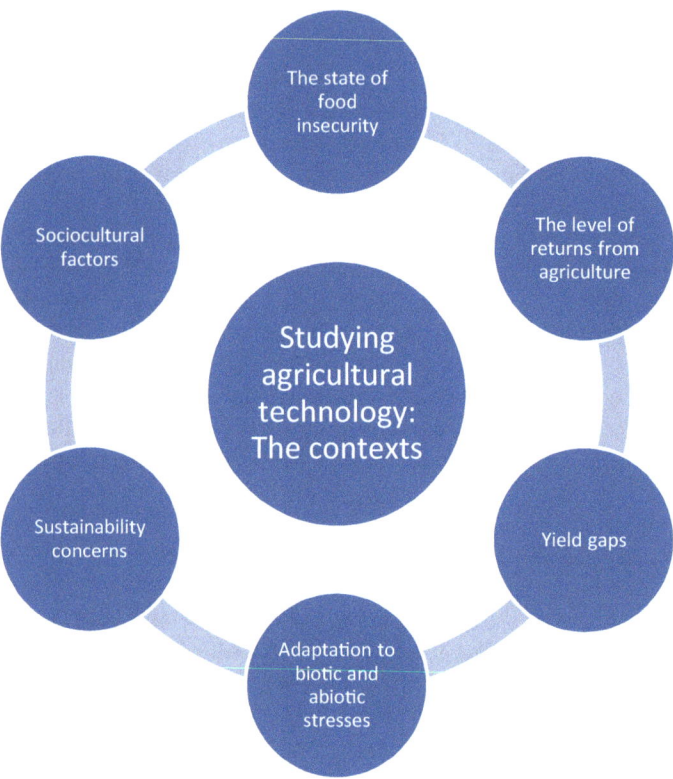

Fig. 24.1 An illustration of contexts in a study of emerging technologies in agriculture

higher productivity is much more resilient than a production system with lower productivity.

Typically, yield gaps are large in countries like India, which are due, to a considerable measure, to unscientific management of farms. Yields may also be low due to different biotic and abiotic stresses, both of which could be exacerbated by the challenges of climate change.

At the same time, the emergence of new technologies is not free from concerns for a social scientist, who would want to understand the *net social value* of changes in technology. The need for sustainable agriculture is an overarching concern, which needs reconciliation based on a reasoned debate. Some new technologies may increase risks in cultivation both from the point of view of the ecosystem and of the well-being of human beings. Scholars might want to follow a three-stage process of assessing these risks: *hazard identification*, *risk measurement* and *risk management* (see Thompson 2005). Hazard identification requires a characterisation of the features and forms of the danger, harm or injury associated with the technology. Risk measurement involves distinct treatments of both hazard (the potential for danger) and exposure (the probability that the hazard would materialise). Risk management is a process of deciding what can be done to reduce the impacts of, if not eliminate, the risks.

In all the three stages of assessing risks, there is a certain interpenetration of the *objective* and the *subjective*. In a characterisation of hazard and exposure, value judgements are inevitable. For instance, there may be cases where the technology is useful to one section but harmful to another section. In such cases, what would be framework of justice and fairness that one would follow in arriving at a consensus? Social scientists, particularly philosophers among them, have written much on these questions of justice and fairness that would come handy in arriving at scientific consensus.

Even if risks are minimum, different sociocultural factors might also determine which technologies are welcome and preferred in certain societies. The best known examples are the opposition to cultivate GM crops in parts of the European Union or the resistance to shift out of traditional varieties of food crops in certain tribal habitats. Religious, spiritual, or historical factors also influence the decisions of societies in choosing technologies (see Conrad and Gabe 1999). Social scientists, particularly social anthropologists, can help in understanding these concerns and provide inputs to science policy.

24.3.2 The Case of Bt-Cotton in India

The case of Bt-Cotton seeds in India can be used to apply the framework discussed above (see Fig. 24.2). Prior to the official introduction of Bt-Cotton in India in 2002, India was a net importer of cotton (see Fig. 24.3). In the late-1990s, Indian cotton imports had accounted for about six per cent of the world's cotton imports (Chaudhary 2005). The reason was that cotton production in India was not catching up with the rising domestic consumption of cotton. The raw material needs of increasing numbers of textile mills in India, which employed lakhs of industrial workers, came to be dependent on imported cotton.

Poor growth of production was due to the low productivity. Cotton yields in India were 190 kg/ha in 2000–01 and 186 kg/ha in 2001–02 (see Fig. 24.4). Cotton plants were also severely infested with bollworms (American Bollworm: *Helicoverpa armigera*) that caused yield losses to the tune of 30 to 60 per cent per year (Kranthi 2012). Incomes from cotton cultivation were also low; low yields on the one hand, and high levels of use of pesticides on the other, were pushing Indian cotton farmers into a hard corner. Large quantities of pesticides were applied in cotton farms to control for the bollworm; studies show that Indian farmers applied about 9400 MT of insecticides in 2001—worth Rs 747 crore—to overcome bollworm infestation (*ibid.*).

The introduction of Bt-Cotton seeds held the promise of reducing bollworm infestations, reducing the attendant pesticide use, raising the production of cotton, and improving farmer's incomes by increasing profitability per unit area. This was the context in which Bt-Cotton seeds were introduced in India.

Fig. 24.2 An illustration of contexts in a study of introduction of Bt-Cotton seeds in India

But the new Bt seeds also gave rise to multiple concerns. Specific groups raised concerns related to biosafety vis-à-vis the impact of the new seeds on soil microbes, reared animals, and the possible presence of Bt in human blood and placenta. Similarly, concerns were raised by a few farmer's groups related to the loss of traditional varieties of cotton and the inability of farmers to save and re-use Bt-Cotton seeds.

Clearly, there was a case for a comprehensive assessment of risks and hazards related to Bt-Cotton seeds. Hazards needed to be identified and characterised; quantification of risks had to be undertaken; and a detailed risk management protocol had to be released. A reliable risk communication strategy also needed to be formulated. Groups with complaints on sociocultural issues had to be engaged to arrive at a consensus view on the new technology.

Our simple framework above allows social scientists to view the introduction of new Bt-Cotton seeds in a more comprehensive perspective. Unfortunately, most critical accounts of the Bt technology were based on narrowly defined perspectives, which ended up skewing the debate in favour of a maximalist usage of the precautionary principle, and towards an irrational amplification of biosafety concerns. As a

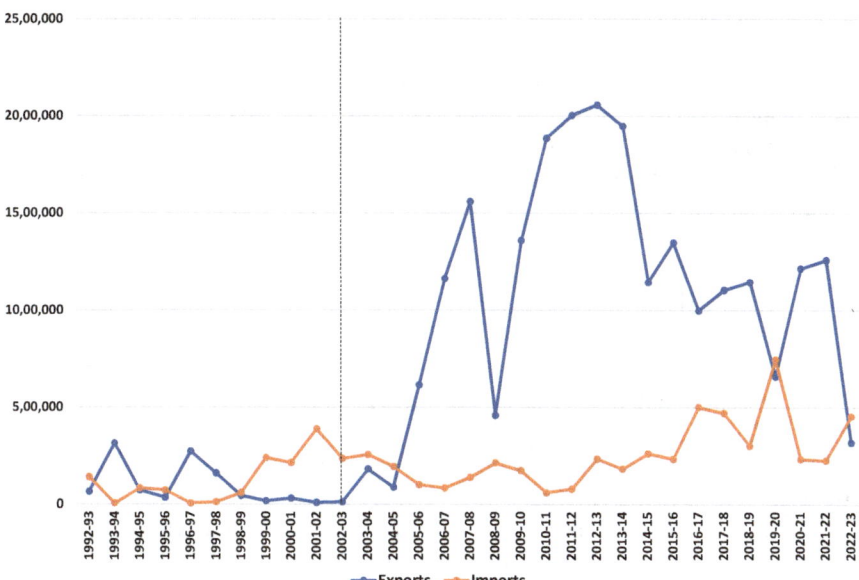

Fig. 24.3 Quantity of exports and imports of cotton, India, 1992–93 to 2022–23, in tons

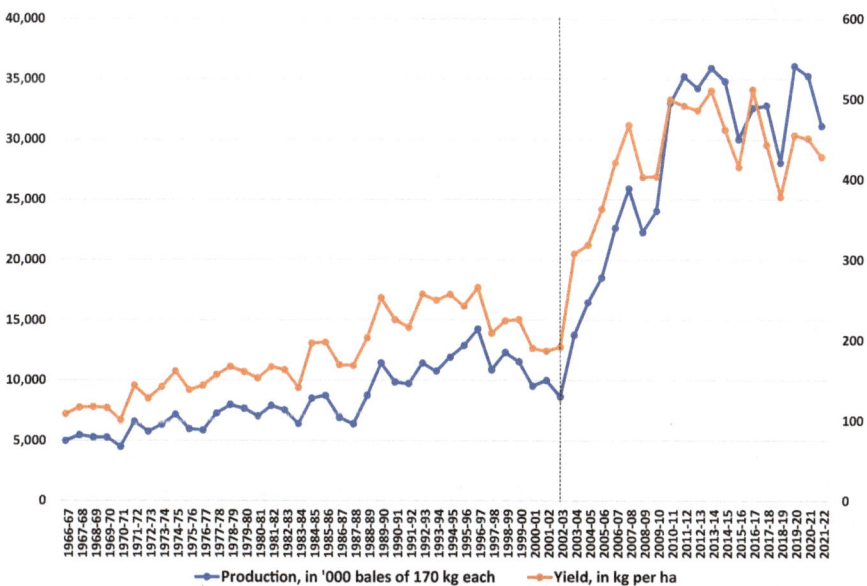

Fig. 24.4 Production and yield of cotton, India, 1966–67 to 2021–22, in '000 bales and kg per ha

result, the regulatory system turned complex and cumbersome. According to Kranthi (2012, p. 4), "while GEAC [Genetic Engineering Appraisal Committee] and RCGM [Review Committee for Genetic Manipulation] should have focused only on bio-safety approval, instead of evaluating and approving the 1128 Bt-hybrids, the identification of appropriate Bt-hybrids or varieties should have been the domain of ICAR (Indian Council of Agricultural Research) and the NARS (National Agricultural Research System)".

24.4 Analysing the Impacts

In this section, I shall attempt to analyse in greater detail the nature of impacts of Bt-Cotton seeds in India. As no new seeds of gene-edited crops have been released in India, any analysis of their impacts would be purely speculative. Hence, our analysis in this section is more generic at one level and purely based on the evidence on Bt-Cotton seeds at another level.

Any impact analysis of new crops in agriculture can be undertaken under three major heads: (a) what were the major economic impacts of the new technology? (b) what are the institutional factors that have guided or restricted the growth of the technology? (c) how have sociocultural factors interfaced with the roll-out of the new technology? (Fig. 24.5). A final assessment must be based on a comprehensive and integrated analysis of all these variables.

24.4.1 Economic Impacts

The economic impacts of a new seed must be primarily seen through the prism of overall net returns. New seeds may be more or less expensive, may increase or decrease the application of inputs like chemicals or water, may increase or decrease the need for hired labour, and may necessitate mechanisation of the farm. If cultivation is risky, the crop may also need to be insured with a regular premium payment. These are aspects of costs of cultivation. These would need to be seen against the gains or losses in the productivity of the new crop, and the movements in output prices. The question to ask would be: do higher levels of productivity more than compensate for a rise in input costs?

In India, the most visible evidence for higher incomes from Bt-Cotton crop than in the past comes from the extraordinary spread in farmer's adoption of the seed. Almost 100 per cent of cotton planted in India today is of the Bt type. Referring to the controversy in India generated by the narrative that Bt-Cotton has "failed", Herring (2006) asks: "rather than asking why there is such a sharp adoption curve of both small and large farmers, and commercial seed firms, across all cotton areas of India, activists continue to declare 'the failure of Bt cotton'" (p. 471).

National averages show that the cotton yields rose from 191 kg/ha in 2002–03 to 499 kg per hectare in 2010–11 and, though it stagnated and fell afterwards, stood at 451 kg/ha in 2020–21 (see Fig. 24.4). Consequently, if India was a cotton importer in

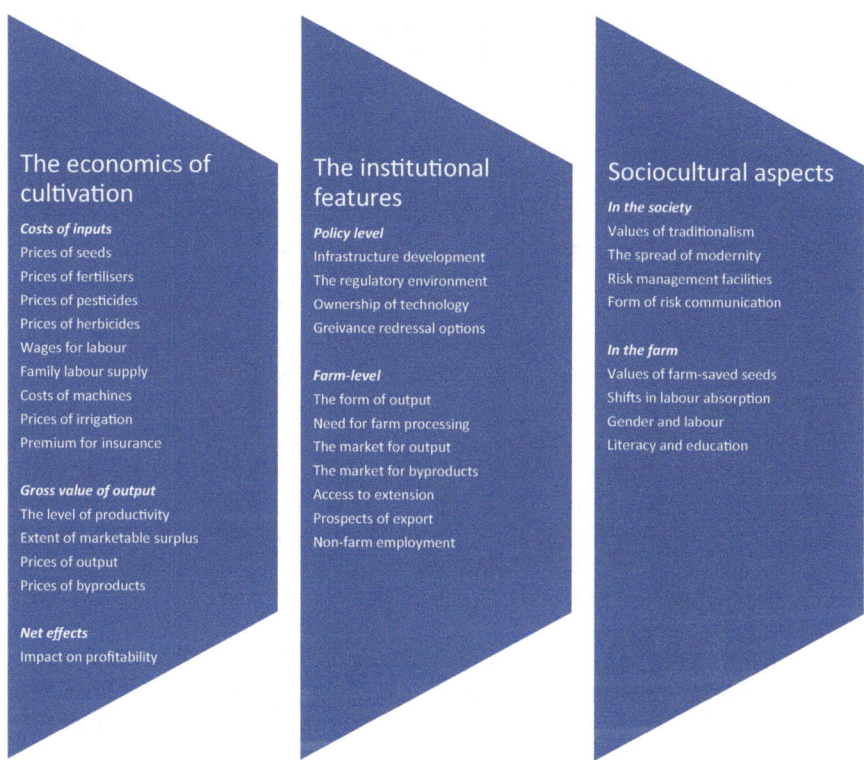

Fig. 24.5 A conceptual framework for a socioeconomic analysis of the impacts of GM crops

the late-1990s, it was transformed into a major cotton exporter to the world after 2003 (see Fig. 24.3). In 2001–02 and 2002–03, India exported just about 8323 tons and 11,475 tons of cotton, respectively. But by 2011–12, India was exporting close to 20 lakh tons of cotton to the world. Exports fell afterwards, and imports rose too, but India still exported close to 13 lakh tons of cotton in 2020–21. The fall of exports was largely due to a moderate fall of domestic production, which meant that domestic production could not catch up adequately with higher domestic demand. I shall return to the reasons for this phenomenon a little later.

To further discuss the question of profitability of Bt-Cotton, I rely on two of my surveys of two cotton-growing villages in Vidarbha in Maharashtra. The two villages were Dongargaon in Akola district and Savali in Buldhana district, and the surveys were organised in 2006–07 and 2013–14 (Dongargaon) and 2009–10 and 2013–14 (Savali). Both were cotton growing villages in the black cotton belt of Maharashtra.

In Dongargaon and Savali, our first round of surveys showed that if *parathi* (non-Bt) cotton plots yielded an average of 2–3 Q/acre, Bt-Cotton plots yielded an average of 6–7 Q/acre (Ramakumar et al. 2009). By the time of our second survey in 2013–14, the average yield in Bt-cotton in the both the villages was higher at around

10 Q/acre. At the same time, costs of cultivation were significantly higher in Bt-Cotton than for *parathi* cotton in both the villages. Yet, net profits in Bt-Cotton plots in 2013–14 were more than double the net profits in Bt-Cotton in 2006–07 (see Ramakumar et al. 2016). In short, on a per acre basis, higher input costs were more than compensated by the value of higher yields.

24.4.1.1 Policy Failure Versus Technological Failure

Let us now return to the question of the decline of cotton production in India after 2013–14. It is a fact that in the 10 years between 2012–13 and 2021–22, there were 6 years where the year-on-year change in area cultivated with cotton was negative. There were also 7 years between 2012–13 and 2021–22 where the year-on-year change in yield was negative. A close perusal of numbers would show that these changes could be explained by (a) the fluctuating receipt of monsoon rains; (b) the spread of Bt-Cotton into rainfed areas, which increased risks of crop failure in the absence of adequate monsoon rains; (c) fall of cotton prices; and (d) the conscious shift of farmers out of cotton into other crops, including maize and soyabean.

In the two villages of Maharashtra that we studied, there were multiple years in which farmers moved out of Bt-Cotton into the cultivation of soyabean or maize. Such shifts of cropping pattern in Vidarbha have often been cited as evidence for the "failure of Bt-Cotton". Much of such inferences are not based on a nuanced understanding of the changes in the economics of cultivation. I attempt a brief clarification below. I shall argue that the shifts in the fortunes of cotton after 2012–13 had more to do with *policy failures* than with *technological failures*.

It may be useful to begin with a quick review of the stance of mainstream economists on such matters. Mainstream economists generally assume that farmers are "risk-averse". The shape of the typical utility function is concave in the case of "risk-aversion", while shape of the typical utility function is convex in the case of "risk-proneness". Let us suppose that the expected income in Choice A is lower than in Choice B. Yet, a risk-averse farmer may choose Choice A if expected future variations are lower. Costs related to "risk" and "uncertainty" are also factored in. A risk has a known probability; but an uncertainty has an unknown probability. The probability of risks is not fully known, and the probability of uncertainties is not fully unknown. So, mainstream economists generally assume that farmers would estimate a subjective probability based on the known probability, which in turn is incorporated into their utility functions prior to a maximisation exercise.

This framework of mainstream economists is ill-equipped to understand the complexity of choices made by farmers in the field. To begin with, there is this old question in agrarian economics of whether farmers are free agents i.e., who are truly free to make their own choices. In fact, many types of constraints prevent them from exercising free choices. Such constraints could be institutional (like tenancy contracts and capital market failures), agroecological (like climate, topography, soil type and irrigation) or infrastructural (such as availability of markets and transport facilities). The mainstream framework fails to comprehensively capture these complexities of constraints that influence the decisions made by farmers. The way out chosen by mainstream economists is typical: what cannot be modelled is often assumed as given.

In my view, a more careful and open-ended mix of qualitative and quantitative tools may need to be employed to address this deficit. In other words, to comprehensively understand the farmer's crop choices, an appreciation of his/her experiences that, in turn, govern his/her choices is essential. In our village surveys, we had multiple rounds of group discussions with farmers where we asked them questions on what drove their crop choices.

24.4.1.2 The Village Case Studies

Building on the literature and our databases, my argument is that the decision-making of farmers cannot be singularly based on the absolute cost/profit differences across crops (an example being: "maize is more profitable than cotton" or that "cotton is less profitable than soyabean"). In reality, farmers undertake a careful analysis of risks in cultivation before they decide on their crop choices. In both Dongargaon and Savali, farmers did not simply compare expected profitability rates across cotton, soyabean and maize. They also asked questions on: (a) the initial current expenditures required for each crop in the context of capital market imperfections; and (b) the probabilities of these initial costs getting sunk in the event of an expected yield-shock, which in turn was a derivative of their own subjective experiences of rainfall variations over the previous years.

Between our first and second rounds of surveys in Dongargaon and Savali, the prices of most farm inputs like seeds, fertilizers, and pesticides had risen sharply. Studies show that such increases in input prices were a result of conscious policy shifts of the Government of India (see Ramakumar 2022). Though input costs rose sharply, there was no commensurate rise in output prices. For instance, the minimum support prices (MSP) for cotton rose at a slower rate than the prices of most inputs.

There were three consequences that followed. To begin with, the sharp rise in input prices forced farmers to incur a large initial current expenditure regardless of the actual level or price of output. But capital markets were not perfect, and banks were not ready to provide farmers with timely and adequate credit. Hence, the real costs of these initial expenditures were burdensome for farmers. Secondly, in this context, if there was a yield-shock due to rainfall variations, there was likely to be an additional financial burden apart from those due to higher input prices and poor credit availability. Thirdly, the inability of the MSPs to rise in accordance with the rise in input costs implied that profitability rates were depressed.

Given this circumstance, a larger risk management strategy informed the farmer's decisions on crop choices. The expected burden of incurring a high current expenditure was compared with the expected loss due to a probable yield shock (or a crop failure). If yield-shocks were perceived to be more frequent, they fell back on a "play-it-safe" strategy of risk minimization. They chose a crop that involved a lesser current expenditure to minimise the losses resulting from possible yield-shocks instead of choosing a relatively risky strategy of incurring higher current expenditures and gambling on a good harvest. More specifically, over occasional seasons, farmers preferred a "low-input and low-output" form of production than a

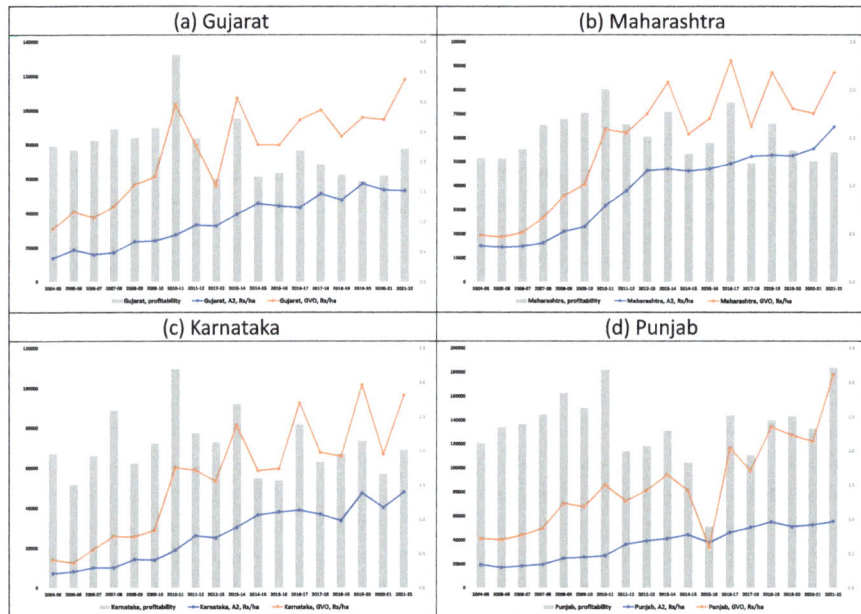

Fig. 24.6 Cost A2, gross value of output (GVO) and profitability rates, Cotton, Gujarat, Maharashtra, Karnataka and Punjab, 2004–05 to 2021–22, in Rs per ha

"high-input and high-output" form of production. Farmers in Dongargaon and Savali employed such a risk-minimising strategy in a strategic combination with an income-maximising strategy, where crops were so chosen as to increase the cropping intensity over the full year (for details, see Ramakumar et al. 2016).

In sum, if many farmers withdrew from Bt-Cotton cultivation in some years, it must not be confused with the claim on the so-called "failure" of Bt-Cotton. These decisions by farmers in parts of India took place within a policy context that led to a rise in input prices, a plateauing of output prices and failures in the capital market. Combined with frequent failures of rainfall, these contextual factors rendered cotton cultivation riskier. In Fig. 24.6, I have provided detailed sets of estimates from the Commission for Agricultural Costs and Prices (CACP) for costs of cultivation (Cost A2), the gross value of output (GVO) and profitability ratios (i.e., GVO/A2) in the four major cotton growing States of Gujarat, Maharashtra, Karnataka, and Punjab. These plots show that while GVO values of cotton have fluctuated over past decade due to price fluctuations and climate variabilities, there is no evidence of a secular fall of GVO reflecting any significant fall in Bt-Cotton yields. On the other hand, what is clear is that the input costs in cotton rose consistently—though at varying rates across States—contributing more to compressed profitability ratios. The fall of area cultivated with cotton, and the shifts from cotton to soyabean or maize must, then, be seen as a response of farmers to higher input costs and the attendant risks they brought along in a period of price and climate uncertainties—and must not be misconstrued as due to technological failure.

24.4.2 Institutional Factors

Several institutional factors at the policy level contribute to the success or failure of a new technology. Primary among them is the presence of markets and infrastructure. The absence of a stable market could be a major disincentive in the adoption of new technologies. Similarly, the presence of a favourable institutional ecosystem would help farmers avoid or overcome multiple market imperfections—both in the input and output markets. An appropriate grievance redressal mechanism would also instil an element of confidence among farmers on adopting new technologies.

Yields stagnated or fell also because cultivation was increasingly spreading into rainfed regions. It is well-known that yields of cotton in rainfed regions can be low under hybrid cultivation, which is more susceptible to moisture stress compared to open-pollinated varieties. Kranthi (2012, p. 34) wrote that:

> Protective and supplemental irrigations for cotton are not possible in...rainfed conditions. Water and nutrient requirement during peak boll formation phase are most critical for high yields. Rainfall starts in June and recedes in September. Boll formation in long duration varieties and hybrids starts in October and reaches a peak in November. Boll formation and retention get negatively affected due to low soil moisture, especially in shallow soils, thus resulting in low yields...[But] farmers often incorrectly attribute such poor performance with Bt-Cotton technology...

The question Kranthi raised takes us to the hybrids *versus* varieties debate in Bt-Cotton; I shall try to elaborate further on that in the next section.

24.4.2.1 The Regulatory Environment

But perhaps the most important among the institutional factors is the regulatory environment. Globally, countries follow two broad models of regulation vis-à-vis GM crops. The first is the *precautionary principle*, followed in the European Union. The Cartagena Protocol on Biosafety, an international agreement, notes thus: "the objective of this Protocol is to contribute to ensuring an adequate level of protection in the field of the safe transfer, handling and use of living modified organisms resulting from modern biotechnology that may have adverse effects on the conservation and sustainable use of biological diversity, taking also into account risks to human health, and specifically focusing on transboundary movements." Essentially, the precautionary principle allows countries to restrict or ban the cultivation or sale of GM crops/seeds where there is a perceived scientific uncertainty regarding its potentially adverse effects on ecology and health. The problem with the precautionary principle is that it could be used to infinitely delay the release of a new seed, effectively amounting to a permanent ban, even if the insistence is only on "an adequate level of protection".

As opposed to the precautionary principle is the principle of *substantial equivalence*, followed in countries like the United States. This principle is based on a comparison of the characteristics of phenotype and composition across the parent crop and the GM crop. If the GM crops are found to be "substantially equivalent" to the non-GM crops, the GM crops are designated as "Generally Recognized as Safe", which does not require a pre-commercialisation approval. In countries where the principle of substantial equivalence is in practice, the regulatory environment is considerably supportive of the release of new GM seeds. The larger number of successful releases of GM crops in the United States, as compared to the European Union, is illustrative in this regard. The more recent decision of the Government of India to allow the commercialisation of GM-Mustard seeds is welcome, but India continues to hold on to a maximalist version of the precautionary principle—as seen during the debates on Bt-Brinjal seeds—that has severely limited the number of GM seeds that can be released for cultivation. Regulatory compliance is extremely costly in India, and this is a serious disincentive for investment in GM-research on crops.

24.4.2.2 The Ownership of Technology

Yet another institutional feature involves the question of ownership of technology. Here, we return to the question of hybrids *versus* varieties that we flagged in the last sub-section. Historically, the role of public agricultural research has been critical in countries like India. However, in the more recent stage of the gene revolution, as opposed to the earlier phase of the green revolution, the role of public institutions has been weakened in GM research, which has provided private companies to monopolize large parts of the seed market. A concern here has been that the choice of techniques in seed production and release may not be decided based on either farmer's welfare or sustainability but purely profits. The case of Bt-Cotton seeds in India is a useful illustration (see Kranthi 2012 for a discussion).

Over the years, the view that cotton varieties would be better suited to India's rainfed conditions than cotton hybrids had emerged as a considered position of the scientists in India's NARS. They argued that Bt varieties had several advantages over Bt hybrids. First, the Cry toxin was in the homozygous condition in a Bt variety, while it was in a hemizygous condition in a Bt hybrid. As the two alleles in the chromosome were in 'AA' or 'aa' condition in a variety, as opposed to the presence of only 'A' or 'a' in a hybrid, homozygous varieties produced far higher levels of Bt toxin compared to hemizygous hybrids. The implication is that the bollworm would be better eliminated in a variety than in a hybrid.

Secondly, because of better toxicity in a variety, pesticide use could more easily be reduced in a variety than in a hybrid. Hybrids were highly susceptible to sucking pests, which was driving up pesticide use even in Bt-Cotton. Immunity to sucking pests could also be better achieved in a Bt variety than a Bt hybrid.

Thirdly, studies conducted at Central Institute of Cotton Research (CICR), Nagpur showed that the Cry toxins of Bt varieties did not segregate easily inside the cotton bolls, while they did segregate easily in Bt hybrids. As a result, the bolls in a hybrid were more vulnerable to worm-feeding than in a variety, as segregation killed the dominant effect of the toxic allele.

Fourthly, the seeds of a Bt variety could be saved and re-used by farmers, thus reducing seed costs and allowing freedom of choice for farmers. Fifthly, a Bt variety allowed about 20–30% higher plant density than in a Bt hybrid, thereby facilitating even higher yields.

For these reasons, agricultural scientists within the NARS argued that public sector science must focus on variety seed development in cotton. However, in India, agricultural research policy has favoured private sector research and has not allocated sufficient funding to (and autonomy to) public sector agricultural research institutions. Private sector was only interested in producing Bt hybrids and not Bt varieties. In contrast, in countries like China, Bt-Cotton varieties, and not hybrids, have driven yield increases (Pray et al. 2011). The Chinese Academy of Agricultural Sciences (CAAS), which pioneered Bt seed research there, concentrated on varieties and shunned hybrids including for the reasons listed above. Bt varieties (as opposed to Bt hybrids) predominated the area sown to cotton in China, while the reverse was true in India.

The inability of science policy in India to promote public and open research, and release more of varieties than hybrids of Bt-Cotton in the public sector must be seen as a failure. The fall of area, production, and yield after 2012–13 is also to be explained by this poor policy choice. It is eminently arguable that the potential for the growth of cotton production in India remained under-attained under a private-led Bt-hybrid regime as compared to a public-led Bt-variety regime.

The above discussion pertained largely to institutional features at the policy level, but there are equally important institutional features at the farm-level. For instance, the form of output matters. The new Bt-seeds released in India were of the long-staple type. The global demand for long-staple cotton (27.5–33 mm) was always higher than the demand for short-staple cotton. Prior to the introduction of the Bt-Cotton seeds in 2002, only about 38 per cent of the Indian cotton production was long-staple; the rest was short-staple cotton, which was considered "inferior" as an export crop (Chaudhary and Gaur 2015). With the universal spread of the Bt-Cotton seeds, the share of long-staple cotton rose to above 80 per cent, which also helped India export more of its cotton to the international market.

24.4.3 Sociocultural Aspects

While yields and exports grew in the long-run, and farmers adopted the seed in large numbers, the dominant public narrative on Bt-Cotton in India remained sceptical. Such a mindset against the Bt seed has much to do with the values of traditionalism deeply embedded in the Indian society. Many groups that argue against Bt-Cotton, for example, constantly harp on the relative superiority of traditional varieties of cotton and how cultivating these varieties were part of the "culture" in rural India. While the constitution of India demands scientific temper from its citizens as a duty, in large part, modern values remain outside the compass of agrarian debates in the contemporary Indian society.

   Given the consistent campaign against Bt-Cotton by various groups, there have been fears among the public on the GM technology itself, which led to a regulatory blockade of the release of new GM seeds in brinjal, mustard and rubber over many years. A part of the problem here has been the inability of the Government of India and the larger scientific community in agriculture to proactively engage with the concerns put forward by various groups in the very public domain in which these are raised. Risk communication strategies, thus, have been a failure that have led to a poor appreciation among the public on the facilities of risk management regarding GM seeds.

   Some groups have also raised concerns on the inability of farmers to save seeds and reuse them in the case of GM crops. While it is important that the right of farmers to save and reuse seeds must be protected, it is equally important to highlight the usefulness of raising seed replacement rates in Indian agriculture. Here again, communication strategies of the scientific community appear to have failed. It is well-known that the overall seed replacement rate in Indian agriculture is just above 20 per cent, and that certified seeds have higher and more stable yields than farm-saved seeds.

   But the skewed debates on the value of farm-saved seeds have led to polarised viewpoints. On their part, private seed companies have preferred policing and criminalisation of farmers to prevent reuse of seeds. Private seed companies fall back on policing because their low-volume, high-value business model is crucially dependent on forcing farmers to buy new seeds every season. On the other hand, an enabling atmosphere to raise seed replacement rates can be generated by a strong presence of public institutions in seed research and production. When public institutions, not motivated by profits, are ready to supply quality seeds at affordable prices, policing becomes redundant, and the purchase of certified seeds could be voluntarily accepted by farmers.

   Claims have also been made that the introduction of Bt-Cotton seeds has reduced employment for agricultural labourers, especially women. Our own field research in Dongargaon and Savali villages show this to be untrue. First, the higher yield of cotton and the larger number of pickings implied that the number of female labour days employed in harvesting was higher in Bt-Cotton. Also, as there was more harvest to be transported, the number of male and female labour days used in harvest and transportation was higher in Bt-Cotton plots. Secondly, seeds of Bt-Cotton were sown in furrows in every plot and not on flat seed beds. This involved the making of furrows, or *zari*, either with bullocks or tractors. Further, no machine or *tiphan* was used in the sowing of Bt-Cotton. Instead, Bt-Cotton seeds were sown carefully by hand by female labourers in every field. This raised the number of female labour days in Bt-Cotton cultivation. Thirdly, a greater number of hand-weedings was undertaken in the Bt-Cotton fields than in the non-Bt fields. This too led to a rise in female labour absorption in cotton cultivation in the village. Overall, the labour absorption in an acre – especially female labour absorption – of Bt-Cotton was higher than in the case of *parathi* cotton.

24.5   Concluding Comments

This article was an effort to undertake two tasks: one, to outline the ways in which social scientists could constructively engage with the scientific community regarding new seed technologies in agriculture; and two, to outline a broad framework within which such an engagement could proceed.

The first point I tried to highlight was the need to understand and appreciate the socioeconomic context in the study of emergence of technologies, including in agriculture. A historical and dialectical approach was argued to be an appropriate and comprehensive framework to undertake this task.

The second point I tried to highlight was that social scientists must adopt a broader and multi-dimensional framework to better appreciate the contexts within which new technologies in agriculture emerge. Narrowly focusing on a few sustainability concerns and sociocultural barriers leads one to poorly appreciate the *net social value* of a technological shift.

The third point I tried to highlight was that social scientists must adopt an integrated multi-disciplinary framework in judging the social value of a technology, such as a GM seed. Apart from studying the economic profitability, they must also study the institutional and sociocultural factors that mould the evolution of a technology on the field. For example, even if a GM seed has a potential to raise yields and profitability, institutional factors and sociocultural barriers could stymie it. Poor policy choices on the techniques to be relied on, policy decisions that lead to counteracting tendencies in the economics of cultivation, and policy preferences for private rather than public research are some of the factors that could undermine the credibility of new technologies and lead to adverse and unexpected outcomes.

References

Burney JA, Davis SJ, Lobell DB (2010) Greenhouse gas mitigation by agricultural intensification. PNAS 107(26):12052–12057

Byres TJ (1988) The agrarian question and the differing forms of capitalist agrarian transition: an essay with reference to Asia. In: Breman J, Mundle S (eds) Rural transformation in Asia. Oxford University Press, New Delhi

Chaudhary B, Gaur K (2015) Biotech cotton in India, 2002 to 2014: adoption, impact, Progress & Future. The International Service for the Acquisition of Agri-biotech Applications (ISAAA), New Delhi

Chaudhary J (2005) The future of Indian cotton supply and demand: implications for the U.-S. Cotton Industry. Ph.D. dissertation, Texas Tech University, Texas

Conrad P, Gabe J (1999) Introduction: sociological perspectives on the new genetics: an overview. Sociol Health Illn 21(5):505–516

Dobb M (1947) Studies in the development of capitalism. Routledge

Engels F (1954) Dialectics of nature. Progress Publishers, Moscow

Fukuda-Parr S (ed) (2007) The gene revolution GM crops and unequal development. Routledge

Gard W (1931) Agriculture's industrial revolution. Curr Hist 34(6):853–857

Hayami Y, Ruttan VW (1985) Agricultural development: an international perspective. John Hopkins University Press, Baltimore

Herring RJ (2006) Why did "Operation Cremate Monsanto" fail? Crit Asian Stud 38(4):467–493

Hille F, Charpentier E (2016) CRISPR-Cas: biology, mechanisms and relevance. Philos Trans R Soc Lond Ser B Biol Sci 371:1707

Hobsbawm E (1975) The age of capital: 1848–1875. Abacus

Hobsbawm E, Rude G (2001) Captain swing. Lawrence and Wishart, London

Kranthi KR (2012) Bt-cotton: questions and answers. Indian Society for Cotton Improvement, Mumbai

Parayil G (2003) Mapping technological trajectories of the green revolution and the gene revolution from modernization to globalization. Res Policy 32

Pray CE, Nagarajan L, Huang J, Hu R, Ramaswami B (2011) The impact of Bt cotton and the potential impact of biotechnology on other crops in China and India. In: Carter CA, Moschini G, Sheldon I (eds) Genetically modified food and global welfare (frontiers of economics and globalization, Vol. 10). Emerald Group Publishing Limited, Bingley, pp 83–114

Ramakumar R (ed) (2022) Distress in the fields: Indian agriculture after economic liberalisation. Tulika Books, New Delhi

Ramakumar R, Raut K, Kamble T (2016) Moving out of cotton: notes from a longitudinal survey in two Vidarbha villages. Rev Agrar Stud 6:3

Ramakumar R, Raut K, Kumar A (2009) Agrarian change in rural Maharashtra: resurveys of selected villages: a resurvey of Dongargaon Village, Akola District, Maharashtra. Research Report, Tata Institute of Social Sciences, Mumbai

Swaminathan MS (1968) The age of Algeny, genetic destruction of yield barriers and agricultural transformation. Presidential Address, Agricultural Science Section, Fifty-fifth Indian Science Congress, Varanasi

Thompson PB (2005) Ecological risks of transgenic plants: a framework for assessment and conceptual issues. *Issues in Environmental Science and Technology*, 21

UNESCO (2000) World conference on science: science for the twenty-first century; a new commitment. United Nations Educational, Scientific and Cultural Organization, Paris

Transforming Agricultural Education for a Sustainable Future

25

R. C. Agrawal and Seema Jaggi

Abstract

The transformation of agricultural education is very much essential in navigating the complex landscape of climate change, food & nutritional security and rural development. With a focus on sustainability, innovation and empowerment, ICAR's initiatives pave the way for a skilled, adaptable and future-ready workforce that can tackle the challenges of a rapidly changing world. By nurturing the seeds of change in agricultural education, we sow the potential for a more resilient, productive, and sustainable agricultural sector. These educated individuals will be the driving force behind innovations that ensure food for all, protect our environment and forge a path toward a brighter and more sustainable future. As ICAR continues to nurture the seeds of change, it contributes significantly to a sustainable and prosperous future for agriculture and beyond. By expanding reach beyond public institutions, it can be ensured that the transformative influence extends across the educational landscape. Through digital transformation, curriculum revamp and collaborations, it is being ensured that agricultural education remains enriching, aspirational, and empowering for students. ICAR envisions aligning its efforts with the Sustainable Development Goals (SDGs) by pioneering research, education and innovation in agriculture, fostering sustainable practices and empowering communities for resilient, inclusive and environmentally-conscious agricultural advancement. Also, the National Agricultural Higher Education Project (NAHEP) has contributed to SDG by providing quality education. Its primary objective is to provide support and strengthen the Agricultural Universities and ICAR in offering more pertinent and superior education to students. By striving to elevate education quality, a

R. C. Agrawal (✉) · S. Jaggi
Indian Council of Agricultural Research (ICAR), New Delhi, India
e-mail: ddg.edu@icar.gov.in; adg.hrd@icar.gov.in

© National Academy of Agricultural Sciences, under exclusive license to Springer
Nature Singapore Pte Ltd. 2023
K. C. Bansal et al. (eds.), *Transformation of Agri-Food Systems*,
https://doi.org/10.1007/978-981-99-8014-7_25

highly skilled workforce capable of perpetually enhancing the productivity of vital sectors, including agriculture can be cultivated.

Keywords

National Education Policy · Agricultural education · Sustainable development goals · Digital education · Skill development · Agripreneurship · Quality education · National Agricultural Higher Education Project (NAHEP)

25.1 Introduction

Agriculture, the cornerstone of human civilization, has undergone a remarkable journey over centuries. From humble beginnings of subsistence farming to the highly sophisticated agribusinesses today, agriculture has always been the bedrock of societies. However, as the world grapples with the challenges of climate change, depleting natural resources and the need for sustainable development, the role of agriculture has evolved. In this transformative landscape, the key to navigating the path towards a sustainable future lies in revolutionizing agricultural education (Tamboli and Nene 2013).

Guiding the trajectory of India's agricultural education is the Indian Council of Agricultural Research (ICAR). The council holds a pivotal role in shaping and adapting the landscape of agriculture in India as it responds to the challenges and opportunities presented by an evolving world. As a premier research and education organization in the field of agriculture, ICAR plays a multifaceted role in guiding agricultural practices, innovations and policies in the changing agricultural scenario while providing quality education.

The creation of trained quality human resources in the agriculture and allied sectors through the establishment of State Agricultural Universities (SAUs), since the year 1960s onwards, and the Deemed Universities (DUs) has ushered in the green revolution, followed by white, yellow, blue and other revolutions. Fuelled by public funding, India boasts an assemblage of 65 SAUs, 4 DUs, 3 Central Agricultural Universities (CAUs) and 4 Central Universities (CUs) featuring Agricultural Faculties, all operating under the aegis of the National Agriculture Research, Education and Extension System (NAREES) overseen by the ICAR. These institutions offer many degree programmes for Undergraduate (UG) to Doctoral levels, encompassing domains like agriculture, horticulture, animal husbandry, fisheries, veterinary sciences, agricultural engineering and more. In addition to imparting education, these establishments engage in critical agricultural research and extend their expertise to farmers and stakeholders. ICAR provides financial support to SAUs, DUs and CUs with Agriculture Faculty throughout the country to strengthen and develop higher agricultural education system. In 2017, ICAR declared all UG courses in agriculture and allied subjects which include Agriculture, Horticulture, Forestry, Sericulture, Community Science, Food Nutrition & Dietetics, Agricultural

Engineering, Dairy Technology, Food Technology, Biotechnology, Fisheries and Veterinary & Animal Sciences as professional degree courses.

ICAR spearheads research in various domains of agriculture, encompassing crop science, animal science, fisheries, agroforestry, agricultural engineering and more. Its extensive network of research institutes and centers conducts groundbreaking research to develop improved crop varieties, drought-resistant seeds, sustainable farming practices and innovative technologies. By fostering innovation, ICAR equips farmers with the tools needed to adapt to changing climate patterns, enhance productivity and reduce environmental impacts. ICAR serves as a bridge between research and practical application. It plays a crucial role in transferring advanced agricultural technologies and knowledge from laboratories to the fields. Through its extension programs, farmer training and workshops, ICAR ensures that the latest research findings and best practices reach the grassroots level, benefiting farmers and stakeholders across the country. ICAR not only advances agricultural research but also shapes agricultural education (Agrawal and Jain 2022). It sets standards for curriculum development and quality assurance in agricultural universities and institutes across the country. By doing so, ICAR ensures that the next generation of agricultural professionals is equipped with the skills, knowledge and mindset needed to address contemporary agricultural challenges. ICAR acts as a driving force that navigates agriculture towards sustainability, resilience and progress, ensuring that India's agriculture remains responsive, competitive and capable of meeting the challenges of the future. The transformation of agricultural education is not just a choice; it's a necessity. As the world seeks solutions to address climate change, food security and rural development, a new breed of agricultural professionals must emerge—individuals who are equipped not only with practical skills but also with a deep understanding of sustainable practices and global challenges.

25.2 The Imperative of Agricultural Education Transformation and ICAR

Agricultural education has a profound role to play in shaping the future of sustainable agriculture. Traditionally, it has focused on imparting practical skills and techniques to ensure productive yields. While these skills remain essential, they must now be integrated with modern knowledge about climate-smart practices, resource conservation and sustainable land management. Agricultural education institutions must take the lead in fostering a deep understanding of agroecology, precision agriculture, biotechnology and renewable energy systems.

The ICAR plays a pivotal role in shaping agricultural education through policy formulation and standard setting. The Agricultural Education Division within the ICAR spearheads advancements in higher agricultural education (Agrawal et al. 2022). The Agriculture Education Portal (https://education.icar.gov.in/) serves as the comprehensive gateway encapsulating all initiatives overseen by the Agricultural Education Division. This division is dedicated to curriculum development, ensuring courses meet rigorous criteria through accreditation procedures, while also offering scholarships to students and faculty and implementing schemes for human resource development programs. Through these efforts, the ICAR ensures that India's agricultural education remains at the forefront of sustainable and transformative progress.

25.3 Nurturing the Seeds of Change in Agricultural Education

Nurturing the seeds of change embodies the essence of transforming agricultural education. Redefining education can sow the seeds of innovation, resilience and sustainability. ICAR is nurturing the seeds of change in agricultural education through its comprehensive initiatives. It's modernizing curricula to integrate sustainable practices, promoting interdisciplinary approaches, fostering hands-on learning through experiential programs and leveraging technology for real-world applications. By encouraging research, innovation and global collaborations, ICAR cultivates a skilled and forward-thinking workforce equipped to address evolving agricultural challenges and drive positive transformation in the sector. A comprehensive, dynamic, vibrant and quality agricultural education system along with development of a strong and effective research and technology, is the fundamental to the national progress and prosperity. Rana et al. (2020) have presented a landscape of higher agricultural education in India. Some of the initiatives for transformation and quality agricultural education by the ICAR are listed below.

25.3.1 Fostering Future Leaders: National Education Policy 2020

The National Education Policy (2020) (NEP-2020) is a pivotal document emphasizing the cultivation of individual creative potential, encompassing cognitive, social, ethical and emotional growth (NEP_Final_English_0.pdf (education. gov.in). Guided by this vision, the Agricultural Education Division has constructed a strategic blueprint to implement NEP-2020 in the agricultural education landscape. This comprehensive plan charts a specific trajectory within designated timelines. This directive aims to ensure multidisciplinary perspectives by aligning with the University Grants Commission (UGC) and other regulatory guidelines. ICAR has been given the responsibility of Professional Standard Setting Body (PSSB) for the Agriculture Education.

Some of the major highlights of NEP-2020 include enhancing the Gross Enrolment Ratio (GER), defining minimum standards for quality agricultural education and ensuring all stakeholders adhere to it, improvement in research contributions, importance of staying relevant and providing placement along with right skills. The multiple exit and entry points into higher education, relaxation of the residential requirements of UG, PG and Ph.D. programmes, restructuring and reformulation of the UG curriculum in accordance with the new system advised by NEP and compliance with Academic Bank of Credits as per the directives of the Ministry of

Education are also some important additions. Various timelines for implementation of NEP by agricultural universities have been defined (Implementation Strategy for National Education Policy – 2020 in Agricultural Education System 2021). In this direction, the Agricultural Universities may start increasing seats by 10% on annual basis until the target is achieved. By 2025–2030 all institutions, located in the same premises, offering either professional or general education may aim to evolve into multi-disciplinary institutions/clusters offering education both seamlessly and in an integrated manner. By 2035, it is targeted to achieve 50% GER in higher agricultural education including vocational education, All higher educational institutions (HEIs) have to aim to become multidisciplinary institutions by 2040.

ICAR-DUs have initiated the process for transforming them into Multidisciplinary Education and Research University (MERU). In this direction, IARI has initiated the regional hubs approach or off campuses to be established for UG/PG/Ph.D. degree programmes, wherein one institute in a particular region will be the nodal institute (planet) and remaining institutes in the region will be the satellite institutes. This shall fulfil the requirement of multidisciplinary faculty and enhanced number of students enrolment. The student intake may vary for different hubs depending on the facilities available.

25.3.2 Attracting Talent in Agricultural Education: ICAR Entrance Examination

The Common University Entrance Test (CUET), conducted by the National Testing Agency (NTA), has been embraced by ICAR for UG courses in Central Universities. This move expands admission opportunities for UG courses in agriculture and allied sciences. Students seeking admission to the prestigious 20% All India Quota seats for 12 UG courses can now embark on their academic journey through CUET (UG) in Agriculture and Allied Sciences. This move encompasses 100% seats in esteemed institutions such as Rani Lakshmi Bai Central Agricultural University, (RLBCAU), Jhansi; Dr. Rajendra Prasad Central Agricultural University (RPCAU), Pusa, Samastipur; ICAR-NDRI, Karnal and ICAR-IARI, New Delhi. This has resulted in a seven-fold increase in the number of registrations over the last year. The All-India Entrance Examination for admission (AIEEA) to M.Sc. in 80 disciplines and the All-India Competitive Examination (AICE) for admission to Ph.D. degree programs remain integral components of ICAR's quest to elevate agricultural education. Additionally, the increase to 30% in All India Quota seats for PG and PhD programs opens avenues for advanced agricultural education.

25.3.3 Revolutionizing Course Curricula for Future-Ready Professionals

The ICAR has been regularly on periodic basis bringing necessary reforms for quality assurance in agricultural education. Council has been appointing Deans'

Committees, which in consultations and deliberations with all stakeholders, have been making recommendations on updating academic norms and standards towards meeting the challenges and opportunities. The 'sixth Deans Committee' is currently working to restructure agricultural and allied sector course curriculum in alignment with NEP-2020 for UG courses. This overhaul includes the integration of vocational courses, fostering skill development, and setting the stage for a new era in the agricultural domain. Since academic session of 2021–2022 the revamped syllabus with new, cutting-edge courses recommended by the committee on 'Broad Subject Matter in Agriculture (BSMA)' has been implemented for PG and PhD programs in 79 disciplines. New courses include genomics (biotechnology), nanotechnology, precision farming, conservation agriculture, secondary agriculture, hi-tech cultivation, specialty agriculture, geographical information system (GIS), artificial intelligence, big data analytics, food quality, safety standards and certification, food storage engineering, renewable energy, mechatronics etc. The courses on personality development, leadership development, yoga practices, life skills, human values and ethics have also been included in the list of non-credit courses.

ICAR has embarked on a significant initiative to integrate natural farming into UG and PG course curricula, contributing to sustainable agricultural practices. A dedicated committee was formed to design a syllabus and curricula on Natural Farming in accordance with NEP-2020's provisions. This endeavour seeks to introduce students to traditional techniques that prioritize biodiversity, soil health and a harmonious relationship with nature. Simultaneously, it equips them with enhanced employability and an entrepreneurial spirit.

25.3.4 Cultivating Practical Wisdom: Student READY (Rural Entrepreneurship Awareness Development Yojana)

The graduates are required to possess professional capabilities to deal with the concerns of sustainable development (productive, profitable and stable) of agriculture in all its aspects. There is need for agricultural graduates to possess knowledge, skills and also entrepreneurship abilities to provide village-based services such as advisories on new innovations, markets and avenues of development assistance for corporate and contract farming. It is essential to develop partnership between industry and universities if the industry has to obtain well-trained agricultural professionals in cutting edge technologies for international competitiveness.

In 2015, the Student READY (Rural Entrepreneurship Awareness Development Yojana) was launched by the Hon'ble Prime Minister for the development of agrientrepreneurs. This was introduced in the UG programme in all the disciplines of agricultural and allied sciences as approved in fifth Deans' Committee. It is a complete one year activity in the last year of the UG programme of Agriculture, Agriculture Engineering, Biotechnology, Community Science (earlier Home Science), Dairy Technology, Food Technology, Forestry, Fisheries, Horticulture and Sericulture. The aim is to develop young agripreneurs for emerging knowledge intensive agriculture. The programme integrates activities to develop skills in project development and execution, decision-making, team coordination, accounting, quality control, marketing and conflict resolutions, etc. with end-to-end approach.

The Rural Agricultural Work Experience (RAWE) program which is a part of Student READY is designed to empower young agri-graduates with practical knowledge, community engagement skills and extension tools. These tools enable them to transfer the latest agricultural technologies to farmers effectively. Also providing opportunity to understand the rural setting, ongoing extension and rural development programs in relation to agriculture and allied activities, imparting diagnostic and remedial knowledge relevant to real field situations through practical training, developing analytical skills using extension teaching methods and exercising strong solutions acumen for ground level agricultural and allied activities.

Strengthening and revisioning the RAWE program is being done to create a framework for actionable, feedback driven outcomes that helps the agricultural ecosystem be better informed. A digital platform is being introduced to streamline reporting, monitoring and data analysis of RAWE activities, ensuring graduates return with knowledge, empathy and innovation, by this means driving sustainable agricultural growth. The system could analyze the information and generate outcomes and outputs on different parameters.

25.3.5 Agripreneurship through Agricultural Education

The changing dynamics of global economy and the increasing demands of growing population require exploring innovative solutions to enhance agricultural productivity, sustainability and profitability. This can be achieved by nurturing agripreneurial ecosystem through strengthening of our agricultural education. Agripreneurship adopts innovative methods, processes, techniques in agriculture or the allied sectors of agriculture for better output and economic earnings. Enterpreneurial mindset needs to be instilled in the students with the necessary skills and knowledge so that the students can think beyond traditional career paths. There are tremendous opportunities in the agriculture and allied sector for promotion of agripreneurship. The allied sectors like, dairy farming, bee-keeping, sericulture, mushroom cultivation, fisheries, etc., have a lot of potential for agripreneurial development. Horticulture, which includes olericulture, pomology and floriculture in itself, provides lots of scope for entrepreneurial ventures. Agri-trading centres, commercial agri-hubs, etc., are some of the prospective entrepreneurial areas that can be extensively tapped on a larger scale.

The higher agricultural education and capacity building programmes are necessary to gain access to employment and self-employment. Besides awarding formal degrees, Agricultural Universities are required to initiate job driven vocational programmes so as to build avenues for off-farm work. In order to promote agripreneurship, the establishment of agribusiness incubation centres at each of the agricultural universities shall give a boost to the start-ups in the agriculture and allied areas. Promotion of creative thinking, problem solving and development of communication skills in students can go a long way in changing the agricultural landscape of the country. Agripreneurship development programs, mentorship initiatives and startup incubation centres need to be organized in all agricultural universities to provide technical support and guidance to the students, the new age entrepreneurs and young minds to transform their innovative ideas into viable business propositions.

25.3.6 Riding the Digital Wave: Embracing Technological Integration in Education

Through the 'Resilient Agricultural Education System (RAES)' subproject under the 'National Agricultural Higher Education Project (NAHEP)', ICAR is embracing digital transformation to augment agricultural education's efficacy. This initiative comprises a robust three-tiered digital framework aimed at bolstering digital infrastructure, enhancing digital capacity and generating relevant digital content for comprehensive consumption. This approach capitalizes on Information and Communication Technology (ICT) to facilitate blended and flipped learning, enhancing collaborative partnerships with industry and government to establish a more market-driven teaching-learning ecosystem. The establishment of state-of-the-art digital infrastructure, equipped with advanced features like smart boards, projection systems and video conferencing facilities, reinforces this transformation across 74 Agricultural Universities. Notable digital initiatives, including 'E-Learning Portal', 'Agri-DIKSHA', 'Virtual Reality Experience Labs' and 'Academic Management System' are designed to enrich the learning experience. A recent addition is the 'Blended Learning Platform,' providing learners and administrators with tools for knowledge retention, engagement and collaboration, aligning with NEP-2020's vision of a hybrid learning ecosystem.

The digital automated pathway provides the step-by-step guide which includes e-learning platforms to meet the demand for formal/informal education, digital access, expanding access to affordable and reliable internet connectivity for education and training institutions, including through public partnerships. Specialization in newer areas viz. database technologies, cloud computing, process automation, software application development, human machine interaction, block chain platforms, geoinformatics, artificial intelligence, machine learning, technical leadership and architecture experience can be created in the agricultural universities in a phased manner. The Massive Open Online Courses (MOOC) are the recent disruptive innovations that provide flexible and affordable system to learn and deliver. It can enable a small group of teachers/mentors to offer learning services to many in the duration of a single course. MOOC can be meaningfully and effectively harnessed for training on a massive scale.

25.3.7 Collaborations and Global Outreach

Collaborations and linkages between universities, ICAR institutes and other institutions like IITs and IIMs helps the teaching - learning process as well as the academic environment in the university/institution. Strong and effective linkages at national and international levels are required for improving and enhancing the quality assurance mechanism and process, faculty and student competence. It is required to promote collaboration through knowledge sharing and partnerships.

A paradigm shift in education, from 'teaching to learning' and expanding the reach and opportunities for learning will necessitate going beyond the current initiatives of upgrading the universities as a global destination for agricultural education. Agricultural research, education, and extension to be mainstreamed into national policies. Restructuring the organisational model of research, from rigid to flexi-program mode and dynamic research consortia led by project leaders on the pattern of international organisations may respond faster to societal demands. Agricultural education is to be harmonized with existing and emerging issues. Agriculture is becoming competitive both price-wise and quality-wise in the entire world. Indian agriculture is no exception, and its objectives have to align with stakeholders' needs, clients' perspective, peer concerns and market vibes.

25.3.8 Empowering Students through Collaborative Initiatives

ICAR has been taking initiatives to collaborate with other organizations to strengthen its educational and other related activities.

ICAR and Heartfulness Education Trust Join Hands ICAR's collaboration with the Heartfulness Education Trust (HET) pioneers an innovative partnership, focusing on agricultural research and education. Students benefit from specialized programs, ranging from mindfulness techniques to leadership development, enhancing both academic and life skills. In tandem, the introduction of an agricultural curriculum in school education fosters an early appreciation for agriculture's significance and role.

Integrating Agricultural Curriculum in School Education Recognizing the significance of agriculture in the Indian context, the Agricultural Education division has taken an important step towards introducing agricultural curriculum in school education. A brainstorming meeting on Mainstreaming Agricultural Curriculum in School Education (MACE) was organized wherein principals of schools, senior teachers, and other experts, including ICAR, Central Board of Secondary Education (CBSE) and National Council NCERT, officials, participated in various sessions to discuss the need for and process of including agriculture as a subject in the school curriculum. As an outcome of MACE, collaborative discussions are underway with the (CBSE) to incorporate agricultural concepts and principles into the school syllabus. This proactive measure will not only cultivate students' interest in agriculture from a young age but also foster an appreciation for the vital role it plays in our lives.

Expanding Horizons: Mainstreaming Agricultural Education in Private Universities ICAR's endeavors extend beyond public institutions, as it collaborates with private universities offering agricultural higher education. This move diversifies the educational landscape and equips a broader student base with agricultural insights.

The Journey Forward: Industry-Institute Partnerships for Global Competitiveness Fostering industry-institute partnerships amplifies students' readiness for the competitive global arena. These collaborations provide practical exposure, access to cutting-edge research, internships, and real-world projects. This bridge between theory and practice equips students for success in the dynamic agricultural sphere.

25.3.9 Strengthening of Infrastructure and Resources

The Agricultural Universities are engaged in complete integration of teaching, research and extension for holistic rural development. But most of the SAUs have not been able to achieve integration of these functions due to limited physical and financial resources and also lack of appropriate human resources. To improve the efficiency of education, research findings and extension activities of the universities and also to inspire academicians and researchers in planning suitable location specific strategies to improve the financial status of their respective universities, the concept of revenue generation or income generation models are need of the hour. In addition to the enhanced government support, the Universities are required to generate their own resources and become self-reliant through innovative programmes, consultancy, enrolment of foreign students, sale of seed/ planting material etc.

25.3.10 Accreditation for Quality Assurance

The assessment and accreditation of higher agricultural education was initially looked after by the UGC. This responsibility was assumed by ICAR in 1965 and further after reorganization of ICAR in 1973-74, the Standing Committee on Agricultural Education was replaced by the Norms and Accreditation Committee (NAC) in 1974 with the responsibility of determining norms for accreditation of AUs. The National Agricultural Education Accreditation Board (NAEAB) was set up in 1996 by ICAR with well defined guidelines to improve and sustain the quality of agricultural education. This activity is now being done by NAEAB and guidelines were framed to make accreditation process more objective. Subsequently, the entire process of submitting documents for accreditation was made online through dedicated accreditation portal. The accreditation process is further being streamlined and simplified by reducing the time gaps in all activities.

ICAR's Model Act, a foundation laid in 1966 and subsequently revised, provides a comprehensive legal framework to guide the implementation of provisions across India's states, fostering unified agricultural education and research. In pursuit of excellence, ICAR has revised the Model Act in response to changing requirements, rectifying anomalies in existing university acts and enhancing provisions to ensure unmatched quality in agricultural and allied sciences disciplines (Model Act for Higher Agricultural Educational Institutions in India 2023).

25.3.11 Agriculture's Ascent in National Institutional Ranking Framework (NIRF)

In the globalizing world and an increase in internationalization of education, the ranking of the universities has become essential and of vital importance as university ranking and the indicators of excellence widen the basis of information for a well-informed university choice. The inclusion of agriculture and allied sectors in the prestigious National Institutional Ranking Framework (NIRF) marks a significant milestone. The parameters identified in NIRF falls in five major heads: (1) Outreach and Inclusivity, (2) Teaching, Learning and Resources, (iii) Research, Professional Practice and Collaborative Performance, (4) Perception, and (5) Graduation Outcomes. The NIRF list, released on June 5, 2023, includes the rankings of 40 AUs under the "Agriculture and Allied Sectors" category. ICAR's four Deemed Universities, including the Indian Agricultural Research Institute, New Delhi; National Dairy Research Institute, Karnal; Indian Veterinary Research Institute, Izatnagar; and Central Institute of Fisheries Education, Mumbai, have secured rankings within the top 10. These accolades underscore their unwavering dedication to academic excellence and innovative practices.

25.4 Cultivating Excellence: Transforming Agricultural Education with NAHEP

The National Agricultural Higher Education Project (NAHEP) of ICAR initiated in 2017–18 aimed to develop resources and mechanism for supporting infrastructure, faculty and student advancement. It also aimed to provide means for better governance and management of Agricultural Universities, to develop a holistic model in order to raise the standard of current agricultural education system that can provide more jobs, entrepreneurship oriented and on par with the global agriculture education standards. The project has addressed the engagement areas of integration, transformation and inclusion. These areas foresee increased agricultural productivity and support quality improvements of higher education so as to create a more skilled workforce that improves the productivity of key sectors, including agriculture. The project has supported Agricultural Universities and ICAR in providing more relevant and higher quality education to students. Quality has been addressed by supporting Agricultural Universities in technically sound and verifiable investments resulting in increased faculty performance, attracting better students to these Agricultural Universities, improving student learning outcomes and raising their prospects for future employability, particularly in the private sector. About 1000 students have undergone international trainings under the project which has given them an exposure to international standards and has resulted in broadening their horizons and fostering a global perspective. This exposure has provided the students learn to conduct themselves in any environment, gain confidence and develop networking skills. The faculty of Agricultural Universities have also been benefitted by undertaking trainings and exposure visits to international organizations. A number of capacity building programmes on various cutting-edge topics have been organized for the faculty and the students. (Annual Report 2022–23, NAHEP).

The project also aligns with the United Nations Sustainable Development Goals (SDGs) by addressing critical facets of education, innovation, climate resilience, gender equality and economic growth within the realm of agriculture (Pathak et al. 2022). The project has contributed to the achievement of following SDGs:

• Inclusive and equitable quality education and promoting lifelong learning opportunities for all—This will promote equal access to affordable vocational training and greater gender and wealth equity through universal access to quality higher education.
• Inclusive and sustainable economic growth, employment, and decent work for all—This seeks higher levels of economic productivity through diversification, technological upgradation and innovation. A stronger innovation culture has been fostered by twinning participating Agricultural Universities with other higher-performing centers of learning (both in India and abroad) and strengthening Agricultural Universities-private sector linkages to better orient student learning toward relevant skill sets.
• Resilient infrastructure, promoting sustainable industrialization and fostering innovation—This would enhance scientific research and increase both the research and development workforce and its associated budget.
• Action to combat climate change and its impacts—This would improve education, awareness-raising and institutional capacity on climate change mitigation, adaptation, impact reduction and early warning. Agricultural Universities have to internalize climate change and resilience in current and future course content and tie this with experiential learning for certificate, undergraduate and post-graduate students for practical career applications.

References

Agrawal RC, Jain V (2022) History of agricultural education in India. Agri Rise Agric Educ Dig 1(1):9–15

Agrawal RC, Pandey PS, Seema J, Vanita J, Agnihotri MK, Sankhyan S, Nidhi V (2022) Achievements in agricultural education in independent India. In: Pathak H, Mishra JP, Mohapatra T (eds) Indian agriculture after Independence. ICAR publication, pp 311–332. ISBN: 978-81-7164-256-4, 425

Annual Report 2022–23. National Agricultural Higher Education Project (NAHEP). Project Implementation Unit, ICAR. https://nahep.icar.gov.in/

Implementation Strategy for National Education Policy – 2020 in Agricultural Education System (2021). *ICAR publication* ISBN:978–81–7164-233-5, Pages 93

Model Act for Higher Agricultural Educational Institutions in India (2023). *ICAR publication*, Pages 59

National Education Policy (2020). Ministry of Human Resource Development, Govt. of India. NEP_Final_English_0.pdf (education.gov.in), Pages 66

Pathak H, Agrawal RC, Tripathi H (2022) Role of National Agricultural Higher Education Project in achieving sustainable development goals. Agri Rise Agric Educ Dig 1(1):22–29

Rana N, Agnihotri MK, Agrawal RC (2020) Landscape of higher agricultural education in India. ICAR publication. ISBN: 978-81-7164-193-2, Pages 74

Tamboli PM, Nene YL (2013) Modernizing higher agricultural education system in India to meet the challenges of 21st century. Asian Agri History 17(3):251–264

Correction to: Transformation of Agri-Food Systems

K. C. Bansal, W. S. Lakra, and Himanshu Pathak

Correction to:
K. C. Bansal et al. (eds.), *Transformation of Agri-Food Systems*,
https://doi.org/10.1007/978-981-99-8014-7

The original version of this book was inadvertently published with errors. The copyright holder name of the chapters has been updated with this erratum.

The updated version of this book can be found at
https://doi.org/10.1007/978-981-99-8014-7